Enabling eBusiness

Enabling eBusiness

INTEGRATING TECHNOLOGIES, ARCHITECTURES AND APPLICATIONS

W.S. Whyte
University of Leeds, UK

JOHN WILEY & SONS LTD

Chichester • New York • Weinheim • Brisbane • Singapore • Toronto

Other Wiley Editorial Offices

John Wiley & Sons, Inc., 605 Third Avenue,
New York, NY 10158-0012, USA

WILEY-VCH Verlag GmbH
Pappelallee 3, D-69469 Weinheim, Germany

John Wiley & Sons Australia, Ltd, 33 Park Road, Milton,
Queensland 4064, Australia

John Wiley & Sons (Canada) Ltd, 22 Worcester Road,
Rexdale, Ontario, M9W 1L1, Canada

John Wiley & Sons (Asia) Pte Ltd, 2 Clementi Loop #02-01,
Jin Xing Distripark, Singapore 129809

Library of Congress Cataloging-in-Publication Data
Whyte, Bill.
Enabling eBusiness: integrating technologies, architectures, and applications /
W.S. Whyte,
  p. cm.
Includes bibliographical references and index.
ISBN 0-471-89941-0 (alk. paper)
  1. Electronic commerce. I, Title.

HF5548.32 .W49 2001
658.8'4–dc21   2001017599

British Library Cataloguing in Publication Data
A catalogue record for this book is available from the British Library

ISBN 0 471 89941 0

Typeset in 10/12 Times by Deerpark Publishing Services Ltd, Shannon
Printed and bound in Great Britain by Antony Rowe, Ltd, Chippenham, Wiltshire
This book is printed on acid-free paper responsibly manufactured from sustainable forestry, in which at least two trees are planted for each one used for paper production.

Contents

Dedication

To Marian, Alasdair and William

Preface

It is the purpose of this book to explain how technology can enable end-to-end eBusiness. It is addressed to a wide audience, not only nut-and-bolts technologists, technical architects and team leaders, but also to functional managers, corporate strategists, instructors and students. All these require to share a common understanding in order to make eBusiness a success, because, as we say many times in the text, the task of understanding is at least as much cultural as technical.

If I am at all successful in this aim, it is because I have been very fortunate in having so many colleagues in universities and industry who can cross those divisions. These people have been patient enough to put up with my incessant asking 'But what *is* so-and-so and how does it work?' Mostly my job has been the management of their knowledge and wisdom onto the pages of this book. The faults will be in the interpretation, not the source!

One question definitely worth asking in this electronic age, is 'What differentiates a book from an on-line document?' and the answer is almost always, 'Good editorship'. Here too, I have been very fortunate in again working with Ann-Marie Halligan and Laura Kempster. They and the rest of the team at Wiley, are truly professional and a pleasure to work with. Finally, I would also specifically like to thank Cliff Morgan of Wiley, for a most productive discussion on meta information – whatever the quality of the content of this book, I now at least know how to describe its electronic structure!

W.S. Whyte
billw@comp.leeds.ac.uk

Introduction to eBusiness

eBusiness is a technical issue. The needs of enterprises and the desires of customers have not changed. Profitability remains the difference between income and costs. But, solely because of technology, major new opportunities for revenue growth and avoidance of expense have become feasible. eBusiness has happened entirely because of a conjunction of improvements in technologies and has yet to reach full speed and realise its potential. These technologies are, at the highest level, only two in number: more powerful computers and faster communication links.

Replacing the physical store by its virtual equivalent will mean that dramatic reduction in the cost of real estate can be achieved and the saving invested on creating new eShops to increase revenue within a global market. In a competitive world, new costs will also be incurred. To handle the increased volume of business or to meet increased customer expectation, these on-line outlets must be backed up by on-line customer service operations which are available on 24 h/7 day terms and these must be integrated with slicker, electronically assisted supply and fulfilment operations.

With this integration in place, it may no longer be necessary, or necessarily desirable, for any one company to do every thing to 'get the melon to the customer'. Instead, a number of different companies will be able to work together in a virtual enterprise, provided information can be interchanged effectively and securely.

1   eBUSINESS AND eCOMMERCE

eBusiness is a broader and more easily defined term than the commonly used *eCommerce*. The latter is sometimes used narrowly to refer to 'shopping on the Internet' but also in a much broader context to include virtually every electronic trading and support activity a business can undertake. In this book we use the term *eCommerce* rather informally to refer to a middle position, that of retailing, principally shopping. We do not restrict the delivery channel for these services to that of the Internet, since there are other channels, of which digital interactive TV and WAP mobile telephony are only two of the examples we shall cover. We shall also use terms such as *electronic retailing*, *on-line shopping/retailing* and *electronic merchants*, more or less as synonyms. Again note that 'on-line' does not necessarily mean 'Internet' and, in the case of TV and mobile radio, the 'line' is conceptual rather than real.

Where the range of activities is extended to cover such areas as logistics, marketing intelligence-gathering, collaborative working and so on, we prefer to use the term *eBusiness*. Sometimes when we want to emphasise the fact that access to the organisation is mainly via a communications network, we shall use the term *virtual business*. Where the business units and their electronic processes operate across a number of geographically separated sites, we shall talk about *distributed business*. If the conditions are such that a number of legally distinct trading entities set up systems that allow them to co-operate across distance, then we call this a *virtual enterprise*.

2 BASIC TECHNOLOGIES

We said that there were only two base-level technologies involved in supporting the eBusiness explosion: computing and communications. But it is reasonable to point out that these have been around for some decades and to ask why they should suddenly have become the engines of dramatic change. The answer is simple: both of them have quite recently 'turned the exponential' in terms of performance and reduction in cost. It is now well-known that performance per dollar in computing systems doubles every 18 months or so, has been doing so for two decades, and is likely to continue to do so for at least 15 years (see the graph on page 65). By the end of this period, computers will be more than 1000 times as powerful as today's machines, for the same price. A similar trend in telecommunications is happening. Twenty years ago, a single, heavy, rigid, expensive to install and maintain coaxial cable could carry a maximum of around 10,000 voice telephone calls or their equivalent in data over a few kilometres. Today, operational systems using individual tiny, flexible optical fibre carry orders of magnitude more, over intercontinental distances. Systems in the laboratory can carry many times more traffic on a single fibre than exists in the world today. The drop in cost is truly dramatic and we are only at the start. The message for business is simple: within a couple of decades, they will have wide access to affordable networks of effectively infinite carrying capacity and to computers with almost unimaginable processing power, and they must plan on this assumption.

3 OPPORTUNITIES (AND THREATS)

The technology developments we have described open up enormous opportunities to businesses and, equally, will enable grave competitive threats:

- *Geographical freedom*: certain parts of a trading organisation are freed from the tyranny of location. An on-line shop-window exists only in 'cyberspace' and is not constrained by rent or availability of real estate. Customers can come from anywhere – which is an advantage and a disadvantage. Trading globally certainly increases potential market size (perhaps particularly for products which would otherwise be niche) but it increases the complexity of product nativization, rules and regulations, language and, perhaps most importantly, how to deliver the goods. There is, in short, a major impact on the complexity of marketing and logistics. That is, eBusiness requires better integration of back-office processes.
- *But no hiding-place*: geographical freedom comes, not just to the vendor, but also to the customer. Customers have the ability to access shop-windows anywhere in the world and, provided the vendor can meet the fulfilment requirements, decide to purchase elsewhere than from their traditional supplier. This puts cost and quality of service pressure on the traditional vendor. It may also give rise to a new breed of organisation which can provide broker functions between multiple suppliers and customers, broking on availability, price, interpretation of requirements and specifications.
- *Temporal freedom*: the *24 h/7 day company* which shifts its operational base around the clock is heralded as an example of the new opportunity for trading on instant gratification and overtime-independent automation. This it is, but it is also a demanding requirement to put on a company's processes, that they can operate flawlessly without any downtime, a condition that is seldom achieved with today's computing systems. There is also a need to provide human cover over a similar span, as an exception handling procedure for where things wrong or simply become too complicated. People must be scheduled in to operate call centres at appropriate skill levels and the centres themselves must load-share the diurnal variations in traffic.
- *Freedom to customise*: To meet a 'giving customers what they want' strategy, there must be a precondition that one knows what they want. Electronic trading provides an almost unique way of acquiring and processing that information. Each transaction with a customer

can be recorded and profiles of customer likes and dislikes acquired. Customers can even be encouraged to construct (virtually) on-line, complex products out of a kit of parts. Technology makes this possible in two ways: first by removing the need for expensive human support and, secondly, by efficiently transferring this specification to the manufacturing arm without failure or other costs.

- *Collaboration in distributed enterprises*: if communications and computing can provide very low cost ways of passing moving images and voices of people who are geographically remote from each other and can do the same with high volumes of the data required to support manual or automated tasks, then companies can concentrate on what they are good at, seamlessly interacting with other organisations with different areas of excellence to create an end-to-end virtual enterprise.
- *Low transaction costs for payments*: automated taking of payment and error-free processing of it will significantly reduce transaction costs to customer and supplier alike.
- *But opportunities for fraud*: criminals should love the on-line society, at least in its present form. It is very insecure and offers great opportunities for low-cost, automated attack.

4   eBUSINESS EVOLUTION

In general, businesses are capital assets and usually managed to a large degree by risk avoidance. Thus, they evolve their working strategies rather than throw away all existing practices in one dramatic act. Not unnaturally, therefore, when trying to capitalise on the opportunities offered by new technology and in defending against its threats, they tend to do the easy, cheap and necessary things first. In terms of the easy and cheap, the obvious case is that of the on-line Web site. This provides a niche channel to market with minimum cost and exposure. This is not on-line marketing, as its proponents would claim. It is simply an on-line market stall and one which, by and large, advertises the goods for sale rather than promises and effects delivery.

On the other hand, the back-office processes of the company have tended to evolve out of necessity to keep down the costs and manpower of individual functional units, without taking much notice of direct marketing to customers, or even internal cross-function operation. Functional-based point solutions have led to the proliferation of incompatible databases and processes. They are often hosted on incompatible hardware and software. Particularly severe has been the divide between voice telecommunications and data networking, with a serious impact on the quality of service offered through help-desks and other interfaces with customers.

On the other hand, by not allowing the world to have any connection to critical processes, the internalised approach to computing has at least protected companies from excessive fraud or malicious damage. Unfortunately this makes organisations unaware of the real risks that will happen when they opens themselves up to the world through on-line trading.

Logistics has resisted much of the automation that has affected the rest of the company. It inhabits an environment that is hostile to technology and a workforce that is resistant to it. Because it involves considerable on-the-road activity there has been until recently no reliable and affordable way of maintaining contact and control.

The message in all of this is clear: successful companies will be those which are early and effective adopters of solutions that are integrated, across *heterogeneous platforms,* and between *organisations, different functional units and applications,* within a *secure environment* that is *customer-facing.* This requires the development of architectures that can work in a distributed and heterogeneous world together with a cultural adjustment where traditional, safe, inwards-looking design and entrepreneurial, rapid development and deployment approaches can work together in mutual respect. The cultural shift may be the more difficult of the two! eBusiness requires technical *integration* and cultural *integration* for its success.

5   OUTLINE OF THE BOOK

Since this is a technology book, we do not attempt to build elaborate business models for eBusiness, but we do need to maintain at least a simple picture of the components of a typical end-to-end business (Figure 1).

No special claims are made for the extreme accuracy of this model, or even for the completeness of the linkages shown, but it probably repre-

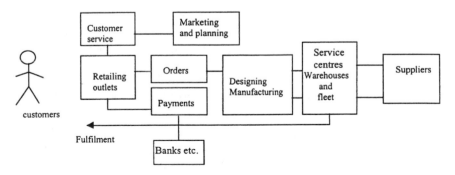

Figure 1   A simplified model of a business

sents the major business units and activities that have been the targets for improvements based on electronic technologies. At the very least, it has the advantage of identifying the areas where major individual initiatives have been centred. As we explained, these have historically been under-taken more or less in isolation from each other, but now require to be considered as an integrated whole, if we are to benefit fully from creating an electronic business.

From this diagram, and looking at events over the last decade or so, we see it is quite logical to follow eBusiness development through a number of strands:

- *Retailing ('electronic commerce')*: this is exemplified by Internet retailing via a Web site, with emphasis on the presentation to customers and less attention to back-office functions. Delivery of this service depends on transmission channels to the home and on the existence of suitable domestic terminals.
- *Enterprise business systems*: the first electronic systems were used to run in-house operations, for a limited range of functional units. Little thought though was initially given to connecting them together and even less to making some of their functionality available on-line. More recent developments have concentrated on *component architec-tures* that achieve this integration without replacing existing systems. Once information exchange becomes desirable across organisations and between organisations and their customers, then consistent defi-nition of terms becomes critical and the analysis of market informa-tion becomes of great value. *Knowledge management* is the buzz-term.
- *Operational security*: exposure to on-line trading soon alerted busi-nesses to the need for better security measures to protect against theft or sabotage.
- *Customer support*: independently of the growth in selling across the Internet, was the rapid increase in call-centre activity. This has been predominately voice-telephony, with computer-hosted scripting of standard procedures very much a point-centred, subsidiary activity.
- *Maintenance operations*: have also followed a traditional path of physi-cal visits, but automated methods are now being proposed.
- *Warehousing/logistics*: these now also need to be integrated into the end-to-end supply chain, having been largely left to their own devices in the past.
- *Marketing*: finally, marketing and sales campaigning are just begin-ning to apply their professional skills to the new on-line sales chan-nels.

Following on from this, it seemed logical to structure the book in terms of the discreet developments above, as they mostly also tend to fall into the domain of individual functional departments. So the parts and chap-ters are organised according to Table 1.

Table 1

Part 1: eRetailing (eCommerce)	Part 2: eBusiness systems	Part 3: Trust, security and eMoney	Part 4: Service, supply and marketing
Retailing principles and models	Systems architectures	Issues of trust	Managing the supply chain
Retail communications networks	Managing on-line knowledge	Security	Customer service
Customer terminals		Payment and electronic money	Marketing
The retail server			

Part 1, *Electronic Retailing*, develops a simple business model for retailing and discusses how far this can be delivered across domestic telecoms links and implemented on a surprisingly large range of home platforms. The basic design of shopping servers is discussed. Part 2, *eBusiness Systems*, is a more technical discussion of *n*-tier architectures for end-to-end business, including the emerging solutions for object-oriented component middleware, intended to interconnect distributed, heterogeneous platforms, with full transactional reliability. It also describes collaborative working tools for virtual enterprises and outlines the current state of systems for corporate knowledge management.

The critical importance of trusted systems design is the subject of Part 3, *Trust, Security and eMoney*. Here, attitude and the development of sensible security policies are seen as important as raw technology. Examples of attacks are given and solutions described, particularly in relation to payment methods. Part 4, *Service, Supply and Marketing*, is application-driven, and describes technologies for tagging, packing and scheduling goods, the evolution of call-centres and electronically-supported maintenance operations. It also covers some aspects of how traditional marketing techniques can be applied in the on-line case. Finally, the *Appendices* contains a bibliography and index, plus some thoughts on the necessary skills and attitudes of the people who will enable the eBusiness. Somewhat reluctantly, for I believe we need to be brave or foolish to make technology and market predictions, I also include some comments, which are no more than intelligent guesswork, as to where eBusiness is going.

Part 1: Electronic Retailing

Chapter 1: The Principles of Electronic Retailing

Standard models of retailing are related to the requirements for on-line retailing services.

Chapter 2: Retailing Network Technologies

eRetailing requires that customers be connected over electronic networks to the retailing servers, but we have to accept that these networks were originally designed for applications other than electronic retailing and it has therefore largely been shoe-horned into existing solutions, notably voice telephony and one-way wireless broadcasting. The problem is most acute at the last mile or so between the domestic customer and the local telephone exchange, where the local loop wires limit the maximum data rate to just a few tens of kilobits per second. This is not really suitable for high quality multimedia. Not even basic rate digital services such as ISDN can solve this problem, but newer developments such as Asynchronous Digital Subscriber Loop (ADSL) and cable modems may alleviate the problem. Radio networks tend to be slower and rather expensive but here too, increases in speed and the development of packet services will improve things.

Chapter 3: Retail Terminals

There are a surprising number of options for retail terminals, including not just the personal computer but also interactive TV, mobile devices, kiosks, airline seatback screens, etc. Although the PC is likely to dominate in the near future, the telephone and a number of telephony network services are the reality behind most of today's home shopping. In the domestic market, cost is seen as a very significant factor and this has a profound effect of the technical design, construction and flexibility of the products. There is controversy surrounding thick versus thin client architectures. Not unrelated is the competition between the PC and interactive TV set-top box, for the home-shopping platform. There is also competition in the mobile market between telephone-based devices, personal organisers and palm-top computers.

Chapter 4: The Retail (eCommerce) Server

There are fewer options for retail servers than for terminals and they are all based on client–server models within standard *n*-tier architectures. The basic principles of Web servers are reviewed, including static and active page presentation, through CGI to Java and active components. Solutions to the problem of statelessness and the construction of shopping carts are discussed. In scaleable applications, Web servers need to interact with separate databases, with ODBC being an example of a reasonably open architecture. The performance of a server and how to measure it, is not a simple task, although most retailers will make use of Internet service providers. The demand for video clips and continuously streaming video-on-demand services requires new technology for high throughput and reliable mass storage. Design issues for these are considered.

1

The Principles of Electronic Retailing

The highest profile activity in *electronic business* is undoubtedly that part which is concerned with selling things to the general public. Sometimes this activity is completely identified with the term *electronic commerce*, which often, in turn, means 'selling things on the Web'. On the other hand, electronic commerce can be extended to cover the entire supply chain, thus giving rise to confusion with *eBusiness*. So, in this book we prefer to use the term *electronic retailing*, to describe those processes necessary to allow someone to view, select and purchase goods remotely, whether by Internet, telephone, interactive TV or other electronic media. Again, we accept that this term is not completely accurate, either, as much of what we shall say concerns business-to-business transactions as well as transactions to end customers, but the emphasis will be on the electronic delivery of the *shopping experience* by a number of channels. Although it does not cover all aspects of electronic business, electronic retailing provides a good introduction to many of the necessary components of electronic business as a whole.

1.1 BUYER BEHAVIOUR

Although this book is predominantly about technology, this must be built on a solid business model, and so we first spend some time trying to define the requirements for this shopping experience. Potential customers, whether electronic or otherwise, must be assumed to have some fairly consistent pattern of behaviour which determines how and why they make a purchase. There are many theoretical models of this process of moving through the purchasing cycle. One simple model, adapted from [1] is given in Figure 1.1.

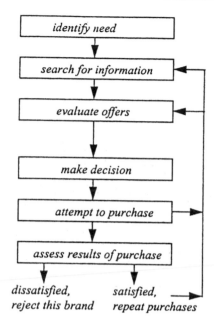

Figure 1.1   The purchasing cycle

This model is a useful skeleton for designing a sales and marketing process that covers each of these stages. It is beyond the scope of this book to cover them in detail. What we can do is relate much of this theory to a very simple model of the interactions that necessarily must be attended to, whenever a supplier wishes to support (or persuade) a customer into making a purchase. This, the 'C-SIT-F' model, proposed elsewhere by the author [2,3] is shown in Figure 1.2.

'C': we require a *channel* via which we can make the customer aware of our products and our brand. This can be a real shop window, an Internet

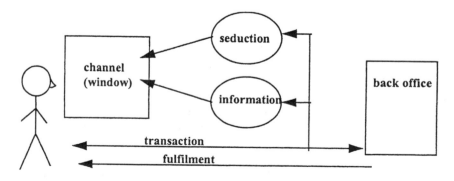

Figure 1.2   C-SIT-F: a simple model of shopping

page, etc. 'S': this channel must practise a level of *seduction*, to attract the initial attention of the customer. 'I': seduction on its own is insufficient. *Information* on price, functionality and availability is required in order to support the purchasing decision. 'T': a mechanism is also required whereby *transaction* of the purchase can be achieved. It may also be required for after-sales services that can differentiate the retailer. 'F': the customer must then receive the goods, through some form of *fulfilment* mechanism.

1.2 RETAIL TYPES AND THE C-SIT-F MODEL

Clearly there is a big difference between the way that 'real' shops display themselves in public market places, not least in the layout and branding of their shop-fronts (Table 1.1).

Banks make heavy use of their own logos and other own-brand identity; their window displays are 'information-based' with a plethora of graphs and figures. Department stores, on the other hand, may give some visibility to manufacture's branding and concentrate more on seductive product displays set within attractive surrounds. Sometimes the only other information given is that of price and size usually on discreet tags. Sometimes, in 'up-market' stores, not even that is on show. White and brown goods (washing machines, TV sets, etc.) are generally accessible to customers, to allow them, as much as possible, to assess the functionality or at least 'feel the quality', a mixture of seduction and information. The architecture and the colour schemes of food-stores are usually heavily self-branded. Inside, they make much use of lighting to present the product in the best possible way. Customers have access to perishable goods, whereby they can assess quality. Sometimes, even, smells, real or synthetic are used to evoke pleasurable responses.

Considering that all of this costs money and employs people with considerable talent, we would be foolish to write it off as irrelevant to electronic shopping, although some of it may be difficult or impossible to replicate. In the chapters that follow, we shall look at the technology components necessary to deliver these shopping experiences. First, however, we outline some approaches to describing the processes that shoppers might carry out, seen in terms of a technical architecture for shopping.

1.3 THE RETAIL PROCESS MODEL: THE DAVIC WISHLIST

Electronic shopping does not stop with the definition of a virtual shop-window. If we are going to create a complete retailing experience, we also need to consider how shoppers and virtual sales assistants can move

Table 1.1 Retail types

	Seduction	Information	Transaction
Retail banking	Low. Mainly for branding and 'lifestyle' presentations	Moderate–high. Facts and figures about money, graphical presentation	Very high. Banking is about transfer of data – sums of money and customer details
Food retailing	Moderate, creating the market-stall image for veg/fruit. Aroma of fresh bread	Low-moderate. Mainly prices and contents lists, increasingly recipes, possibly a need to 'feel the tomato'	Very high, low unit-priced items, regular price changes, loyalty point details
Consumer goods, do-it-yourself	Moderate. Life-style presentations (e.g. for sports goods)	Moderate–high. Information leaflets, demonstrations	Moderate, larger ticket items
Fashion	Very high	Very little text but need to feel the quality, try on the garment	Relatively low, moderate ticket
Entertainment, leisure	Very/extremely high. Largely it is the seduction of the product that is being sold. For music, quality of sound is critical	Often high, particularly in travel products where there is no possibility to carry out real inspections	Varies over the whole range. Holiday bookings are highly interactive; CD/tape purchases, etc. involve only payment transactions

through the store and interact with each other, the products and the transaction services that underpin everything. A lot happens in a real shop and there is a need to codify this behaviour before it can be implemented on a distributed computer environment. Some of this codification, particularly in the case of Internet-PC instantiations often appears to have been created on-the-fly; an alternative, more thoughtful approach has been worked out by the Digital Audio Visual Council (DAVIC), [4], an industry body concerned with specifications for interactive television. Whatever one thinks of the eCommerce potential of interactive TV versus the on-line personal computer, we can at least thank the promoters of the former for having laid-down a specification (or at least, a wish-list) for a set of properties they expect from an on-line shopping service (Table 1.2).

The DAVIC specifications are intended to cover the most complex as well as the simpler tele-shopping applications. Some of these are quite demanding and we have to remember that DAVIC comes at things, at least originally, from an interactive TV viewpoint, rather than from an Internet/Web approach. Interactive TV may lie on the seductive side of the SIT model. At least its roots are, based on the need to entertain and the rather lavish production budgets of film and television. (So, incidentally,

Table 1.2   DAVIC specification

DAVIC specification of tele-shopping functionalities

1. The system should permit a content provider to create a virtual store
2. The system should enable a content provider to determine the layout of the 'virtual store'
3. The content provider should be able to assign products to 'virtual departments'
4. The system should permit multiple items to be displayed simultaneously (e.g. for comparative choice)
5. The user should be able to place selections in a 'virtual shopping basket' prior to committing to purchase these items, maintain a record of total cost, and be able to adjust contents as better alternatives are found in other 'stores' or 'departments'
6. The system should enable a transaction to take place between a user and a product supplier
7. A user, within a tele-shopping environment, should be able to request exchange or return of goods
8. The user should be able to store/readily retrieve product information from one 'store' for comparison with offers found elsewhere (a virtual shopping list)
9. The user should be able to commit to purchase items in a 'virtual shopping basket' using a choice of methods of payment
10. The user should be able to amend an order already placed, or enquire of the status of an existing order
11. The system should enable an order placed by a user to be processed, and for the status of the order to be reported
12. The system should permit collaborative (group) shopping
13. The system should facilitate the use of intelligent agents (aware of user preferences and parameters) to locate items matching needs

is fashion-retailing, to which it bears more than a passing similarity.) DAVIC is also standards based and with a strong European 'thoughtful' approach, intended to cover all eventualities. The Web approach is, at present, anyway, pragmatic and sees HTTP/HTML/browser architectures as the way forward. If it cannot be done with this, then the application will not be done at all. Some of the DAVIC items do look rather difficult to achieve satisfactorily via the HTTP/HTML/browser approach.

Nevertheless, the DAVIC list does have the benefit of clarifying a set of aspirations for the shopping experience that we can offer up against the realities of the shopping platforms currently available or planned.

1.4 PLATFORM ARCHITECTURE

Although a number of technical solutions have been proposed in order to provide eServices that satisfy the shopping models described above or elsewhere, it is possible to represent nearly all of them in a simple high level 'architecture', as shown in Figure 1.3.

This breaks the platform into three components:

*   an *access terminal* that the customer can use in order to interact with the virtual store. This can be on the customer's premises, in a public place, inside a shop and so on;
*   a *merchant server* under the control of the vendor;
*   *network technologies* capable of transporting messages from both of these across the intervening distance.

Each of these three components imposes constraints on what can be

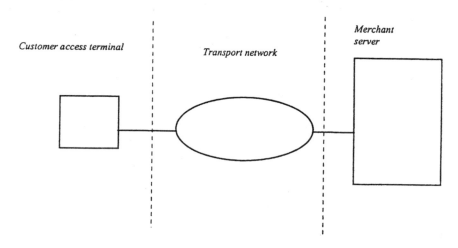

Figure 1.3.   A high level view of shopping architecture

provided – quality of picture, cost of service, reliability, etc. Some examples from each component will be natural partners with examples of the other components; some will be incompatible or at least a poor match. In chapters that follow we shall review a number of these individually and in combination. Notice that we described them as 'almost all' of the necessary components. They are highlighted because they represent the current high profile model of on-line shopping as essentially a product-display and purchase activity, over the Internet. We must not, however, forget that the real world of shopping – and the competitive thrust that will drive the electronic shopping revolution – is as much an issue of providing support services such as maintenance, help-desk and repair, as well as, of course, fulfilment and handling of returned goods. With that in mind, and as a subject for later chapters, let us now examine the components of Figure 1.3.

2

Retailing Network Technologies

The single most characteristic aspect of the eCommerce environment is distance: distance between each part of the supply chain, 'from melon to customer', and the single solution to this, is electronic transmission – how we connect the customer to the supplier. If we could do this, instantly, passing infinite rates of data, in both directions, at zero cost, then there would not be any more to say. But of course, we cannot and, as a consequence, this has given rise to a number of alternatives, each of which imposes constraints on what we can do, or afford to do. This in turn leads to a number of different services which have been proposed for eBusinesses in general and eRetailing in particular. Among these are basic wired telephony and computer networks and radio systems such as digital TV and mobile telephones. We have to accept that these technologies were originally designed for applications other than electronic retailing and it has therefore largely been shoehorned into existing solutions, notably voice telephony and one-way broadcasting. The future is brighter, however, and we shall explore some new methods.

Retail customers are markedly different from business ones in at least one way: they are intensely interested in price and, in particular, what they think they are paying, on a service-by-service basis. Businesses may be aware of the costs of their internal networks and their telecommunications charges, but their customers are not generally aware of how much of this is passed onto them within the overall price of the products they buy. But at the retail end, the connection between electronic shop and end customer, is paid for, at least in part, directly by the customer and becomes part of the latter's discretionary spend and subject to detailed scrutiny. Attitudes to these costs are also affected by fine details of the tariff arrangements: the use of a TV set, for example, is generally 'free' once the subscription or licence is paid, whereas telephony charges based on duration of call, are seen as a good reason to make as few calls as possible. These marketing decisions have a major effect on technology developments. For example, one of the strong drivers for developing networks on *packet* technology is that this is believed to be a much cheaper way of providing service than traditional *switched telephony*. Another example is *digital interactive TV*, which again is seen to be a less expensive way of customising programming to the needs of individuals. Imaginative service and pricing developments by traditional telecoms companies and new entrants are likely to be a dynamic activity in the near future and will have a lot of influence in the rate of growth of electronic shopping.

Noting therefore that the retail end of the transmission network has some definite peculiarities, this will be the main subject of this chapter. We shall leave inter and intra-business networks for later chapters, with one exception, that of networks for small suppliers, which share many of the limited capacity and low-cost characteristics of end-customer services. Also, some of what we say here is applicable to inter-business and, it also has to be remembered that, in the end-to-end delivery of a service, the performance will be determined by the summation of the performance of the individual links. Clearly, any overall limitation in service will be principally determined by the weakest links. As we shall see, often this will be the immediate connection from the customer

2.1   BASIC REQUIREMENTS AND LIMITATIONS

In terms of user requirements, we must first ask of our networks that they be reliable and 'secure enough' for our purpose. Given these essentials, we can then begin to negotiate about price, performance and quality of service, the last two generally being inversely related to the first. Defining performance and quality of service is not easy to do, and there are horses for courses, but we can try to summarise a number of generally applicable parameters:

- *Speed of access:* when we want to see or hear something, do we get it straight away, or do we have to wait several seconds for it to be delivered?
- *Interactive or one-way:* does everything come from the far-end without any possibility to interact with it? Do you press buttons to get data back to you? Can you interact, e.g. by voice, with someone at the other end without significant delay interposed in the conversation?
- *Quality of the signal:* is the sound quality clear and easy to understand? Is any print easy to read? Are images clear and of sufficient size?
- *Richness of signal:* is the signal text only or is it accompanied by audio and video? Is sound up to broadcast quality? Stereophonic? Of good

dynamic range? Is the video simply static images, a succession of static images, flickering and discontinuous 'moving images', TV quality images, three-dimensional?

- *Security:* how secure is the network? Not only against eavesdropping but also against people impersonating us or corrupting our data.
- *Mobility:* are we constrained to operate from one point or can we connect to the network at a number of points? Can we switch from one application to another without having to hang-up and reconnect?
- *How much does it cost:* and are costs at a flat rate or time/data volume dependent?

As we shall see, compromises on these requirements may need to be made and today's solutions may require a veneer of integration of a number of different solutions. It is not possible to set a single specification for a complete eRetailing customer interaction that covers the range from viewing the product to negotiating a purchase or reporting a problem to a help desk. In the first case it is likely that we would want high quality images but not be too concerned with high-speed interaction; in the second scenario, the situation would probably be reversed. However, experimentation and analysis would be the wise course of action in any real situation.

2.2   BIT-RATE AND LATENCY

In technical terms, the network requirements described in the previous paragraph really boil down to two factors: we must be able to pass data correctly at a sufficiently fast rate, (the *bit-rate* requirement), and, for many applications, the time taken to go from source to destination (the *latency*) must be below a certain limit. Both of these parameters must be truly end-to-end between the system providing the service and the eye, ear or mouth of the customer. Any time taken, for example to code up and decode video images, must be taken into account, and any overhead on the data, in the form of error-correcting codes (used over noisy transmission channels) must be deducted from the raw data rate. Table 2.1 contains an approximate specification for the *multimedia* requirements for a range of services.

The most demanding requirements for multimedia are those needed for transmitting real-time audio and video in both directions. Adequate speech quality can be obtained at a few tens of kbit/s. The bandwidth for video is higher: tens of Mbit/s are required for high quality TV, although *video compression* techniques can produce a quality comparable to decoded satellite broadcasts at 2 Mbit/s. This is also around the lowest acceptable rate for video conferencing (and we need this 2 Mbit/s for transmitting pictures from *each* conference studio). Smaller size video

Table 2.1   Multimedia requirements

Service	Bit-rate	Latency (delay)
Web page of text or small still image	A few kilobits/ second	A few seconds to retrieve a new page
Real-time audio		
Basically understandable (e.g. basic telephone quality)	A few kbit/s in each direction	Less than 200–300 ms
Seductive quality ('hi-fi')	Several tens of kbits/s to 1 Mbit/s	As above, if two-way, not critical if one-way
Animated graphics	A few kbit/s	Not really applicable
Basic moving images for information	0.1 Mbit/s	Not really applicable
Basic moving images for two-way video telephony	0.1 Mbit/s each way	Less than 200–300 ms
High quality seductive moving images	2 Mbit/s	Not really applicable

images (a window within a personal computer screen for example) that are a little jerky, with poorer definition and which tend to break-up if there is too much movement, can be delivered using as little as 64 kbit/s. A succession of still pictures, refreshed in less than a second, can be sent by domestic modem connection across the Internet, under certain circumstances.

Because of the need to conserve bandwidth, all practical coding methods for video, and some for speech, introduce a delay between the capturing of the picture by the TV camera and the generation of the bit stream. There is a further time taken to pass across the network and, finally, a further delay between the reception of the data and its conversion to a picture on a screen (Figure 2.1).

Some of this delay is simply due to the time taken for the coding algorithm to work, but some is also due to the fact that the algorithm involves

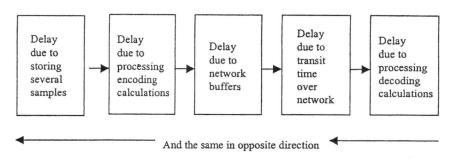

Figure 2.1   End-to-end delay

taking the difference between successive picture frames and only transmitting a signal representing the difference between the frames. This latter delay is obviously inherent in the system. Coding and decoding delay is often not a problem in the case of one-way reception (for example, in watching a TV film), provided, of course, the sound is delayed in synchronism, although delay of this nature can be, literally, fatal, if one were carrying out remote surgery, for example. A greater problem occurs when conducting two-way (or multiway) conferencing: the introduction of any delay beyond a few milliseconds in duration creates an artificial, stilted interaction.

In general, the strictest requirement for entertainment multimedia is in the need for a sufficient bit-rate to cope with high quality video and sound. Delay is not an issue because there is no need to get the content exactly when it happens. In interactive conferencing, the requirement is exactly the opposite: limited bandwidth is acceptable, but long delay is not.

Another difference between entertainment services and peer-to-peer bi-directional applications is in the relative costs of coders and decoders. Entertainment services can afford expensive encoding equipment for the relatively few coding centres, but require cheap decoders for the multiplicity of receivers (e.g. TV sets) that are bought in a very price sensitive market. Conferencing however, obviously requires symmetrical coding and decoding.

These differing requirements explain why today's solutions for entertainment and for interactive communication have tended to take different paths in trying to fit their signals into the very limited capabilities of today's networks.

2.3   LAYERED MODELS OF DISTRIBUTED SERVICES

When discussing distributed services, it is often convenient and informative to use a layered description which separates out the complete system into a number of levels or 'layers', that range from the most concrete voltages and signal shapes, etc. to the most abstract applications that are carried. The International Standards Organisation (ISO) has defined a set of *layers* for this purpose [5]. Their approach has been criticised by some as being too theoretical and too late, but it does provide a reasonable starting point. Confusingly, this is known, not as the ISO model but rather as the *Open Systems Interconnection (OSI)* Reference Model. The OSI layers are shown on the left-hand side of Figure 2.2.

The ISO layers are numbered from 1 to 7, as follows:

- Layer 1 *physical*: wire, fibre, etc. technology and the capacity of the system to carry data.

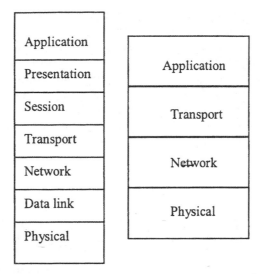

Figure 2.2 Two-layered models for distributed services

- Layer 2 *data:* the detailed structure of the data units as they cross a single link in the transmission path.
- Layer 3 *network*: the way that data is steered (connected) through the network.
- Layer 4 *transport:* the steps taken to make sure that the data is reliably transmitted once the route has been set up.
- Layer 5 *session:* the rules that set-up and close the data transmission.
- Layer 6 *presentation:* how the information is formatted together with any data syntax rules.
- Layer 7 *application:* the rules, procedures, interfaces that allow applications to interact with lower layers.

Notice that the right hand side of the figure also shows an alternative way of looking at things. People with a computer, rather than telecommunications, bias often prefer this model. Both views are tenable and an approximate mapping can be made between them, without too much trouble, although it is not exact. As this is a network chapter, we tend to prefer to use the OSI seven-layer model, but the four-layer version might be considered more suitable for the chapters that deal with computer platforms.

2.4   PHYSICAL LAYER OPTIONS AT THE CUSTOMER END

It is primarily the physical layer characteristics of any specific transmission system that sets the limits for the 'signal quality' we can achieve, in

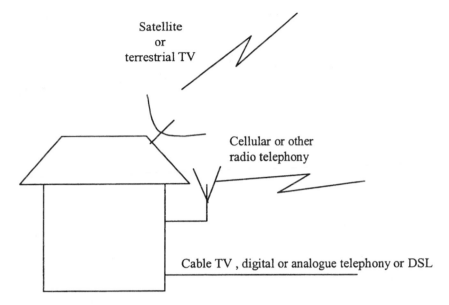

Satellite
or
terrestrial TV

Cellular or other
radio telephony

Cable TV , digital or analogue telephony or DSL

Figure 2.3   Network options for the domestic customer

the transmission of text, graphics, audio and video. The principal channels, which can be used to connect to a user of an eRetailing service, are
shown in Figure 2.3.

We see that these consist of:

- Telephone lines, (including *digital subscriber loop (DSL)* technology).
- Cable TV.
- Broadcast radio/TV.
- Cellular or other radio telephone networks.

Here we are concentrating on domestic users. Notice that we have not
made any mention of 'computer networks' as a way of delivering retail
information to customers. The fact of the matter is that, unlike the inter
and intra-business cases, the domestic customer end is still primarily
based on networks that either provide telephony or entertainment
services. There is no real 'computer network' at the physical layer.
Computer data is fitted into these using a variety of expedients, as we
shall see.

It is important to realise that within these categories there are a number
of variants and it is possible to use a mixture within any one instance of
delivery to a single customer (for example, broadcast a shopping channel
and receive orders over the telephone network). Also, it may well be
possible – and desirable to the retailer – to be able to deliver the same
service over more than one physical channel, for example to a PC and to a
digital TV set.

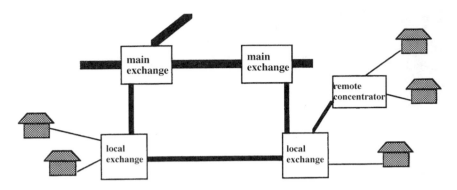

Figure 2.4 'Tapered star' telephony network

2.5 TELEPHONE LINES

Although we mentioned earlier that we need to examine the communication channel from end to end, in the case of the wired telephone network the principal bottleneck at the physical level is in the part between the customer and the local exchange. Figure 2.4 gives a simplified picture of a typical wide-area telecommunications network.

Conceptually the design is that of a *tapered star*. The *main exchanges*, the relatively small in number, big, central points that connect long-distance trunk cables together, also connect to more numerous *local exchanges*, which connect directly to customers (or, sometimes, to remote concentrators which are simplified exchange units). The cables that connect local exchanges to main exchanges, and the cables between main exchanges, are much fewer in number than the cables that connect individual houses. They also connect to fewer pieces of exchange equipment. It is therefore feasible to make them capable of better performance, in terms of greater data carrying capacity, than the local cables. Increasingly, the inter-exchange 'cables' are actually optical fibres. Of course, there is an economic limit on the performance one can afford to pay for on the long-distance portion of the connection, but, in terms of technical principle, the long range circuits are capable of transmitting many megabits of data every second. Moreover, as there are relatively few long-range circuits it is possible to upgrade them in line with new technologies much more easily than it is with the many millions of customer connections. Thus, telecoms network performance tapers from the centre out to the periphery.

The overall performance, including the data rate, provided to each individual household is therefore in many ways dependent on the *local loop*, the 'last mile' or so between the local exchange and the customer premises. In the sections that follow, we shall discuss the various ways that signals can be carried over this local loop.

2.6 ANALOGUE TELEPHONE LINES

The vast majority of domestic Internet users connect to the World Wide Web via their pre-existing telephone lines. These use copper cables whose performance is severely limited in terms of attenuation and noise at high data rates. The lines are optimised for analogue speech transmission. *Analogue* implies that the electrical signal on the line is an exact replica of the continuous changes in air pressure produced at the mouth of a speaker, rather than for the on/off signalling of discrete, *digital* bits of information. It is a reasonable approximation to say that the digital data from a computer is converted (*modulated*) into a series of analogue tones of various pitch and amplitude before being transmitted over the telephone network. The piece of equipment that does this (and converts it back at the other end) is, of course, the *modem (modulator-demodulator)*. By using fairly complex transmission schemes and also making use of the fact that most data streams contain recurring patterns which can be efficiently re-coded, it is possible to achieve data rates of up to about 56 kbit/s. This means, for example, that it is really not possible to send high quality moving images or high quality audio over standard telephony modems.

Telephony services in many countries are also provided by cable TV companies. Note that cable TV companies do not use their TV coaxial cable in order to distribute telephony. Instead, they offer it on conventional telephony wires that are usually installed at the same time as the coaxial cable is pulled into each subscribing household. Thus, telecoms networks provided by cable companies operate to the same specification as the telecoms networks described above.

2.7 CABLE TV NETWORKS

Cable TV networks, on the other hand, were of course designed specifically to permit the distribution of moving images into the home. Cable TV is distributed into domestic premises using flexible *coaxial cable*, which is quite similar to the cable that connects a TV set to an aerial or a satellite dish. The cable can handle a very high bit-rate, at least over the short distances involved.

Originally cable TV comprised simple one-way transmissions of multiple channels of TV identically into every home (Figure 2.5).

This is, essentially, 'TV broadcasting in a pipe', all the TV channels being broadcast from a central hub onto a master, rigid coaxial cable that was broken out into smaller, more flexible cables, with one-way amplifiers blocking off the possibility of return messages.

However, these systems are rapidly being replaced by new ones: a street of houses is served by an optical fibre which carries all the program-

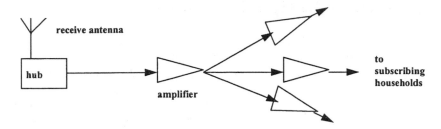

Figure 2.5 Early cable TV scheme

ming channels. A backward path is also provided on this fibre, or on a separate one. The fibre is terminated in a street cabinet, which holds electro-optical equipment that splits out the signals destined for each individual house. This individual traffic is carried into each house on a separate 'tail' of flexible coaxial cable. As well as having enough band-width to carry in the TV channels, the coaxial cable also can handle data to and from a 'cable modem'. This allows the householder to have very fast Internet access, using relatively cheap modems. One service is advertised as offering a speed at least five times faster than a telephone connection at a cost £169 for a modem and Ethernet card. It would seem quite feasible to increase that speed even further in the future. There are a number of complex business and technical reasons why cable companies have been slow to develop this market, but in the longer term, their cable networks do pose a potential threat to telecoms companies.

2.8 DIGITAL TRANSMISSION IN THE TELEPHONE LOCAL LOOP

To counter the threat from cable, telecoms companies have been interested in increasing the speed of data transmission over the telephony local loop. There are a number of different strategies that can be employed. The simplest is obvious: use more than one pair of wires, and this is exactly what used to be done by businesses which expected to handle a large number of calls. This technique has, however, been obsolete for a number of years because digital services have been developed that allow the simultaneous transmission of multiple voice circuits channels (typically around 30) on pairs of rather better quality cable, into business premises. (These cables have often, today, been replaced by optical fibre.) Essentially we can treat the business telephone network as being subsumed into the main *integrated services digital network (ISDN)* – the local loop essentially disappearing for large businesses. The data rates available to them can be as large as they can pay for.

For smaller businesses and, possibly, domestic customers, there is also a digital option, *basic rate ISDN (BRI)*. This offers customers a two-way, digital communication over their existing single pair of copper wires. Prices vary from country to country but the cost is approximately twice that of analogue telephony. For this, one gets the ability to set up two independent, fully duplex (i.e. working simultaneously in both directions) 64 kbit/s channels. These behave very similarly to conventional telephone calls: they have dial-up numbers and are switched through the network in identical fashion, with low end-to-end delay and guaranteed performance for the duration of the call. Call set-up and take-down is actually quicker than analogue telephony, of the order of half a second. Thus they can be used for quick 'click-through' operations, from an on-line retail service, to summon a live *pop-up operator*, to deal with a sales query, for example. Although each of the two 64 kbit/s channels can be used separately to make calls to independent called parties, they can also be made to a single party, with both channels being synchronised to deliver data within a specified maximum relative delay. This is convenient for a number of applications, such as higher quality video at 128 kbit/s. A variant, $N \times 64$ kbit/s, can extend this synchronisation to cover, typically, 6×64 kbit/s, channels. Moving video at this rate (384 kbit/s), whilst not exactly to full broadcast standard can be very good, even for cluttered, rapidly changing images.

As part of the ISDN standard, and supported in most countries, there is also an 8 kbit/s data channel. This offers a reasonably fast *packet-data* service, which may be tariffed by the telecoms company on the basis of number of bits sent, thus providing a lower-cost solution than a full 64 kbit/s channel. A possible application is for retail *point-of-sale* (POS) terminals. In the domestic case, it may be of more interest for telemetry (e.g. utility meter reading).

The growth in the use of ISDN has been steady, rather than spectacular in many countries. In the US, for example, there have been problems with different main-network data rates and from competing technologies. In the UK, there has been rather a waiting game played out between equipment manufactures and telecoms companies, with applications and attractive pricing both holding off before the other moved. Until quite recently, ISDN was marketed almost entirely as a business solution, with little attention given to its appropriateness for domestic customers. In the last couple of years, its potential as an access service for relatively high-speed Internet connection (perhaps four times faster than analogue modems) has been promoted. It is, however, under increasing threat from the new technology of *digital subscriber loop*.

2.9 DIGITAL SUBSCRIBER LOOP

Naturally, telecommunications companies have been concerned to counter the threat from cable companies who are not just selling access to
CATV, but are also frequently packaging it with telephony. Cable modems
for Internet delivery (see page 28) and interactive TV services are two
obvious markets that benefit from the higher bit-rates available over
cable. One solution, *Asymmetric Digital Subscriber Loop (ADSL)* was originally developed as a system for sending *Video-on-demand (VOD)* TV quality
images over the existing copper pairs of wires that carry domestic telephony. It has more recently been seen as a way of providing fast, very high
quality (e.g. moving images) access to the Internet.

There are two problems that have to be tackled if we want to send
higher speed data down the copper wire used for carrying local loop
traffic. Firstly, there is a limit to the bit-rate it can handle without the
signal dying away as the distance from the source increases. A signal at
1000 bit/s, is hardly attenuated at all after travelling a distance of 1 kilometre. However, a 2 Mbit/s signal, (such as would be needed in order to
deliver good quality moving images), will be reduced to one hundredth of
its size, or even less, over the same distance. This in itself would not be a
problem. It is easy enough to design amplifiers which can restore such
signals to their original level, but this is where the second problem with
copper pairs comes in: induced noise, particularly noise that is induced
close to the receiver and therefore not attenuated with distance. Amplification of distant faint signals is no longer the answer: the noise is amplified by the same amount and therefore still drowns it out. Copper pairs
are rather good at picking up noise from other sources, particularly those
on other pairs within the same multi-pair cable. This problem can be
reduced somewhat by twisting the pair of wires together (Figure 2.6),
but it cannot be entirely eliminated and unfortunately, it gets worse as
the bit-rate increases. Our 1000 bit/s signal will induce a version of itself
into an adjacent pair of wires, that is only a few thousandths of itself,
whereas a 2 Mbit/s signal is transferred across as several parts in one
hundred.

Looking again at Figure 2.6, we see receiver B situated at some distance
from transmitter A:

- B receives a much attenuated 2 Mbit/s from A.
- Close to B is transmitter C and the wires from C run alongside the
  wires from B.
- The signals in C's wires radiate slightly and induce a signal into B's
  wires.
- Because C is close to B, its interfering signal (called *near-end cross-talk*)
  will not be attenuated much and therefore completely drowns out the
  faint signal from A.

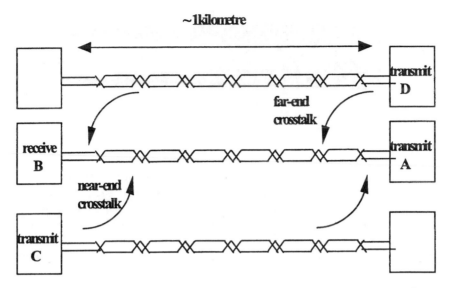

Figure 2.6   Near-end and far-end cross-talk

- Amplification of the signal received at B will therefore only produce a louder but no less noisy signal.

But consider transmitter D:

- It too is generating a small cross-talk signal into the pair connecting A and B.
- However, it is as far away from B as is A, and its signal will suffer the same attenuation as the signal from A, by the time it reaches B.
- So, the signal received at B will always consist mainly of A's signal and only an insignificant level of D's. (The noise from distant sources is known as *far-end cross-talk*.)
- In this case, amplifying the received signal at B will be effective as a way to receive A's signal.

ADSL (at least in its original design) relies on the fact that the signal traffic is always one-way – from the service provider to the domestic customer. In this case, as shown in Figure 2.7, the centrally transmitted signals and the cross-talk both die-off at the same rate.

Typically, bit streams of up to 2 Mbit/s, corresponding to an individual channel of TV quality, can be sent over a distance of several kilometres, covering the vast majority of connection links between local telephone exchanges and the customers they serve.

We can consider the signals on the wires to occupy certain *frequency bands* within the *frequency spectrum*, just like the channels employed by radio broadcasting. It is possible to use transmission techniques that shift

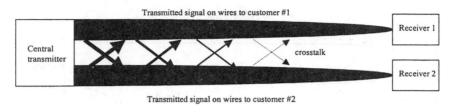

Figure 2.7 Principle of ADSL

the basic data signals into selected parts of the spectrum available on local wires, to meet our requirements. As shown in Figure 2.8, there is no interference from the two-way telephony signal, because the video signal operates within a frequency band above that of the speech. Similarly, a narrow-band two-way control channel, provided to allow signalling messages between client and server, is also out of band with both. (Two-way transmission is possible in this channel because it is at lower frequency than the high-speed channel and therefore does not suffer the same degree of attenuation and cross-talk.)

ADSL has been successfully piloted in a number of countries and its success has led to other configurations which permit higher bit-rates in both directions (either over shorter ranges or by using different cables). Telecoms companies see ADSL as giving them a low-cost, rapidly deployable solution to competition from cable companies for TV services and, perhaps now more importantly, a way of providing much faster Internet connections. (We discuss the ADSL/Internet server in *The Retail Server* chapter.) They are also looking at a variant of the cable company hybrid fibre-coax solution – hybrid fibre-copper – where there is a street feed over fibre, with the existing copper pair, rather than coax, completing the final

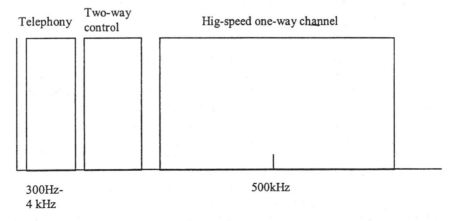

Figure 2.8 The frequency spectrum for ADSL, control and basic telephony

delivery to the premises. Standards for bit-rates of up to 51 Mbit/s are currently being defined.

2.10   BEYOND THE LOCAL LOOP: CONNECTING TO THE INTERNET

We said that the data bottleneck in domestic telephony was in the local loop and it is true to say that higher bit-rates can be provided in the main, inter-exchange networks. But this comes at a high price and long-range data connections are rapidly moving off conventional telephony-based solutions onto those involving the Internet. What we mean by 'the Internet' in terms of our layered model is explained more in *Part 2, eBusiness Systems Architecture*, where we discuss inter-business networks, but we need to consider some of the aspects here, if we are going to understand the connectivity between an eCommerce site and an end customer.

In that chapter we explain the way that, under the layer 3 (Transport Layer – see also page 24) *Internet Protocol (IP)*, data is split up into virtually autonomous *packets*, each carrying the addresses of the sending and receiving *hosts* (computers). The packets separately find their way across heterogeneous interconnected sub-networks. Each sub-network, perhaps a LAN, or even a bigger network is connected to another via routers.(Figure 2.9).

The routers manage the naming conventions for the individual hosts connected to the networks so that they are unique across the whole of the Internet.

This arrangement has to be modified in the case of a domestic user connected to an *Internet Service Provider (ISP)*via a telephone line and a modem because, in this case, the home device is not part of a LAN sub-net. Instead, as shown in Figure 2.10, the user is connected to a bank of modems in the service provider's premises which, together with associated software, can handle the connection as a standard telephone call.

For the duration of the call, the service provider assigns a unique, but temporary, IP address to the user's computer and informs it of that address, which is used by the user's browser in any communication

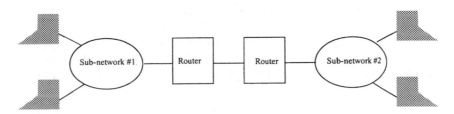

Figure 2.9   Connecting sub-networks via routers

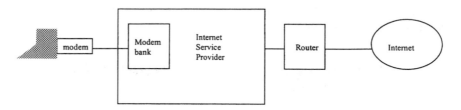

Figure 2.10   Connecting users to the Internet via ISP

with the Internet, for the duration of the telephone connection. Typically this temporary binding of an IP address to a host is done using the so-called *Point-to-Point Protocol (PPP)*. Some service providers provide an option whereby users can bind this IP address permanently to their account numbers; other providers do it as a matter of course. But it is not otherwise guaranteed. There is no certainty that a user, coming in via an ISP, will always present the same IP address to an Internet server. Clearly this has some significance with regard to the collection of server statistics and to customer management (see Part 4 *Electronic Marketing*).

2.11   RADIO NETWORKS

Although there have been attempts to introduce point-to-point wireless services in direct competition to fixed lines, these have not, at least to date, had a particularly significant impact even though they could, in principle, offer high-speed, say 2 Mbit/s, delivery. The basic principle is simple: small radio dishes are fixed to the houses of individual domestic customers. These dishes communicate bi-directionally over short range with antennae installed at local telephone exchanges or other points where access to wide-area communications can be made. A UK government auction of radio spectrum in the 28 GHz for this purpose had just begun at the time of writing and may re-invigorate the market for such services. Some companies in the US are also beginning to re-enter the market.

But today, radio-based domestic telephony is almost entirely dominated by cellular mobile phones. The number of Internet users has until recently attracted the greater amount of publicity; however, the market is beginning to realise that the growth in mobile telephony has been equally spectacular. By the end of 1999, about 37% of Western European residences had mobile phones, more than twice as much as Internet penetration.

The earlier mobile phones were analogue and used only for voice. Even today, data is less than 10% of the voice traffic. However, in Europe at least, the development of the digital *GSM* service offers the potential for integrated voice-data solutions that should have a major impact on eBbusiness. GSM in its original form follows conventional telephony principles, apart from the obvious enhancement of mobility [6]. Telephone calls are switched through the network, in the same way as line-based services – in fact, for most of their distance, they make use of the line-based networks. Indeed, for most eBusiness voice services (e.g. in call centres) mobile and fixed services can be treated as identical, apart from tariff differences, and these are likely to converge.

Data was initially carried across GSM in the voice channel, at a maximum rate of 9.6 kbit/s. Recently, with the launch by some mobile service operators of an improved radio transmission scheme, the *HSCSD service* has increased this to 28 kbit/s, and with data compression, this can be extended to 57.8 kbit/s. This still uses the telephony channel and incurs 'telephone-style' call charges.

2.12 PACKET-BASED RADIO

There is considerable work underway intended to bring to market alternatives to the rather expensive switched-telephony approach to mobile data. In particular, there is a desire to base tariffs on the amount of data sent rather than connect time. An early solution is the GSM *short message service*, which as its name implies, transmits short messages keyed in using the normal mobile keys or by a full alphanumeric keyboard. It has become much favoured by school children as a way of passing messages between friends, which goes to show that user-reinvention may provide new uses for products, beyond the imagination of their inventors.

Radio data services will be much extended in the near future with the introduction of the *General Packet Radio Service GPRS* [7]. Figure 2.11 demonstrates the principle of operation between the mobile units and the radio base-station.

The base-station makes available to all users a single bunch of eight radio channels previously used for voice calls. To the user, these channels behave as a single channel of eight times the original capacity, giving a maximum total capacity of around 114 kbit/s at any one time. Any user within radio range can send traffic to this channel (there is an identical and separate receive channel as well).

However, the most significant difference between this approach and that of existing GSM is in the use of packet transmission. Users send and receive data packets to this virtual channel as and when they wish, up to the total capacity of the channel. It thus behaves rather like a slow Ethernet in the sky. Mobiles clearly need to be specifically designed for transmitting and receiving on this channel and of course, the radio base-

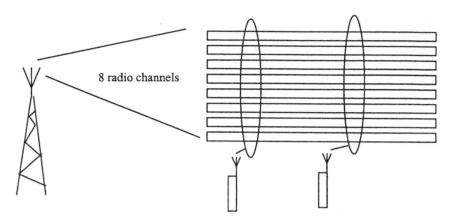

Figure 2.11 GPRS radio principles

stations have to be modified, but relatively little else has to change in the existing mobile network, apart from providing an IP backbone network (Figure 2.12).

The location of the user is found as with conventional mobile telephony: mobiles continually send their whereabouts to a *home location register* which can be queried in order to route the packets through the central networks.

By the beginning of 2002, with the installation of new transmitters and receivers, GPRS should be enhanced into *EDGE*, with a theoretical potential of up to 384 kbit/s, but this will be location and service-supplier specific. Since these are packet services, (as explained previously), Quality of Service will vary depending on instantaneous demand. GPRS is intended to meet the needs of customers who require reliable data

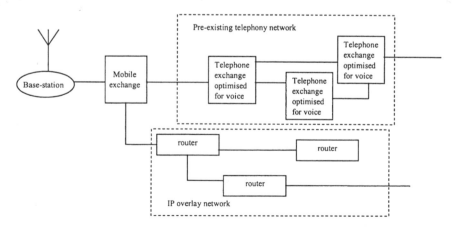

Figure 2.12 Providing mobile packet services

services at minimum cost. It is primarily adapted to small bursts of data of a few hundred bytes occurring several times per minute and to less frequent data blocks of several kilobytes. These data patterns are based on the assumption that its main applications are as wireless PC office activities, electronic money transfer, traffic reports, fleet management and *Service* [8].

GPRS has some interesting service features, including:

- Point-to-multipoint, which broadcasts to all, or a selected group, within an entire area;
- Group call, sent to a selected group of users, wherever they are.

There are clear opportunities for this as a field-force control mechanism, or for the rapid and simultaneous broadcast of premium information to a group of subscribers. (Stock market, re-routing service in case of road obstructions, etc.) It also might offer a simple way to up-date pricing on remote vending machines or utility meters.

Since information is sent as *packet-data*, transmitted in bursts, the cost is cheaper than that of a full telephone call. We discuss details of packet protocols in the next few sections and in the chapter on *Retail Terminals* where we examine the particularly important *Wireless Application Protocol, (WAP)*.

2.13 UNIVERSAL MOBILE TELEPHONY SERVICE (UMTS) AND MOBILE DATA

The ultimate goal of today's mobile telephony strategists is the *Universal Mobile Telephony Service (UMTS)*,which is intended to shift mobile telephony from a connection-based, telephone architecture to a computer-network based one, using extensions to the *IP*. It has to be said that there are some uncertainties in the future direction and timeliness of UMTS, not in the least because of the parallel development of GSM [9]. One thing is certain, the terminals for UMTS will not be limited to simple telephones, but will also include laptop computers and other devices.

Not unnaturally, UMTS is very concerned with the radio or *air interface* aspects, in order to achieve high data rates and effective hand-over from one radio cell to another as the mobile moves. This has involved a great deal of standardisation effort across the world, further accounting for relatively slow progress [10].

Indeed, the issue of providing effective mobile data services is proving quite complex. There is even a problem with definition of the term. For example, the mobile IP working group of the Internet engineering task force distinguishes between simple *nomadicity* and *true mobility*.

In the case of nomadicity, all that is required is the ability for someone located other than at their home site, to be able to re-start all applications at this site and then close them down before moving. This can simply be achieved today, by remote log-in to ones home server or Information Service Provider. True mobility, on the other hand, requires the ability to cater for some additional requirements:

- Providing a way whereby a mobile terminal can continue to communicate with other terminals and servers although it has moved its point of 'attachment'. (Attachment can be by physical plug-in to a wired network or by setting up a radio path to a new base-station.)
- Allowing the terminal to keep a single, unchanged IP address (its *home address*) even though it moves about and still carries on the session with others.
- Communication should be possible with static units that use IP but do not use mobile IP.
- Avoiding introducing additional security weaknesses.

Currently, the only truly mobile networks in terms of the IETF are 'telephone' ones, based on telephony networks that provide data paths within the voice channel. However, the IETF have begun to create a specification for how IP would work in a mobile environment. They admit that solving the mobile IP problem is only part of providing a highly satisfactory solution to the true mobility problem: for example, there are also benefits to be gained by improving TCP and other Internet protocols concerned with maintaining individual parts of the link. However, they believe IP to be the critical area and have made good progress. The details are fairly complex and their full description would not be appropriate here. They are fully and clearly explained in an excellent book by James D Solomon, the co-chair of the IETF working group [11]. We can only hope to give the briefest outline of the problems and the proposed solution. The approach is to enhance the performance of the mobile's *home router* (notionally, mobiles have a static home base where they can be registered), so that it takes any packets sent via it and destined for the mobile unit, and encapsulates them in another packet which is routed to the current address of the mobile (Figure 2.13).

The home router is constantly kept up-to-date of the mobile's position, the latter transmits this information to the router each time it shifts to another link. (The mobile knows which link it is on because, there are routers on the link which constantly broadcast the name of the link.) In fact, the process of identification and routing is very similar to that used by switched mobile voice telephony, (except of course that no actual connection is permanently established and control messages and data do not travel by separate paths).

Finally, we should mention that the aspiration of UMTS is to offer data rates of up to 2 Mbit/s, although it is probably best to consider mobile services as being relatively narrow-band and not intended, for some time

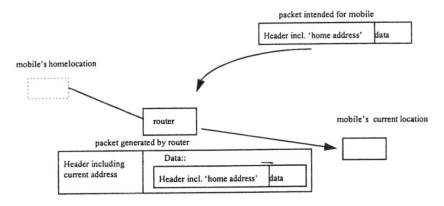

Figure 2.13 Principles of mobile data routing

to come, to be suitable for very high-speed Internet services, or, all the more so, for moving video. Their great strength is indeed their ability to provide services anywhere and on the move.

2.14 BLUETOOTH

The solutions mentioned above are designed for radio communication across distances of up to several miles, or even tens of miles, and with the signals thereafter transmitted across conventional line-based telecommunications networks. It has been felt that a number of valuable applications would also benefit from short-range communications between hand-held devices and other terminals (Figure 2.14).

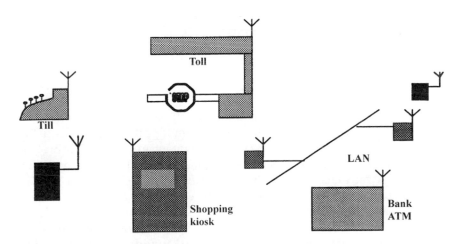

Figure 2.14 Some applications for short-range radio

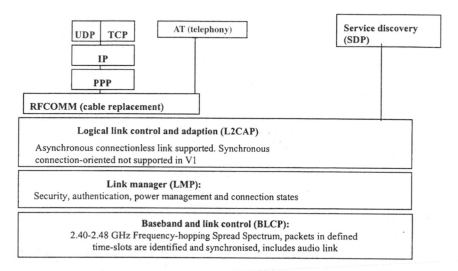

Figure 2.15 Part of Bluetooth and associated protocols

For example, we can instance the transfer of electronic cash from a smartcard-based mobile phone to a dealers radio-enabled cash register. For this and other applications, the industry has developed *Bluetooth* [12]. This comprises radio standards and protocols covering several layers of the ISO stack.

At the physical layer, Bluetooth operates in frequency bands of 2.400–2.4835 GHz (US, most Europe), 2.445-2.475 GHz (Spain), and 2.4465–2.4835 GHz (France). It has a range of tens of metres and a data rate of around 1 Mbit/s. Figure 2.15 shows how Bluetooth is layered and how it relates to some existing protocols.

A particularly significant aspect of Bluetooth will be its ability to carry WAP services (see Part 1, Chapter 3: *Retail Terminals*).

2.15 BROADCAST TV

Television is already used as an eCommerce channel into the home, in the form of advertising. Clearly, it offers a medium for providing high quality, seductive vision and sound, certainly equivalent to at least 2 Mbit/s in terms of information rate. What has been lacking until quite recently, has been a backward channel which allows rapid conversion into a transaction and perhaps a sale. However, the growth of, first analogue satellite channels, and then digital, satellite and terrestrial broadcasting, has led to development of subscription or pay-per-view. Because this requires a mechanism for payment extraction, service providers have encrypted their offerings and provided the customer with some form of decryption equipment (usually within a *set-top box*), which, conveniently can act as a host for creating a backward channel. The simplest way to use such a box is to enable it to make telephone calls to a payment centre for pay-per-view purposes, but other alternatives are also available. We look further at the set-top box in the chapter on *Retail Terminals*. What is important to note here, is that interactive television may provide a serious rival to the home computer in the eCommerce environment, being cheaper, more user-friendly and possessing a high-speed download channel.

2.16 SATELLITE BROADCASTING AND NARROWCASTING

Television services are provided to domestic customers by terrestrial and satellite transmitters, with negligible difference in performance. Virtually all domestic radio telephony is, however, only distributed using earth-bound transmitters and receivers. In recent years there have been attempts to change this by using *Low Earth Orbiting Satellites (LEOs)*, of various degrees of complexity, with aspirations of providing interactive data as well as voice [3], with service characteristics similar to that of terrestrial mobile systems. None of these have yet proved to be cost-effective. Part of the problem with LEOs is that the satellites are not *geostationary*, that is, they do not maintain a fixed position overhead. Thus adding an additional complexity to mobile operation in that traffic has to be switched on a regular basis between satellites whether or not the communicating parties are moving. Other problems include the relatively large size of the mobile units and some doubts about performance during heavy rainfall.

Partly because of the emergence of LEO systems, there has been renewed interest in extending the capability of high altitude *geostationary satellites (GEOs)*. In the past, these have been predominately for broadcast purposes or for intercontinental trunk telecommunications but some people believe they could also be used for interactive or Web-TV. One problem with GEOs is with the transmission delay, of the order of 0.5 seconds there and back. This limits the data rate of TCP/IP connections to around 200 kbit/s where hosts are set to their default options [3]. This is not insurmountable, but makes GEOs less than an ideal, off-the-shelf solution for high-speed Internet services.

2.17 SUMMARY OF TRANSMISSION OPTIONS

We have seen that a number of options exist for connecting up retail

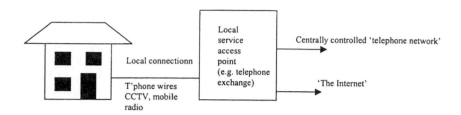

Figure 2.16 Local loop and wide area networks

customers. We now try to summarise these in terms of the various service requirements for eCommerce. It is important to remember that the trans-mission network has short distance (local loop) and long distance (wide-area or trunk) components (Figure 2.16).

There are many options for the local connection, which tends to be the bottleneck for overall data rate. In terms of service quality, the long-range part of the network has two principal alternatives: a well-specified, centrally controlled telephony-based network which can provide guaran-teed data rates with very low end-to-end delay, at a price, or a much cheaper packet-based Internet option, whose overall speed and end to end delay are much less rigidly defined. Thus, although we have main-tained that the theoretical limitation to data-rate, etc. lies mainly in the local loop, we must not neglect the ad hoc impairments and restrictions imposed by heavily loaded wide-area Internet paths. The overall perfor-mance of these paths will be determined by the performance of all ISP's involved in the chain between customer's terminal and on-line merchant's server. These are in turn determined by commercial factors, which then rely on the cost of providing adequate circuit and routing capacity. This will increasingly be an issue once high-speed delivery over the local loop, using cable modems, ADSL or wideband radio links, becomes a reality. In practice today, it is usually the speed of the local connection using analogue modem techniques that limits the aver-age speed in retrieving data, but the variable delay in the long-distance Internet connection (combined with local loop slowness) that rules out interactive voice and video.

Interactive voice/video services are today still best met by end-to-end use of telephony networks. In the case of voice-only applications, fixed line and radio mobile provide broadly equivalent performance. The slow bit-rates available from mobile services really precludes their use for video telephony, for which fixed line, basic rate ISDN is probably still the best option. Packet-based voice telephony, which relies on uncongested passage across IP networks by service providers is still of poor quality, but may ultimately be the way to provide interaction between humans in an on-line transaction.

Seductive multimedia has its benchmark standard in broadcast TV, which

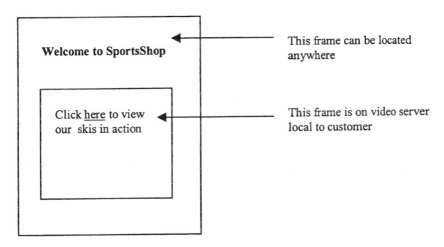

Figure 2.17   Inserting video-on-demand clip in a Web page

today predominately uses terrestrial/satellite radio broadcast or cable transmission. In the near future, telecoms companies hope to be able to provide equivalent performance using ADSL technology over existing telecoms wiring in the local loop. Video-on-demand services, where each user can select their own viewing choice and time of viewing, will, of course, require storage of the programme material at or near the local loop of the customer, to avoid the need to shift large volumes of data across the wide-area in real time.

Web services (which are generally seen as the major growth area) are more difficult to define and therefore their transmission options more varied: if an interactive service, such as on-line voice is required, then consideration must be given to end-to-end delay. If all that is required is to pull down pages of text and static images, then all the options mentioned in this chapter can be used, the major performance metric being data rate. Where high quality multimedia is required, then this might need to be *cached* locally on a *video-on-demand server* (see the *Retail Server* chapter). Figure 2.17 shows how this could be done in a framed Web page.

The 'welcome page' is hosted on the vendor's server, which could be anywhere in the world. It does not require too fast a data rate to supply basic text and graphics. However, the inset frame is intended to give a seductive movie which requires high-speed data transmission to the customer, in order to give the desired effect. Clicking on the '*here*' link connects the customer to a local video server which supplies the video data over an ADSL or CCTV local connection.

3

Retail Terminals

All trade requires some form of physical outlet where customers can see the goods set in an appropriate setting, can find out information about them and conduct a purchasing transaction. This is true even for electronic shopping. What is surprising, perhaps, is the range of customer access platforms or *terminals* available. In this chapter, we shall look at traditional methods of eShopping based on the telephone and fax machine, the emerging market for kiosk and personal computer terminals and the prospects for interactive television and mobile terminals.

3.1 THE TELEPHONE

In the past, mail-order was the mechanism for remote shopping, it is likely that a large part of its future lies with the Internet, but today the majority of remote shopping transactions are carried out over the telephone. Measured against our SIT model of shopping (page 12), telephony provides customers with a very convenient way of conducting *transactions*. It is not so good for rapidly gaining *information* and 'persuasion' rather than *seduction* is its main outbound sales technique. However, paper catalogue or magazine advertising can supplement telephony's deficiencies in information and seduction and, as we shall see, telephony can easily be combined with TV to provide a highly effective channel.

Today, catalogue shopping provides most of the reality behind the term 'home shopping', providing around 5% of all retail sales in the US and UK, exclusive of automobile sales. Until quite recently the ordering transaction was based on mail services and conducted via an agent, who usually made a commission on these sales. In recent years, the agents have virtually disappeared and telephoned orders make up the majority of purchases. People have taken to direct placement of their orders, whereby they are able to speak to a human agent at the other end of the

Table 3.1 Questionable use of multitone calling

Bear-faced robbery?

This ingenious, although ethically questionable, sales method was reported as
happening in the US, a few years ago: the scene consists of a small child watching
television, unattended. An advertisement promotes a lovely, cuddly teddy-bear.
'You can have this bear', a voice croons. 'All you have to do is pick up the
telephone and hold it near the TV'

The child does so and, after a few moments, the advertisement plays out the
multitone dialling code of the supplier. This sets up the call to the supplier and,
since most US telephone systems have calling line identity, the identity of the
caller is passed to the supplier. The supplier now uses an inverse directory to
convert the number to an address and mails off the bear, with a 'return if not
satisfied,' invoice enclosed. How easy will it be to part the bear from the child?

telephone. They are also increasingly happy to use voice response
systems which, perhaps combined with the tone dialling keypad of the
telephone, allow them to place orders without human operator intervention. A rather sharp use of multitone dialling was reported from America,
a few years ago, Table 3.1.

The medium for handling telephone orders is, of course, the *call centre*,
which we discuss in Part 4, *Service and Support*.

3.2 TELEPHONY SERVICES FOR HOME SHOPPING

Over the last two decades, telecoms companies have begun providing a
number of features and Value Added Network Services (VANS), many of
which are directly applicable to home shopping:

- *Multitone dialling:* this was mentioned above. Most developed countries now have a significant penetration of tone signalling from the
  telephone for dialling. This can be used, once the call is completed, as
  a way of controlling equipment such as the order-placing system
  described about. However, many countries in Europe and other
  parts of the world still also have a significant number of so-called
  'loop-disconnect' dialling telephones, which operate by opening
  and closing a switch which interrupts the electric current in the caller's telephone line. This interruption is recognised at the caller's local
  telephone exchange but, unfortunately, is not passed all the way to the
  called party's line. All that can be heard are a set of 'clicks' which
  cannot be reliably decoded into a set message.
- *Calling line identity (CLI):* this is a potentially very useful network
  feature which can provide the vendor with immediate information

on the caller's telephone number. Again unfortunately, it is not available in many countries and in some countries, for example the UK, it is not passed between the different telecom companies' networks for technical and, possibly, commercial reasons. It is also supplied at the caller's discretion – callers can withhold it by prefixing their dialling with an agreed code. Consequently, any eCommerce design that makes use of CLI must also be capable of exception-handling the cases where it is not present. Furthermore, for reasons that are probably accidental rather than planned, CLI is more frequently provided from domestic, or at least direct-exchange-line, customers, rather than from the Private Branch Exchanges (PBXs) of large companies. This is the case where connections are over the 'PSTN' not where an 'ISDN' end-to-end connection (see page 29 for explanation of the difference) is involved. In the latter case, the caller's identity can be passed on. Finally, before leaving the subject, we can note that the technology used for sending the CLI data can also be used to carry out other functions such as telemetry. We shall look at uses for this in Part 4 *Service and support*, when we discuss customer support.

- *Free-call, low-call and premium rate services:* these services have been provided by telecoms companies as an overlay to their existing networks and will be integrated into the emerging development in telecommunications as part of the *intelligent network* concept. We discuss the intelligent network in more detail, in Part 4 *Service and support*, in the context of call centres. These services are quite simple in principle although providing them nationally, perhaps internationally, is a major upgrading task for the telecoms company. A customer dials the special service number (800 for free-call, etc.) and the remainder of the digits. At a point quite close to the caller's local exchange, the dialling is recognised and the call is routed onto a series of paths through the network to arrive at a destination agreed between the telecoms company and the company who purchased the service. It is possible to arrange for this destination to be varied according to time of day, location of caller, etc. perhaps to assist with load handling across the company's call-centre agents. It is also possible to route calls made to different service numbers to a single destination, Figure 3.1.

When the call arrives at the single destination, it is accompanied by information provided by the network, concerning the *called* (not the caller) number. This allows the answering service to know which company representative the caller wants to contact, and the answering service can then take on that role – double-glazing, travel-agent, whatever.

Telecoms companies also have a lucrative market in selling easy-to-remember numbers, particularly when they are evocative of the seller's brand. In the US, there is the advantage of standardised positioning of

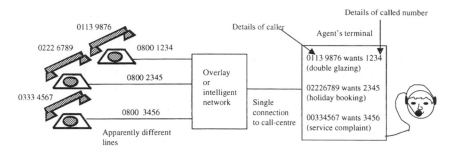

Figure 3.1 Call discrimination facilities

letters of the alphabet against the numerical keypad on telephones. In the UK, the standardisation was allowed to lapse many years ago, when the increase in the number of telephone exchanges meant that they could no longer be given names rather than numbers. Since that time, telephones have been sold in the UK with varying arrangements of letters against the numbers and the opportunity to use consistent alphabetic dialling may have been lost for ever. Companies must therefore try to make do with numbers. Some have been extremely successful in this. The Forte hotel chain successfully improved brand recall and total number of calls to their reservation service, by positioning it behind a single, free-call number: 0800 40 40 40. (Oh eight hundred, forte, forte, forte.)

Another aspect of special rate services is how the tariffing rate can be used to say something about the service, and even as a means of collecting revenue. Basically, there are three rates: 'free', 'local' and 'premium'. In the mind of the customer, they may evoke the following impressions:

- *'free'* – this is an advertisement or an obligation on behalf of the retailer (e.g. a 'within-guarantee period' help-line).
- *'local'* – a factual information point, or perhaps a concession, from the vendor to geographically distant customers, making them feel valued as much as local customers. The retailer is trying to cover its cost rather than selling an added-value service.
- *'premium'* – customers paying at premium rates for this service expect to be provided with something of value.

Vendors need to position their offerings in accordance with these customer perceptions. It is generally recognised that information services have got to be just that – not thinly disguised advertisements [13]. However, it may be worth exploring customer attitudes to individual offerings through focus-groups or test-marketing: there are instances of users being prepared to pay premium rates in order to get product information, particularly where different models are compared. Interestingly enough, this does not appear to be true for Internet services, which are expected to be free. Handy hints and do-it-yourself repair information can further enhance this sense of value, whilst at the same time reducing the vendor's cost of answering queries. An on-line service, telephone or otherwise is much easier to sell to the board if it can be positioned as revenue earning (or at least cost-recovery) rather than as a pure advertising loss.

A significant added value of premium rate services to the retailer is as a mechanism for collecting revenue. The telecom provider handles all the charging and billing for use of the service. In past years for technical, regulatory and business inertial reasons, the charging mechanism was inflexible. The premium rate was set at a single, call-duration rate above that of a basic call and the excess revenue thus gained was split between the telecoms company and the vendor. This was not ideal for the selling of information, particularly valuable material, because it meant that charging was based on length of call, rather than on value imparted. This led to a number of stratagems for increasing call connection time: lengthy 'teasers' and convoluted navigation through multiple selections using touch-tone keying; in the case of information retrieval via fax machines, artificially slowing down the transmission rate. This was condemned by one telecoms regulator, (wrongly, in the author's opinion). In recent years, the three inhibitory factors have been eased and a more flexible charging regime has emerged. However, there still remains in the minds of some customers that some of these are scam practices used to promote tacky services.

Although it is not a pure 'telephone" service, we also should mention *Internet callback applications*. This is a method which combines the advantages of Internet services and telephony: on-line customers accessing a company's Web site and wishing to be contacted, can complete a form which includes their telephone number. The form is routed by the company's server to an automated but manned reception point, for example a call-centre. The call-centre agent can then make an outbound call to the customer. This is actually cheaper to the company than having an 800 freephone number available to customers. The sales dynamics of this approach are slightly different from in-bound 800 calling. The customer is relieved of the need to go to the telephone – and from the spontaneous impulse of a phoning act with instant response – but eased into a less stressing act that might elicit more information from the form. Market testing is the only way to decide whether this is an effective approach.

We also should note that premium rate numbers can be used for charging for Web-based services. As we mentioned in the chapter on *Retailing network technologies*, 'Internet' connections from private customers usually use the conventional local telephone network. The telephone connection charges that are charged by Internet Service Providers are usually set as low as possible, in order to encourage users to access the service, but it is also possible to use higher rates as a way of collecting payment for added value services. One typical example is in browsing on-line maps. Customers access the Map-provider's home page using the normal low-cost tariff. Once they have made a selection however, a piece of software is automatically down-loaded into their PC. This closes the current connection over the telephone network and re-dials a new connection, this time on a premium-rate number. Thus, while the user is examining the on-line map, they are paying a telephone connection charge at a higher rate. The excess charge is the fee for using the service and is provided to the service provider by the telecoms company, which bills the customer as part of their standard telephone bill.

3.3   SCREEN AND KEYBOARD TELEPHONES

There have been several attempts to launch telephones which have screens and/or keyboards. None of these attempts have, so far, met with much success. In the 1980s, some telecom companies marketed a product which was essentially a remote 'dumb' keyboard which could be used to input data to a remote computer, using an integrated modem. Usually the products also had a simple VDU screen for reception. (Providing a printer was too expensive.) The one notable exception to the general failure, was the French Minitel approach, where telephone customers were provided with a free keyboard/text-only display. As a free service, this has proved to be quite successful.

The development of low-cost liquid-crystal and other text-only displays has also led to trials of screen-phones, built to a price (typically below £50), for a range of possible applications, notably telebanking. Such telephones were, for example, used for Mondex electronic cash trials (see page 290). Again, the take-up was disappointing, probably because the applications on offer were not sufficiently valued to persuade the customer to purchase a screen-telephone at double the price of an ordinary one. The rise of eMail, as a valued application, has tempted telecoms companies to try again with marketing keyboard/display telephones which can act as (very) thin clients to an eMail service. These products currently retail at around £70. This eMail solution may be seen as low-cost in the customer's eyes when compared with the alternative of a PC, and its ease-of-use may be also an attractive feature. However, its functionality is very restricted compared to a PC, and mass-market eMail has yet to arrive as a killer application. The jury is still out as regards to its market potential. If it does become accepted, then the obvious retailing opportunities are in transaction-intensive, rather than seductive, services or even information services requiring display of volume data: banking, ordering goods not advertised on Web pages, regular shopping lists, etc. Interestingly enough, mobile telephones are increasingly acquiring data and keyboards services (as we shall see later in this chapter) and already have small screens. Whether the success of these will create market pull for fixed line screen and keypad telephones is an open issue.

3.4 FAX MACHINES, TELEX

Interactive voice is not the only service supported by the fixed telephone network. Typically, five to 10 percent of network revenue comes from facsimile traffic. There are over one million fax machines in the UK alone; they therefore have sufficient penetration to be an effective platform for eCommerce and, indeed, they are used as such in a large number of inter-business transactions. There is even a large network of world trading still carried out using the switched telegraph (*Telex*) network, which only provides basic text, but usually very reliably in parts of the world otherwise lacking infrastructure. The market for truly domestic use of fax is very small and most of what follows is really concerned with inter-business or small business use.

'Junk-fax' is a pejorative term for a rather generally unwanted service, and covers the practice of faxing out unsolicited information, generally basic advertising material, to prospects who have unwittingly had their fax numbers disclosed. Improperly managed, fax-marketing can lead to a large number of customer complaints, but, used in a sensible way it also can be a very effective, low-cost service to administer and deliver. Modern fax machines can be programmed with a string of numbers, to allow the batch mailing out faxes automatically, at low tariff, off-peak times. Machines will make repeated tries to each number and produce logs that record the success or otherwise of the call.

An increasingly adopted alternative is to use a computer to generate the fax message and then a fax modem card, either in an individual PC or on a LAN network server, to dial the call into the telephone network and transmit the data. This is an attractive method for dealing with a range of suppliers, some who have PCs and others who only have fax machines. This is a common situation with the global purchasing of small ticket items where a network of relatively small individual suppliers is involved. It is also a very convenient way of trading with suppliers whose languages require unusual character sets. Indeed, in dealing with some countries where the packers cannot be assumed to be literate, a faxed picture of the goods required is the only way to ensure correct completion of an order.

Food retailers, in particular, have the problem of trading with a wide range of suppliers, with varying degrees of sophistication. One option that has been adopted by a number of them is to use software that acts as a gateway between combinations of *Electronic Data Interchange (EDI)*, telex and fax. Standard EDI purchase requests are generated by the retailer's computers and sent directly to those of its larger, more modern suppliers or converted by the software into formats and transmission protocols suitable for the telex or fax machines of the smaller suppliers.

Paradoxically, although in the longer term this application of fax use is clearly under threat from computerisation at the purchasing end, in the short-term it is likely to benefit more from the growth in Internet services than be threatened by them. Since fax messages do not need to be sent in continuous, real-time manner as is provided by the expensive telephone network, they can instead be sent via an Internet connection at considerably reduced cost.

A slightly different use of fax is in serving in-bound requests for product information and order taking. Here the selling company maintains a computer-based catalogue of product information. By one of several means, (advertisement, targeted mail-shot, leaflet from sales representative, etc.), a potential customer is given a telephone number they can contact for information. The contact is achieved by the customer's fax machine, having first been set to *polling mode*, initiating the call. It is then in a position to receive the relevant information, rather than itself sending a fax. In effect, it behaves like a printer, with the selling company's computer behaving like a sending fax machine. Selection of the required material (a specific product brochure for instance) can be achieved by the very simple expedient of providing different dial-in numbers for each, where the selection is not very extensive. Where a more sophisticated facility is required and the data can be hosted on a computer, then making use of the touch-tone keys of the fax machine, perhaps in conjunction with a voice dialogue, can effect a very flexible solution.

It is even possible to fax computer-readable information from a fax machine to a computer, for example, a completed order form (Figure 3.2).

The selling company's computer transmits, on request, an order form to the customer. The customer can write on this form as instructed either making a simple mark across a point in the fax or by writing text within a defined area. When this is faxed back to the seller, its computer can locate the marked box and also carry out a simple character recognition algo-

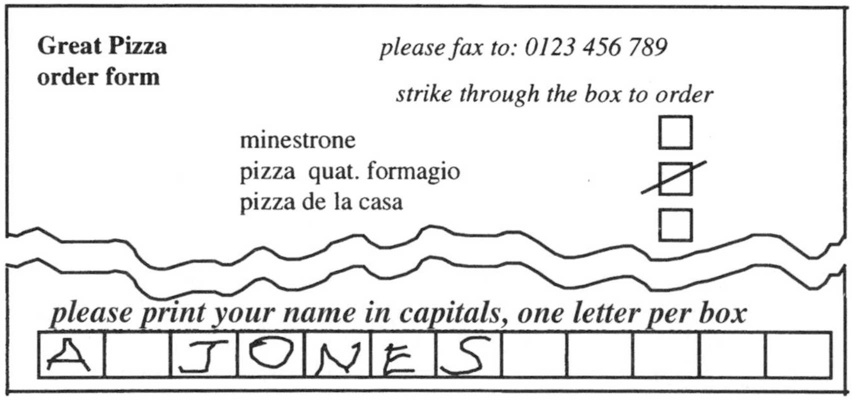

Figure 3.2 Example of a machine-readable fax

rithm, which is helped considerably by constraining text to a specified location of known size.

Applications for this include situations where the person at the ordering end is not computer literate or does not possess a computer, or someone who communicates in a language written in an alphabet which is not easily available on a keyboard – again something that happens in food retailing.

3.5 VIDEO TELEPHONES

The marketing experiences with text and keyboard telephones are more or less recapitulated in the case of videophones, at least in the domestic context. AT&T trialed analogue videophones in the 1970s; even when these were provided free, the user responses were disappointing, probably because of picture quality and, possibly, because there were very few other people to call. A decade later, ISDN videophones began to emerge from the laboratory. These were initially very expensive and, although much of the image-processing circuitry is now available on large-scale integrated circuits, their integration into stand-alone videophones still results in products costing in the range of several hundred pounds. The installed market for these is, so far, exclusively for business use (*desk-top conferencing*) and it is unlikely that the domestic market is ready to take off. A slightly cheaper option is the video card for incorporation within a personal computer, where it can make use of associated electronics, power supply and housing. A low cost camera is fixed, usually, on the top of the PC. Cards similar to these can also be installed in public kiosks (see below), where they can provide interactive video between customer and a sales or help desk. As we mentioned, the transmission method for which they have been designed, is ISDN, using one channel at 64 kbit/s, two at 128 kbit/s, or, perhaps, more channels up to a maximum of six. The quality of the pictures at 64/128 kbit/s is 'adequate' rather than excellent, significantly poorer than business conferencing systems, which typically operate at 2 Mbit/s. The performance is not just limited in that it lacks resolution, but also in not being able to cope with a significant amount of motion. Although adequate for viewing the head and shoulders of a person with whom one is having a business meeting, their specification might seriously degrade a multimedia selling performance.

Indeed, there is a serious deficiency in our knowledge of the psychological effects of end-to-end delay and of camera positioning, in cases involving emotional transactions – for example, on-line face-to-face selling or complaint handling (Figure 3.3).

The codecs used in videophones introduce a noticeable, fraction of a second, end-to-end delay into the video connection. Combining this effect with a camera in its typical location above the monitor, results in you

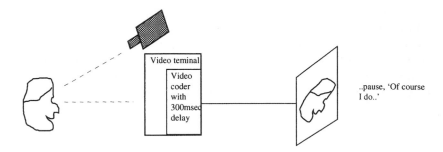

Figure 3.3   Psychological issues of video terminals

interacting with a person at the other end who hesitates before answering you and won't look you in the eye!

ISDN videophones do, however, today offer the only feasible way of bringing real-time interactive video into the average domestic premises. The alternative, of using a computer modem and Internet protocols is much worse. Individual image quality is poor and it is not possible to provide a stream of images at a sufficient rate to give the impression of continuous movement. This situation may change with the introduction of video-on-demand services, although it should be remembered that the market priority is to provide fast services only from server to client, not in the other direction. From an eCommerce point of view, however, this may not be a significant disadvantage, as most scenarios would involve video in that direction, for example a 'sincere' help-desk agent, or an on-line help manual with moving video. A review of interactive video standards can be found in [14].

3.6   OTHER VIDEO DEVICES

The particular characteristic that distinguishes videophone instantiations is that they are real-time interactive, that is, two-way moving pictures more or less as they happen. This low latency, symmetrical operation demands quite a lot from current networks. For many applications, this is not required. For instance, images need not be moving or they may not need to be coded and decoded with minimal delay. In these cases, already the Internet can cope with images download into PCs, in the form of JPEG, gif, etc. files, as well as RealVideo, etc. moving images. More recently on the market are consumer devices specifically designed for handling network imaging. Digital cameras are available at a range of performance specifications and prices. Cameras at the top end of the market are only an order of magnitude or so below that of photographic image, with tens of *megapixels* ('picture points') resolution. Some have the ability to record a few minutes of moving video. They have obvious applications for electronic business in making it easy for vendors to capture and display images of products for sale (real estate agencies and second-hand auto sales, being cases in point) and for remote diagnosis of damage by repair companies and insurance assessors. But they also open up the possibility for domestic users, either in selling their goods at on-line auctions, or in supplying on-line retailers with details for which they might offer a product. For example, one's own figure and features could be sent to fashion retailers to allow them to offer clothes, makeup, hairstyle, etc.

3.7   MULTIMEDIA KIOSKS

Some of the earliest schemes for 'shopless' shopping, have been developed around the concept of the multimedia kiosks that can be located in any public or semi-public place. Kiosks are commonly built out of a number of standard computer components, and are housed within a ruggedised enclosure. Frequently a touch-screen is deployed rather than a keyboard. It is common practice to provide seductive presentations using high-quality images delivered onto a standard VDU screen, and sometimes sound, stored locally on compact discs or hard disk drives, in preference to providing them over the network.

Recent developments in transmission: cable modems, digital subscriber loop technology and radio local loop, (all of which are discussed elsewhere), have challenged the assumption that local storage is necessary. However, this earlier limitation has led to a useful understanding that it is a sensible design practice to separate out the information and seduction components. The latter are expensive to produce, if high quality is required, but, unlike the information components of price, range, availability, etc. they do not require regular updating. It would be very bad practice if one were to embed the price of a garment within the image that displays it.

There is a potential to provide a communications facility that allows the user to transact with the retailer. (Although many such services have not yet been made available.) Analogue telephony or basic rate ISDN can be used to talk to a live assistant. If ISDN is used, then a low definition video connection can be made. This can provide a 'pop-up' assistant, who could be summoned on request to deal with difficult transactions. A variation on this is to use a camera at the kiosk end, to provide a security feature, such as customer identity for banking applications. Kiosks pre-dated the popularisation of Internet/Web transmission and presentation of information, but this is changing and kiosk/payphone installations are now being trialed that are based on standard Web access, with a transaction facility that uses HTML forms.

Kiosks can even provide a degree of direct fulfilment: they can be equipped with printers so that they can issue tickets or vouchers once the goods have been chosen from the screen. In fact, the printer is often the most troublesome component, liable to jam and requiring to be reloaded with paper. The use of smart-cards in place of paper tickets would enable the use of a much more reliable card interface on the kiosk.

In some countries, notably Japan, direct selling of physical goods from kiosks is very common. The kiosk can be integrated into a business's standard service and supply chain, provided a network connection is made to the kiosk. It thus becomes possible to monitor remotely the quantities of stock held within the kiosk, use remote alarms to protect against vandalism and detect system faults, and adjust prices to cover seasonal demand and special offers.

Another development has been in terminals that take the form of seat-back devices in aircraft. Business class passengers in particular, represent a high income, bored and captive population for on-line games, messaging and shopping. Many airlines have introduced such facilities which generally utilise a small display screen, say five inches in diagonal, and a simple control panel similar to a TV remote control, which can be detached from a bracket in the seat arm-rest, to which it is hard-wired by a short cable. Designing for these systems requires careful consideration of several aspects. Aircraft-borne computers and associated hardware are a specialist field, operating under stringent regulations regarding safety, particularly in terms of electromagnetic compatibility with aircraft control systems; consequently the computers use proprietary designs which may be unfamiliar to a design team used to a Unix or Microsoft/PC environment. There is also the constraint of designing within the limits of the small remote control unit. This does not fit easily with the 'traditional' Windows environment of standard Web browsers. Finally, where communication with the ground is concerned, either for messaging or surfing, this has to be carried out within the limits of airborne communication services. Even for short-haul flights, where the radio path is directly between air and ground, this can be of rather limited speed; in the case of long-haul, communication is frequently via geostationary satellites which, because of their distance from the earth, introduce a significant end-to-end delay into the signal. This, combined with possible fading and noise, which require messages to be repeated, can seriously impair transmissions, especially these based on standard protocols such as TCP/TP.

All of these aspects have made in-flight systems adopt proprietary browser, transmission and server architectures. Thus, although they are an important revenue stream for eCommerce, the design of airborne sales systems is a specialist area.

3.8 PERSONAL COMPUTERS

Not unnaturally, the PC has become associated in most people's minds as *the* platform for accessing eCommerce services: eCommerce began as a technology-oriented service, with the early adopters being IT proficient and the majority of purchases being computer products or software. (This is still true.) Whatever the longer term trend (and, as we shall see, there are alternatives) the PC is likely to remain the main eCommerce platform for the next few years, based on using one or other of the industry-standard 'Web browser' software that runs on the PC 'client' and communicates with the vendor server over the Internet, via an *Internet Service Provider.*

Although we are talking about PCs, we must not assume that we are concerned with a single model or homogeneous provision of functionality. The business market for PCs is geared to a rapid churn-rate of hardware and software, rapidly bringing every user up to the latest standard, but this may not be the case with domestic users who have to purchase the equipment themselves. Of the residentially located up-market machines, many are bought by companies for staff who work at home, or by self-employed people using tax concessions. On the other hand, 'true payers', apart from computing fans, are less likely to have the latest state-of-the art machines.

Indeed, many end-customers will not be accessing an eCommerce site from home: either they will be doing so as part of their work, as buyers, for example, or simply using the company's facilities for private use. It is a fact that there are a much higher proportion of people with PCs at work than in the home. According to Forrester Research, in 1998, only 6.8% of European homes possessed Internet connection [15]. Virtually every desk-worker in the US has access to a machine, and the rest of the 'developed' world also has a high level of provision (Figure 3.4).

Furthermore, as we said, these machines will probably be to a higher specification than domestic ones; almost certainly they will be in most cases connected to the Internet via a higher speed connection.

Of course, there are ethical issues concerning using company machines for private use. There may even be tax implications. But eCommerce vendors might be able to cooperate with the companies, for example on the grounds that families with both partners working may find it an incentive to be able to pre-order shopping in their lunch break. In any case, Internet users will tend to have a higher income than average and ought to be reasonably comfortable with IT. This then represents a potentially valuable market segment for advanced, sophisticated services.

PCs per deskworker

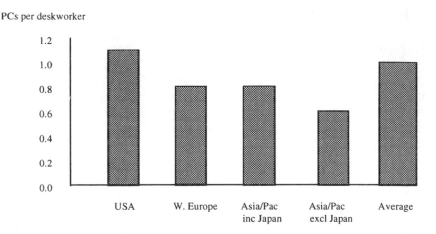

Figure 3.4 PCs per desk worker, 1997 (source: 1997 IDC Global IT Survey)

3.9 PERSONAL COMPUTER SOFTWARE

A simplified model of the software contents of a computer might look like that shown in Figure 3.5.

But the reality of the situation as seen and purchased by the user is much less 'clean' and more complex. Today the PC market is overwhelmingly dominated by Microsoft's Windows environment, with a minority presence of Apple products making up most of what remains. Neither of these is really an operating system, in the exact sense of the word, rather more a loosely bundled collection of user-interface, utilities, application

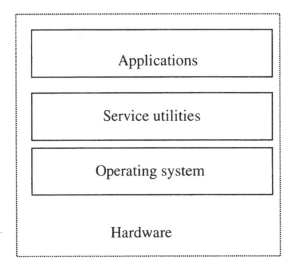

Figure 3.5 Simplified computer model

and operating system components, running on 'real' operating systems kernels of some vintage. Except for very recent versions of PC software, perhaps not even for them, Internet capability has been bolted on rather than designed in.

In all cases, Internet access is achieved using an additional utility, a *browser*. This is a much understated name, implying as it does only the ability to wander (*browse*) fairly arbitrarily through a file structure (in this case, the files available on all computers connected to the Internet). Browsers actually do more than provide a passive file-manager service for the user; they are almost equivalent to a complete Windows or Apple environment, except for an operating kernel.

As we see in the chapter on *The retail server*, where we discuss the client-server operation of on-line retailing, Web services did indeed begin essentially as a simple means of simple remote file access, but they have evolved into more complex processes, including the ability to run programs remotely. Today, the users' PCs often download program (*executable*) code from the retailer's server and run it locally, as part of a browser operation. Originally, (when PC owners were predominately computer-fluent), there was a belief that loading down this locally running code would be done deliberately and consciously by users, under control of the browser. This approach, with the code module known as a *plug-in*, is indeed still adopted for many applications that are 'advanced' at any point in time. Currently, music decoders are a case in point. However, the plug-in process appears to be rather too complex, or at least annoying, to the general user. Instead, the executable code is often imported and run by the browser directly, without requiring manual intervention except possibly by clicking on labelled buttons which are presented within the browser screen.

Java, an open programming language, and Microsoft's ActiveX exemplify two distinct types of client programming.ActiveX is not so much a programming language as an integration of a number of distributed computing components, together with a development environment. Of most significance to our current discussion are *ActiveX controls*, a set of components typically for designing on-screen controls, such as sliders, buttons, pull-down menus, and so on, that can be loaded down into browsers along with the basic Web page, to enhance user interaction. ActiveX standard libraries are widely produced, but only for the Windows environment. On the other hand, *Java* has been specifically designed to be an open language and one which can, in theory, run on most operating systems. (Albeit with performance penalties.) Java also operates in a rather untraditional manner, in that it is both *compiled* and *interpreted*. This is unlike many languages which follow just one or other choice. In interpreted languages, the code is limited to a number of basic operations which can be translated more or less line by line by an *Interpreter* programme at run-time. The interpreter has control over data typing and memory allocation. Compiled code is generally more flexible and, potentially at least, more dangerous, since it is translated, en-masse, as an off-line operation into code which can not so easily be tested for compliance with the specific computer on which it is to run. In general, interpreted code is thus safer. It is much easier to ensure that all operations are strictly limited so that their consequences, for any values whatsoever that they are fed with at run-time (e.g. user-input to a form), are strictly contained within a 'safe' area in memory. They cannot access critical operational functions of the browser or the operating system, nor can they run any process which might lock-up the system.

In the case of Java, the original code produced by the programmer is compiled, to a degree, before being loaded onto a server. The compilation results in a set of instructions in *byte-code*. However, when this code, a *Java applet*, is loaded down into a client computer along with the basic Web page, it has to undergo a further translation, this time an interpretation, before being executed. This interpretation which binds the still variable elements of the code into fixed values (for instance, assigning physical places in memory) takes place within a *Java virtual machine* (Figure 3.6).

This virtual machine is really a piece of software residing within the

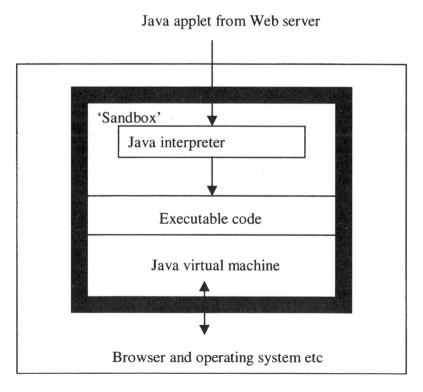

Java applet from Web server

'Sandbox'

Java interpreter

Executable code

Java virtual machine

Browser and operating system etc

Figure 3.6 Providing a safe environment for Java applets

browser. The code is constrained to run within a safe memory space and executable instructions to operating parts of the computer are checked and controlled via the virtual machine's interface with the rest of the computer. Thus, all of the running of the applet is contained within a controlled environment, the *sandbox*. Not only does the virtual machine concept provide a good (though not complete) level of security against malicious operation, it also provides a standard design environment for applets. The application designer does not need to know details of the browser or of the operating system, only the specification of the virtual machine, which is part of the Java standard. Applets once designed, can be reasonably sure of operating on new versions of software.

Microsoft's ActiveX employs a completely different approach to correct operation. No attempt is made to constrain the ActiveX code or its effects to within a safe domain. There is no sandbox. Instead, ActiveX relies on a *trusted code* concept: code is produced and digitally signed by organisations that the user is expected to trust. Using the encryption techniques outlined in Part 3 *Security*, it is possible to provide extremely good evidence that code was indeed developed by the organisation that signed it, and that the code has not been corrupted accidentally or deliberately thereafter.

3.10   PC eCOMMERCE APPLICATION SOFTWARE

What then is this downloaded client software used for? The most obvious is for dynamic graphics. It is possible to use basic HTML code to display some form of movement, by downloading a succession of pages for example, but this is slow and cumbersome. Using the ability of Java and ActiveX to run programs on the client machine, it is possible to generate the graphics locally, limited only by the power of the PC, not by the network. This way, seductive presentation can be achieved, even over slow modem links. Another common use is for form checking: basic browsers allow clients to pull down forms from a server, complete them and send them back. The browser, however, does not check the form to see that the data (for example, a date field) is within feasible limits. On the other hand active browsers can download and run forms which include a piece of executable code which can carry out this checking, within the PC, and signal to the user, in case of error. As Web pages move from HTML to XML (see page 179), where the latter page generation language contains better checking facilities, this will become even more powerful.

Clients PCs *pull* information down from the server. About 2 years ago, market analysts were predicting a move towards *push technology*, where the server actively supplied information to the client without the user necessarily requesting it. The name is rather a misnomer: the client actually had to ask periodically for an update from the server, using a piece of software that has been previously downloaded. Server-side software allows the information provider to configure the server as a number of channels, analogous to TV channels, which contain services such as 'sport', 'shopping', etc. that clients have been configured to receive [16]. The users also select the frequency of updates, thus instructing the client to check the server at that rate and download any changes. The client software also provides a 'tuner' control that allows the user to select the appropriate channel. Despite the market predictions, push has not been as vigorously accepted as expected. The channel concept has, however, been modestly adopted by some information providers, particularly in the news media sector. This is discussed further in Part 4, *Marketing*. Incidentally, the one exception to the lukewarm reception of push, has been eMail. Not so much in the retailing market, but certainly in the information retrieval environment, there has been a steady growth in eMail push of reports and news updates that are sent automatically to a list to which a user has subscribed. We discuss some of the applications of such user-groups in the chapters on marketing and knowledge management.

3.11 PROBLEMS OF PC OWNERSHIP

For general users, the PC is not an easy consumer product to own or borrow, or even to purchase. Away from the office environment where there may be a computer support unit and there probably are colleagues who can help to fix problems, the home user has several difficulties, which may mean there is a need to provide alternative solutions.

Ensuring that the machine is equipped with the latest software, even if it is free and can be downloaded over the Internet, is a chore that many people avoid. Although new versions of Internet browsers are regularly released, any Web service will have to be designed to cope with several versions, some of considerable vintage, if it is to provide wide access.

Another issue with equipment for domestic use is its basic reliability: PC software and hardware are probably not good enough for critical processes in the home, where there is no access to expert support facilities or the possibility of swapping a machine over in the event of a break-down. PC hardware faults are not uncommon; one source [17] reports a typical availability at figures two orders of magnitude worse than a telephone exchange and the standard PC is not considered acceptable by telecom authorities for use as a home telephone exchange, on these grounds. Software problems are even more frequent: Hatton [18] claims that defects occurred every 42 min on systems using Windows '95 + Professional Office, 28% of which required reboots.

PCs are relatively expensive items of consumer equipment: if we were to use a common business accounting practice of amortising their purchase over three years and make a modest estimate of £1000 per machine, we find a cost of over £300 per year. Compare this with a TV set at a typical price of £300 and an average lifetime of 10 years; that is, £30. Under these assumptions, a PC costs ten times that of a TV set.

PCs are also quite bulky, something that we do not always appreciate when we use them at work or in a separate study, but has been an issue in home-working trials for operatives who have not had the benefits of reserved accommodation. If PCs are to become a truly mainstream mass-market terminal, they may have to fit into a living room or a kitchen, and they may have to withstand a degree of physical abuse. This may mean a reduction in size, perhaps the integration of the processor unit, the monitor and the keyboard/trackerball. It may also mean that printers and suchlike ancillaries are banned.

These issues of maintenance and cost have led some commentators to argue [19] that the next trends in the PC market will not be in the direction of increased functionality, rather they will be towards a stripped-down machine which relies on a network connection to a remote server. This is the concept of the *Network Computer (NC)* or *PC-Lite* as it is variously called. In terms of a client-server model of computing, network compu-ters represent the 'thin client' option, with most of the power resident on the server. Some of this power will remain there; some of it, application-based software, for instance, could be loaded down into the NC, as and when required. This will reduce the need for large amounts of memory in the NC and also ensure that the latest versions are available. IDC Corp. [19] have predicted a slow take-up of network computers initially – and these will be primarily for business – but forecast that the home market for them will overtake that of the standard PC, by 2005.

Is this a realistic scenario? In order to generate an informed opinion, it is necessary first to look at the construction of a PC and try to gain some insight into how it is likely to evolve.

3.12   PCS – TODAY AND TOMORROW

As a piece of machinery, the PC is not actually particularly complicated, compared say with a motor car, or even perhaps a television set – this is by no means the least of the reasons why it has become so successful. In order to understand this, and to attempt to chart its likely progress, it is impor-tant to look at this in some detail. In Figure 3.7 we show the principal components of a PC.

The top three components in the diagram represent the usual ways of providing interaction between the user and the computer (and the services behind it). Screens are typically about 15 inches in diagonal size and the best quality of image is still given by the venerable cathode ray tube display. Although liquid crystal and other displays are making some inroads into this dominance, it is unlikely to lose its dominant

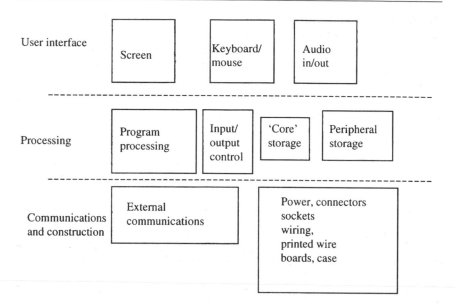

Figure 3.7 Sub-units of a personal computer

position for some years yet, and we can assume that the quality of the shopping experience will be similarly restricted for this period. Keyboard and mouse (or tracker-ball) are the dominant input media and it is probably not too risky to say that they too will remain so for a similar period. There is some interest in voice input-output for mobile applications, but this is still not very reliable and, in any case, may be annoying to others in the vicinity. Thus, there is unlikely to be any radical change to their input/ output control features shown in the next level down.

However, entirely the opposite is true for the programme processing capability of the computer, its 'intelligence'. The heart (or perhaps the brain) of any PC is the processor chip, an integrated circuit comprising upwards of a million or so active electronic components. The processing power of a computer is approximately related to the number of these active circuits multiplied by their speed of operation. What has been astonishing, is that over the last three decades the increase in the number of such components available on a single integrated circuit; what is breathtaking, is the further potential.

The graph shown in Figure 3.8 demonstrates the extraordinary phenomenon known as 'Moore's Law' (after Gerald Moore one of the founders of Intel), that the number of transistors per chip (or, equivalently, the cost per unit of processing power), more than doubles every two years. What is more, is that many experts believe that this trend will continue for at least one more decade. If this prediction is correct, computers in the year 2010 will be about one thousand times as powerful as they are today. Accepting the prediction, we can cast around for something to use this massive power for. One application comes immediately to mind – better quality graphics. Although we are still limited to basic visual display units, we can imagine them being driven by more powerful software that allows the creation of virtual reality interfaces of extremely high quality, and thereby highly seductive for computer games or for the creation of on-line stores.

This trend in processing power, as indicated in Figure 3.8, also extends to 'core memory'. The term originally referred to the construction mechanism for this memory (literally, tiny toroids of magnetic material). The term more often used today is 'random access memory' or RAM, but we have preferred to use the earlier term here: it reminds us that core memory is 'core' to the processing elements of the computer, compared with disk drives and the like which are peripherals.

Core is required when software is to be run quickly; the program to be executed must sit within the core, rather than on the peripheral, because the latter is much slower to access. In the case of operating systems, for example, some kernel activities such as control of basic input and output, always remain within the core, whilst less time-critical or less frequently used utilities are held on the disks and only loaded into core when required.

Core is also used when we want to carry out software implementations of video and audio coding and decoding. Coding schemes such as the video *MPEG* standards, which, need to store successive frames of a video stream so that they can transmit only the changes, require fast access memory. Often it is convenient to replicate the screen layout with a set of 'video-plane' memory locations, which can hold the entire data for a single video frame, plus another plane for text and graphics that will be

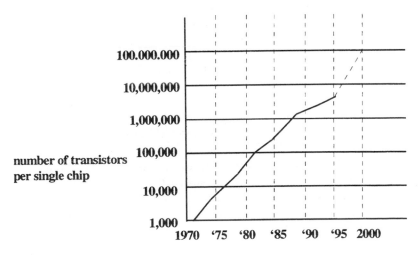

Figure 3.8 Moore's Law

overlaid on the basic image. (This is a common design for TV set-top boxes. See for example [20].)

All of these requirements – increased processing power, video display and coding – provide a reason to be cautious about the claims that a PC-lite/network computer might be a stripped-down version of today's PC. It is quite likely that what we are seeing is a slowing down, rather than a reversal, in the processor and core memory specifications. Designers probably ought not to think in terms of trying to design to a very much restricted hardware specification. For instance, prototype developments in TV set-top boxes (which are, to almost all intents and purposes, PCs) have revealed that it is a bad thing to restrict oneself to a small size of memory for image planes.

If core memory and processing speed are subjects of dramatic improvement, not so much can be said about peripheral memories and other devices. Peripheral memory capacity and access speeds have grown at a steadier rate, broadly linear with time. Marching in time with them has been the storage requirement for complete operating systems and utilities. For instance, this book is being written on a PC running Windows 98 with Office 95 and a few other utility programs. This consumes a total of around 800 Mbytes of disk storage. Windows 98, the operating system, itself consumes more than 300 Mbytes.

Here again we have an issue for the network computer. Are we going to economise on equipment cost and maintenance problems by removing the disk drive and instead, load down parts of the operating system as and when required, from a central server? If so, we are either going to have a rather constrained operating system, utilities and applications, or a very fast channel to load them down from, if we are going to provide an adequate service. Up until quite recently, this network requirement might have been the biggest stumbling block in the path of a domestic market for the network PC. Where were we going to get a fast enough downloading capability? In the business and, perhaps, educational environments, it might be possible for the server containing the program fragments to share a local area network with the client PC and have the ability, over that network, of delivering data at rates well in excess of 1 Mbyte. But what about the domestic user, stuck at the far end of a telephone line offering data rates over the analogue telephone network of less than 8 kbytes/s, or over the more expensive ISDN digital network of only four or five times that? (See Part 1, *Retailing Network Technologies*.) But, as explained in that chapter, recent developments in cable modems and digital subscriber loop may have enhanced the case for the network PC, by promising at least, data rates approaching 250 kbytes/s. The network PC debate is live again and may significantly affect the evolution of the home shopping platform.

3.13 LOW-COST DESIGN OF THE eCOMMERCE HOME PLATFORM

As they stand at present, PCs are not really consumer products, certainly in terms of construction technology. They have been designed to a marketing formula which gives too much emphasis to functionality, evolution and flexibility and not enough to price. In the business market, where computers began, choice was based on *product differentiation* – comparing of one computer with another, in terms of their price, certainly, but also their flexibility, expandability, evolution, even the whim of the IT department or the user (neither of whom had to pay directly for it out of their own money). In the domestic, consumer goods market, purchasing decisions are dominated by *substitution purchasing* – not simply choosing between one computer and another, but across a plethora of wildly different choices: do I want a computer, a camcorder, a hi-fi unit or even a weekend holiday or some smart clothes? In these cases, pricing becomes a critical issue: there is a sharp *price point* above which goods become much less attractive.

In the light of the much tighter price-pointing of the domestic market, we look again at Figure 3.7 and consider how the configuration of its components can affect price. One thing needs to be considered straight away: it is not necessarily the most exotic and complex components which control the cost. It is certainly not the software, for instance. Nor is it now the very large scale integrated circuit. This certainly has made computers into products that are much more affordable than they have ever been; indeed, without such integration they would not even have been feasible. But once this integration has been done, the opportunities for cost reduction lie with the more mundane elements such as cases, connectors, printed wire boards, etc. Indeed, apart from one or two electronic items such as the processor and some other specialised silicon, PCs have not really been constructed as mass-market devices. They consist of individual printed wire-boards, each fitted with expensive wiring connectors, sometimes even with rugged edge-connectors capable of handling the weight and bending moments of the large circuit boards, within a chassis which is assembled from individual components. Usually there are expansion 'slots' – sockets and card guides that allow for the insertion of additional printed wire boards (video accelerator cards, additional memory devices, etc.) – as well as sockets for attachment of external devices.

All of this is indicative of medium volume production, leaving open the options to add functionality over and above the original design. All of this costs money.

Compare this with the design principles for the manufacture of consumer goods. Cases and the structures that support the active components are often 'formed' (e.g. by plastic injection moulding) rather than assembled, the printed boards themselves are often supported at multiple points within the case and permanently held in place by low-cost fastenings. The placing of active components is determined by cost of construction rather than by partitioning that enables functional enhancements: the number of inter-board connectors is minimised, special purpose integrated circuits with largely fixed functionality are deployed and expansion slots are avoided. Although these methods are essentially for cost reduction, they also result in a rather more reliable and resilient product that can be used without problems in the somewhat more hostile consumer environment.

In parallel with this preference for cost reduction over flexibility is a desire to simplify the user interface and restrict the number of options. (Sometimes it is questionable whether this actually achieves its aim as most users of video cassette recorders might attest, but few have the flexibility or confusing complexity of a PC interface.)

3.14 GAMES-MACHINES, PERSONAL SOUND SYSTEMS AND OTHER CONSUMER PRODUCTS

Nowhere is the quest for affordable functionality and corresponding design techniques pursued more vigorously than in the consumer markets for games and personal sound systems. However, this approach may have gone too far for either of these to become serious rivals to the PC as eCommerce platforms, in the near future, but they cannot be totally discounted as they both can benefit from networked connections.

Looking at games, we note an emerging market for interactive, multi-user games. This may lead to the design of very low cost modems at reasonable speed for conveying information between players and, as a bonus to the vendor, a potential mechanism for selling new games that can be down-loaded over the network. From this, it is only a short step to opening up the selling channel to any eShopping transaction. One problem, which we shall come across again when we discuss mobile platforms, is the incompatibility between the display and control keys of a typical games machine and the range of functionality needed to display typical Web pages currently used by on-line shops, based on PC display and keyboard. A chicken-and-egg situation arises: if most eShops require PCs to view them, then games platforms may be inadequate, but no-one will develop an alternative, more suitable display repertoire for their Web pages unless games platforms are equipped with Web page capability. There are at least two possible evolutions that may get round this: firstly, that the multi-user games market is so successful that Web viewing rides on the back of it, initially as a secondary service at secondary cost and functionality; secondly, the games machine becomes a low-cost eMail platform, for recreational use (as is happening with the use of the GSM mobile telephone short message service by school-children) and again the Web viewing capability initially emerges as a secondary use. Around the time of writing, Sony have announced that their new *Dreamcast* games console will include an Internet browser, eMail service and a fast, upgradeable modem which only adds £20 pounds to the basic £200 platform.

Personal sound systems, such as Sony's 'Walkman' products, may approach Internet access from a rather different angle: access to downloadable audio. Audio has, of course, been networked since the early days of radio. PCs are now routinely supplied with sound cards and software which convert them into wired-radio sets. On the other hand, the recorded music industry has tended to treat its playback units as non-networked devices, until quite recently. What has stirred this up is the development of audio coding algorithms which make it feasible to download near CD-quality music files across basic modem connections, within sensible time limits. Developments in re-useable disk technology and the development of audio coding and compression standards, notably the *MPEG3* standard [32] now make it possible to download high quality music and speech signals via an Internet connection using low-cost modems and basic analogue telephone lines, at an affordable price and speed.

An area to watch in this regard is the development of *mass storage devices*. To date, the main mechanisms for cheaply storing data at volumes greater than a Mbyte or so have been magnetic tape or optical CDROM. Tape has the problem that it is a serial medium and needs to be spooled to the appropriate place before it can be read. CDROM has been predominately a replay-only medium, although more expensive, read/write versions are available. It is speed-limited (although it can give reasonable video quality) and it is rather bulky for small portable devices such as mobile telephones and PDAs. Quite recently, the market has seen the emergence of *high storage memory cards*, which use solid-state memory technology, which is fast, compact and economical with power. Storage capacities at time of writing are confined to relatively modest figures of 64 Mbyte, but a development consortium comprising Panasonic, Toshiba and SanDisk have claimed they are confident that a 1 Gbyte card will be available in 2002. The impact of this technology on eCommerce could be quite significant, obviously in the specific case of selling music or video, but also on the seductive component of retailing in general, allowing high quality catalogue information to be downloaded and replayed at leisure.

More conventionally, fixed magnetic disk drives are now offering storage capacity of the order of 40–60 Gbytes of data. These can be used as digital jukeboxes for downloaded audio, or boxed and sold as solid-state video recorders. For example, *TiVo*, built by Thomson, has 40 Gbytes of memory and costs around £400. For an additional one-off fee of £200, an intelligent programming interface that be programmed to record programme material at a variety of recording qualities, giving between 14 and 40 h of content. The system also has a pause control, which can allow replay of live material. This might be one way to include shopping information into a life-style channel, for instance.

Finally, we should mention the new possibilities for content-generation offered by low-cost *TV cameras* and *digital cameras*. The rise of the video telephone has long been predicted but never realised as a marketable proposition. Part of this is undoubtedly due to lack of bandwidth in the local loop, but this restriction may be removed in the near future by the installation of some of the technologies we mention in the chapter on *Retailing network technologies*. Precisely where videophone, from the customer's end, fits into the eCommerce model is difficult to predict, although there are possibilities both for selling (scanning in your image for fashion purchases, for example) and for after-sales support. ('Now show me the fault in action, please'.) The problems of producing good quality video are, however, well-known. Lighting and camera-handling skills are among the issues. Perhaps, still photography is a simpler solution. There are many affordable digital cameras on the market today and prices are falling as image quality and storage rises. The digital camera, in association with a PC, can act as a relatively low-cost way of sending images across the Internet, to sell houses, send assembly instructions, and so on.

3.15 INTERACTIVE TV

Perhaps a much more serious contender to the dominance of the PC as the home shopping platform, at least in the short-term, is interactive digital television. There are only about 100 million PC owners, but approximately two billion domestic TV users [4]. Industry analyst Dresdner Kleinwort Benson predicts that sales from TV home shopping will top £one billion per year in 2003, rising to £12 billion by 2009. Currently TV advertising presents a seductive view of products into the home, but seduction is virtually all that you get. Prime-time advertising is too expensive to permit delivery of detailed product information, and the transaction method is telephone, post or a visit to a shop. Today's TV lacks personalisation and interaction. This is all due for a change, with the development of *digital interactive television*. The term 'digital' does not mean that the basic picture on the screen will undergo a significant change, but the signal arriving at the set will. The output from the studio will be converted to a digital signal using picture coding techniques which can reduce the bit-rate required to about 2 Mbit/s for picture and sound channel. The quality will be at least as good as current decrypted analogue satellite channels The TV signal may be broadcast over the air using terrestrial aerials or satellites, or may be delivered via cable networks. As we explain in Part 1, *Retailing Network Technologies*, it can also be delivered over standard telephone wiring using the new transmission technique of xDSL.

The programming material may simply be Web pages, accessed using browsers similar, or even identical to those on a PC, – it may even be received on a PC. Programmers may, however, wish to distinguish themselves from PC-Internet shopping services by capitalising on the TV-quality of its productions and image quality. It may be delivered through a user-friendly consumer-electronics product – the TV set.

3.16 THE SET-TOP BOX

We said that the TV set was not 'digital'. The consumer electronics industry is very wary of upsetting customers by introducing dramatic changes to products, that are not obviously of immediate exceptional value, and so the TV set itself will remain unchanged in the medium term. (Not the least because consumers expect to get 10 years' life from a TV set, with each household possessing on average two working sets, and purchasing a new set every five years or so.) If we ignore the relatively few early adopters who may buy sets with integrated digital electronics, the medium term development will be through separate 'set-top boxes'.

Set-top boxes (which, perversely, are often located *beneath* the set) are designed around general purpose PCs, minus, of course, the display and, usually, the keyboard (Figure 3.9).

Set-top boxes have initially been designed to receive signals from off-air transmitters, but common designs have been agreed that will also allow the same box to operate with digital cable-TV or telecoms 'video-on-demand' (see page 30). There is also a telephony interface, currently for analogue telephony, but this could be digital. The digital signal is decoded by a special purpose digital decoder chip, under control of the processor. There are a number of standard coding/decoding algorithms for digital

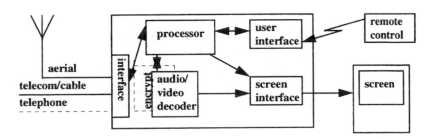

Figure 3.9 TV set-top box design

television signals available, depending on the quality required, but the
industry is moving towards convergence on the *MPEG2* standard. Notice
also the dashed box labelled 'encrypt'. The TV signal may be encrypted to
make it unviewable without purchasing a key. One way to enable such a
key is to allow someone to purchase it over the network. This is one
possible use for the simple telephone interface. Obviously, this payment
mechanism need not be restricted to purchasing TV pictures: it could be
used as a secure means of buying any type of goods.

The picture and sound are then converted to analogue signals and
passed to the TV set for delivery in the normal way. Users communicate
with the set-top box using, typically, a hand-held infra-red remote controller. Navigation menus need to able to operate within any restricted keyset
provided on the controller. The implementation of the set-top box may be
achievable using as little as four VLSI chips: processor, media processing,
media access and communications [21]. This would bode well for low cost
integration within the TV set itself at a later date. The set-top box software
structure also conforms to general purpose computer principles set out in
Table 3.2.

The software is partitioned into a number of layers, each communicating principally with the layer immediately above or below it. Layers near
the top are specific to the current application, whereas layers lower down
are concerned with the operation of the hardware across a generic range
of uses.

For example, in a home shopping application, the user can browse
through the 'content': a multimedia catalogue containing a number of
still and moving images and sound together with product information
and order form. The application designer has generated a set of control
information that links the various parts of the screen images. For example,
selection, by means of the remote control, of a particular number that is
shown on the screen to represent 'next page', ' order form', etc. will result
in a command to the control software to retrieve and display the corre-

Table 3.2 Set-top box software

Software layer	Residence
'Content', for example TV programme or film	Download on request
Control information specific to individual content, e.g. how to navigate through a home-shopping catalogue	Download on request
Generic environment software	Download during setting up of connection
General purpose, multimedia computer, operating system	Permanently resident
Hardware device drivers and network interface control	Permanently resident

sponding result. The content and the application control software are downloaded into the set-top box as and when required, during a session with the application.

At the very bottom of the layered structure are the network control software and a basic level operating system; the latter directly controls the hardware elements of the computer (memory, input-output, the running of programs) and provides, to the next higher layer, a set of instructions that make it easier to control these hardware elements through symbolic instructions such as 'read', 'write' and so on.

There is also a layer between the operating system and the specific item of content: the 'generic environment'. The best way to define this is to say that it is the software that distinguishes the set-top box from a purely general purpose computer and, which, at the same time, allows the box to take on a variety of functions or to allow it to be updated to take into account new functionality. For instance, suppose that some time after you bought your set-top box, you hear that a new service has been developed that can automatically learn your interests in fiction and negotiate with on-line libraries and bookshops to identify books that meet this profile. Because your set-top box has been designed with a standardised generic environment, you can download this application into the box, using very simple, foolproof commands, and thereafter run it automatically. The software may even be intelligent enough to update itself with the names of the participating libraries, by periodically downloading a directory.

This approach, whereby software is not always loaded into the equipment at the initial time of purchase, but is instead swapped in and out and downloaded across the network, is particularly attractive to consumer goods, because it can be arranged so as to minimise the user's involvement in configuring and maintaining the system. In other words, the set-top box is one example of the network PC architecture mentioned earlier. By using consumer-electronics construction techniques, the set-top box can be sold at much lower prices than standard PCs, and, eventually, it is intended that the functionality will be integrated into the TV set itself, obviating the need for a separate box, power-supply, etc. further reducing the cost significantly.

Already set-top boxes are being sold in significant numbers and enhanced versions are also available. These include the ability to alert you to incoming telephone calls, including calling-line identity display. Either as part of the set-top box or as a stand-alone attachment, eMail devices based on using the TV as a monitor are emerging. At least one STB vendor offers an integrated hard disk drive, allowing up to 10 h of video to be recorded.

3.17   STAGGER-CAST AND VIDEO-ON-DEMAND

The least radical variant of digital broadcasting is where the current analogue broadcasting practice is replaced by one where signals are broadcast in digital format. But, there are other more radical options. Since the set-top box is essentially acting as a client in a conventional client-server architecture, it can be used to retrieve individually, items of programming that are unique to the viewer, at the time the viewer chooses, i.e. *on-demand*. That is, provided there also exists a transmission channel that can be dedicated to the individual.

This is not really possible, in the case of broadcast transmissions; the best they can do is to repeat the same content on different broadcast channels, staggered in time. In this way, you, or your video recorder, do not have only one chance to see a programme, nor do you miss out if two programmes of interest start at the same time, as you do with current programming. Instead, it will often be possible to schedule one's viewing better, by selecting the 'staggered' broadcast that meets your needs. The reason that staggercast becomes possible with digital broadcasting is that it requires less radio spectrum per channel than current analogue broadcasting; at least five times as many digital channels can replace one analogue one.

Cable systems, and those delivered over telephone lines by ADSL, can do much better: each cable or pair of wires is a separate channel between the customer and the server. Consequently, there is nothing in principle to stop each and every viewer selecting different programmes and starting them at arbitrary times. The operation is simply that of a file request to a high performance, multi-user file server. The server architectures are discussed in Part 1, *The Retail Server*.

3.18   STANDARDISATION FOR DIGITAL TV – DIGITAL AUDIO-VISUAL COUNCIL (DAVIC)

In 1994 a number of industry players set out to create a non-profit-making organisation 'with the aim of promoting the success of interactive digital audio-visual applications and services by promulgating specifications of open interfaces and protocols that maximise interoperability, not only across geographical boundaries but also across diverse applications, services and industries'.

This organisation, the Digital Audio-visual Council, DAVIC, now has around 160 company members in nearly 30 countries. [4]. Membership includes computer, consumer electronics and telecommunications equipment manufacturers, service providers and government organisations and research organisations.

Table 3.3 DAVIC standards decisions, from (22)

Functionality	Selected standard
Video coding	MPEG-2
High-quality sound	Dolby AC-3
Set-top application programming interface (API)	MHEG, JAVA
Max lines for PC High Density TV	1080
Security API	DVB Common Interface

DAVIC proclaims 'a vision of an audio-visual world where producers of multimedia content can reach the widest possible audience, where users are protected from obsolescence and have seamless access to information and communication, carriers can offer effective transport, and manufacturers can provide hardware and software to support unrestricted production, flow and use of information.' DAVIC is firmly targeted on standards for mass-market TV, rather than for the home computer. This tends to be reflected in its membership and its approach.

DAVIC has been very active in the standards debate and has been quite decisive in selecting a restricted set from a large basket of possibilities. Some examples are shown in Table 3.3.

3.19 I-TV, WEB-TV AND eCOMMERCE

DAVIC is likely to achieve its aim of setting standards for digital TV. What is less certain is the precise nature of the eCommerce channel on a digital TV network. One option, which is almost certainly to be tried is the provision of an enhanced version of today's TV shopping channels. The goods will be sold in a TV show environment, where visual and audio seduction are achieved by 'glitz' – it is difficult to find an alternative, non showbiz term to describe the phenomenon – perhaps supported by a modest amount of information, and certainly backed up by a real-time ordering procedure, either a phone number or an on-line order request. Remember, it is trivially easy to programme the set-top box with account details, including credit card, and the programming encryption system can also be used for secure financial transaction. 'If you like this, press 'yes" should be sufficient to place an order.

One option, proposed by the author and colleagues, several years ago, demonstrated the possibility of a 'stop me and buy one' service linked to Video on Demand. Because each household accesses a separate viewing of a programme, then they can pause it without affecting any other viewers. The idea of 'stop me and buy one' is that you can halt the action at any time to purchase items that you have seen on the screen. The (rather fanciful) example we demonstrated was buying the tie worn by an actor.

What is not yet clear about interactive TV is the nature of the majority of content that will be viewed. A few years back, there would have been little argument: it would be enhanced television programmes, with enhancement primarily being an increase in the number of channels and the addition of some level of 'interactivity' with the programme or the associated advertising material. Things are not quite so clear today; there is a school of thought which believes that digital TV is a way of delivering access, not to broadcasters, not even to 'narrowcasters' but to the content available on the Internet. That is, to a much, much wider spectrum of content, on a much wider range of subjects, unmediated by broadcasters or convention, by and large poorly produced and highly variable in quality, generally poor but with gems that we may find personally of great value. This is the domain of 'Web-TV'. In other words, the digital TV set becomes a terminal for accessing the Web.

Implicit in this viewpoint, and relevant specifically to this chapter, are the technical implications that this poses. At first sight, there is no reason why there should be any problems. After all, we have said that a set-top box is simply a computer without a screen or keyboard. That is true, but here lie a number of problems: without a keyboard, how are we going to control the interactivity? Is it possible to provide acceptable access to the Web, simply by using a TV remote control? Alternatively, is it necessary to provide a keyboard, and what format should this take, bearing in mind that it will add to the price, which, in the consumer market, may be a significant issue? In any case, what level of control do we provide? This raises the question of the appropriate browser for TV set. In marketing terms, a TV set is *not* a computer, however similar the interior technology. It is a consumer product with a customer expectation regarding, among other things, stability and ease of operation, by unskilled users. Browsers churn regularly and are incompatible. This is not only true for browsers; it also applies to plug-ins for sound and moving images, as well as other applications. We do not know whether users will happily use their TV sets in two modes: simple, entertainment-based and also, complex IT-based.

One further problem: screen size. The author was involved some years ago on video-on-demand services. We punctiliously developed our applications so that they ran, not on computer screens, but on standard TV sets, so that, in our estimation, we would not display anything on the screen that could not be viewed comfortably. We found that text at anything like the density it is usually displayed on a Web page, is extremely difficult to see. So (rather smugly) we demonstrated large-print text to our marketing people to show them we were customer-friendly. They were nearly satisfied, but they asked us to lay on a demonstration on a real TV set in a real room. The results were disappointing. We had developed 'on-the-bench', close up to the set. At practical distances, a TV set subtends a very much narrower angle than does a PC screen. Text on the latter can often be completely illegible on the former.

This is not to say that Web viewing will never become a part of digital TV; it almost certainly will, but there are significant issues that need to be thought about before assuming that the Web will easily conquer the TV set.

3.20   THE MOBILE TERMINAL

One of the wildest wild card in eCommerce is the role of the mobile terminal. The growth in the number of mobile telephones, for example, truly questions the Internet's claim to be the unchallenged champion of exponential growth (Figure 3.10).

Although the main use of mobile telephony is for voice, the establishment of the *short messaging service* and the *GPRS standard* (both of which we discuss further in Part 1, *Retailing Network Technologies*), are showing that a strong demand exists for mobile messaging and browsing services. The next stage in the evolution of these on-line services is at a bifurcation point (Figure 3.11). Increases in terminal power and network performance may mean that the shopping experience becomes richer in terms of its ability to seduce through vivid, high quality images. Alternatively, there may be more benefit to be had from providing unimproved media quality (perhaps even poor quality compared with what is currently available from fixed networks) but available everywhere and on the move.

Both of the directions shown in the figure are being followed today and future development is likely to be rapid. It is important for eBusinesses to position their offerings correctly with respect to both of them: does the business wish to trade exclusively as a wired or on-air service, provide access to these methods separately, or consider integrating both? Indeed,

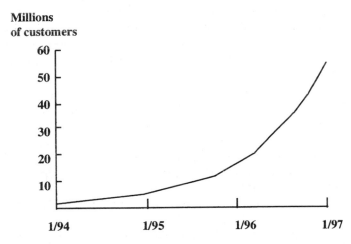

Figure 3.10   Growth in GSM telephony

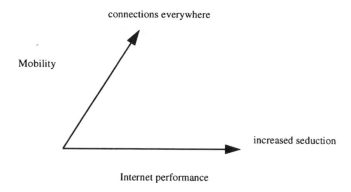

Figure 3.11   Alternative evolutions

what are the differences between the two alternatives, in business terms and in technical ones?

Looking first at business aspects, we can see that mobility has certain significant characteristics:

- The user is away from their normal place of abode or office: we assume that the user has their own terminal and simply wants to carry out normal on-line activities, except in a remote manner. In this case, the important issue is the unfamiliarity with the surroundings and the corresponding lack of support in the event of any problems. The requirement is for uncomplicated and unconditional 'plug-and-play', end-to-end from user terminal, via transmission, to the server.
- The user is truly mobile, perhaps in a car, an aircraft or temporarily stationary but without any access fixed power or data connection points: users may be prepared to sacrifice functionality for portability of the terminal, longer battery life or reliability of connection.
- There is a specifically geographic component to the mobility: users may want to have one-way or interactive communication with some local service – book a hotel room, visit a restaurant, contact a local garage, and so on.
- In all cases there is an implicit requirement for the applications to run unsupervised, perhaps when the terminal is unconnected to the network, perhaps to save batteries, perhaps because it is temporarily out of communication range, switched off in accordance to regulations (e.g. on an aircraft in flight), and so on. When reconnection has been established, it would be preferable for the application to be restored without problems.
- The user will require a 'proactive' relationship with the terminal: in the office, we are usually prepared to accept that our eMail becomes accessible only when we log-on; we do not accept this from our tele-

phones. The latter generally have the right to interrupt our routines at any time. (We may use call screening via the loudspeaker on an answering machine, but we still accept this interrupt.) This convention allows us to service urgent business from distant parties. The same functionality is probably necessary in a mobile terminal.

These are only some of the possible reasons why specifically mobile offerings may be required. One general principle seems to emerge in all cases: if there has to be a trade-off between seductive multimedia and the convenience of a portable and reliable terminal, then so be it. This tolerance is rather fortunate, because, with the current state of technology, this does appear to be necessary.

The first problem is in construction of the mobile terminal: in order to make it light enough, there has to be a severe constraint on the size of its batteries, and hence, on the performance of power-hungry components such the screen and disk storage. Compare, for example the peak requirements of a GSM telephone at 1 W, with that of a portable PC at 15–20 W. We should also add that integrated circuit ('silicon-chip') technologies designed for low power inevitably operate at lower speeds. It is unrealistic to expect applications to run at the same speed as on top-of-the-range PC models.

Furthermore, portability sets higher standards for ruggedness than is required by the desktop computer: case design has to be stronger and the means of connecting the various component parts together has to be better able to withstand vibration. PCs are assembled from a number of printed wire boards onto which the individual electronic components are soldered, and then the boards are pushed into connection-ports which carry the wiring between the boards. The normal range of vibration that a portable is subjected to, is often sufficient to cause these cards to be jolted from their connectors. Thus, portables generally eschew plug-in cards and go for a more fully integrated construction. The printed wire cards are rigidly fixed to the case and interconnection wiring either soldered to the card or connected via lightweight connectors with insertion forces that are very high compared to any inertial force likely to be generated by vibration. One added bonus from this approach is a reduced cost of production; a corresponding disadvantage is that no spare card-slots are left free for future expansion of functionality through the inclusion of a new printed wire board, as is commonly the case with PCs.

A further consequence of the constraints placed on hardware by limited power, space and reliability, is the knock-on effect this has in software. Most operating systems and the languages and scripts that run on top of them have been developed assuming at least the minimum performance and memory size of today's PCs. A great deal of the code is intended to be swapped between hard disk and RAM. Most mobile devices will not possess the former and have severe restrictions on the size of the latter.

Code execution may require that firmware instruction sets and routines are permanently 'burned-into' read-only memory (ROM). Special environments are also required for code development. The target device will be emulated on a different, development machine with higher functionality but capable of testing the code to ensure that it will run satisfactorily on the target. This approach is quite standard practice for engineers involved in the development of so-called *embedded systems*, but may be rather unfamiliar ground for traditional systems programmers. A good description of what is involved in designing for such systems, particularly in the case of Java but also of general relevance, is given in [23].

Consequently, it is sensible to consider most mobile terminals to be similar to consumer products in construction, price and flexibility, rather than to computers.

Alongside this technical similarity, mobiles also tend to share the product marketing traditions of consumer products: high reliability without pro-active maintenance and a relatively slow rate of issue of new versions, again very unlike the PC market. There is a very intriguing issue in the battle for the de facto standard for the user interface. There are three principal contenders, coming from the domains of mobile phone, personal organiser and PC, (Figure 3.12).

If mobile systems are seen as an extension of PCs, then Microsoft is the industry leader, with Windows CE, as its proposed solution for programmable consumer devices. In a joint venture with Qualcomm, they offer an integration of Windows CE with Microsoft BackOffice and Microsoft Commercial Internet System, MCIS, and this is being taken up by some major US mobile phone service providers.

Although some examples of personal customer use are cited, for example Radiant Systems offer a touch screen kiosk that customers can use to place orders within a restaurant, Microsoft's approach is heavily directed towards business and 'professional' use of portable devices which it sees

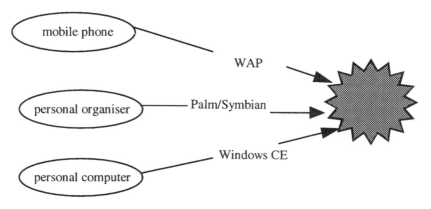

Figure 3.12 Contenders for user interface for mobile terminals

as an extension of its core market for networked business IT. It tends to characterise the mobile terminal as a form of miniaturised PC, either a *laptop,* which has most of the features found on a desk-top PC, or a *hand-held or pocket PC* which handles quick access to personal diary and sche-duling information, eMail or note taking, sacrificing to immediacy, the full functionality of office systems.

Windows CE claims to provide a subset of Microsoft's Win32 API (application programming interface) from the Windows NT operating system widely used for business applications. Making Windows CE in to a pure subset would mean that it should be easy to provide compat-ibility with NT, simply screening off, or providing customised alternatives for, functionality that is not possible in the more restricted footprint of the mobile terminal. It is also possible for developers to use the same devel-opment tools to create Windows CE applications. This should make it easier and faster for companies to undertake development, as they can either use existing staff trained in Windows software, or purchase them from a widely available skill-base. The software operates via a thin layer of code residing between hardware and the operating system kernel. This layer allows customisation for specific hardware devices.

Given that the terminals are intended to be fully integrated with an office process, it is also desirable to provide synchronisation between data held on more than one system. Windows CE provides a means whereby calendars, contacts, eMail, etc. can be managed and backed-up from terminal to one's main desk machine.

Technical characteristics of Windows CE include recognition of the need for a real-time system that embraces hardware, operating system and applications. In particular, it is claimed that the operating system must be multithreaded and pre-emptive, support thread priority (unlike other Windows software, CE interrupt priorities are essentially fixed), a system of priority inheritance and predictable thread synchronisation. Importantly, the information on the real-time performance (maximum interrupt masking, etc.) should be made clearly available to developers.

As can be expected from Microsoft, technical and sales-related docu-mentation is extensive. If the current direction of Windows CE – the 'professional' terminal user, is maintained, it would seem that its biggest eCommerce application would not be in the area of mobile, on-line stores with users being domestic customers. Instead, it might be more rapidly taken up by mobile sales and maintenance staff, and we shall discuss this use further in Part 4, *Service and Support.* However, a number of shopping applications do come to mind: we have mentioned the Radiant Systems kiosk: as we noted, kiosks are essentially ruggedised PCs with special user interface characteristics, such a keyboard-less access via touch screen. This would seem to be exactly the place where one might see Windows CE fit in perfectly. Intelligent vending machines and trans-port-information terminals, for example at bus-stops would be another.

A further use might be for in-car terminals: apart from providing traffic and route information, these devices are seen as a possible selling-tool for local services. You are hungry? It can advise you on local restaurants and how to get there. Some time to spare? A guide to nearby places of interest can be provided. For many years it has been possible to get route maps from motoring organisations. Tomorrow it may be possible to get the on-line guided tour as well.

Coming from the opposite end of terminal functionality are a number of mobile telecoms and personal organiser companies led by Ericsson, Motorola, Nokia and Phone.com. They declare the intention to bring Internet content and advanced data services to wireless phones and other wireless terminals, by creating a global wireless protocol based, wherever possible, on existing standards and technology. Their solution, *Wireless Application Protocol (WAP)* [24], includes a microbrowser specifically designed for the limited capability of the screens and controls of mobile equipment. WAP and HTML are not immediately compatible: pages for a WAP-enabled terminal have to written in *Wireless Mark-up Language (WML)*, and gateways are required between Internet resources and the mobile phone market. WAP is not a radio transmission standard; instead, it is a set of layered protocols (six layers in all) that sit above the level of existing and emerging radio transmission standards and is thus intended, at least in principal, to work over all air interfaces (Figure 3.13).

WAP is tolerant of bandwidth restrictions and long, end-to-end delay ('latency').It does not foresee an early lifting of restriction on battery power or screen size/resolution. In the light of these constraints, WAP

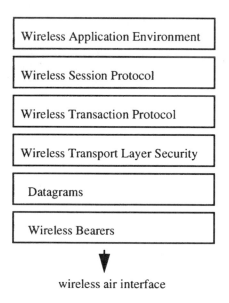

wireless air interface

Figure 3.13 WAP protocol stack

specifies a lightweight session protocol that can allow sessions to be suspended and resumed without the need to set the application up from scratch. Another refinement, intended to save on transmission time, is to transmit the HTTP plain text headers in binary.

Input will also continue to be through a limited functionality keyboard, perhaps without full QWERTY capability. The WML, is claimed to be based on XML principles, and allows a layered document structure, rather than the flat structure of HTML pages. A basic document unit is a 'card' and services are created by allowing users to move back and forth between cards. A user-interface specification has been created that allows the cards to be presented to the user.

Rather than use a full TCP stack in the phone, which would require too much memory, WAP uses a *Wireless Transmission Protocol* that provides a reliable datagram service. It is claimed that these measures reduce the standard HTTP/TCP/IP requirements by more than half.

All this means, of course, that WAP cannot work directly to a standard Web server. There is a need to go via gateway proxy, which can translate between the codes and protocols, provide Domain Name Services (to convert from URLs to IP addresses), optionally manage subscription services and handle access and security. (WAP also has a *Wireless Transport Layer Security* service.)

Although it is described as a gateway service, with the option of auto-mated translation between standard HTML Web pages, it is accepted that, in most cases, it is better to write documents directly in WML, if they are to be displayed effectively on the terminal's limited interface.

There is also a variant of WAP, the *Shared Wireless Access Protocol*, which is intended to provide cordless (i.e. short-range) network access, for hand-held data products, including use in the home. (But also see *Bluetooth*, in Part 1, *Retailing network technologies*.) Wireless home systems are interest-ing, because of the cost of wiring a domestic dwelling for communica-tions, can run into several hundreds of pounds.

WAP is particularly intended to take note of the constraints surround-ing the low-end terminals, some of which we have mentioned earlier. It is particularly concerned to specify that mass-market handsets must price-in at under $149, but notes that market size (one billion subscribers by 2005, according to Nokia), will demand and get optimised solutions. They do not believe that their terminals will be used to 'surf the Net': users will have specific goals that they want to achieve, quickly and everywhere. As described further in Part 4, *Marketing*, WAP also offers a *push service* which allows the server to alert a terminal client to any news items.

In July 1999, Ericsson and Reuters launched one of the first applications intended to demonstrate the utility of WAP for delivering financial market news to the latest generation mobile phones. This would seem to be an example of precisely where one would expect this technology to find a place: the data transmitted does not generally require much seductive multimedia component, if any; instead it is information-intensive and requires very timely delivery.

Although the WAP White Paper claims that '[telecoms] carriers representing more than 100 million subscribers world-wide have joined the WAP Forum', there are some doubts in the market regarding WAP as a long-term provider of a complete application stack from basic radio-networking technology though to screen-based browser. Whereas WAP is being quite kindly received, some believe that a higher functionality, and, indeed, full HTML compatibility, is required. It is possible that WAP will remain, but become invisible, underneath a richer user-interface. In this regard it is obviously significant that, in December 1999, Microsoft and Ericsson announced a 'strategic partnership', whereby they will form a joint company to market eMail solutions, based on an Ericsson WAP stack working into Microsoft Mobile Explorer. This, it is claimed, is intended to allow access to WAP and HTML-enabled servers, although the initial emphasis appears to be specifically targeted on eMail. However, the eventual outcome will be decided by the relative merits of compact, cheap, personal mobility, versus integration with the more-business oriented, HTML/PC market.

Somewhere in between the fully functioned PC and the much more constrained hand-held mobile phone, is the domain of the personal organiser/palmtop. Users of these would prefer to have, at least, good eMail capability and the ability to view most Web pages. Psion, the brand-leader for personal organisers, is one of founders of 'Symbian' [25], a joint venture with Ericsson, and Nokia (and shares memoranda of understanding with Motorola and 3-com). Unlike the developers of WAP, Symbian have developed the *EPOC* product set which is intended to handle standard HTML. EPOC is intended to fit into the software and hardware fingerprints of smart telephones and handheld computer-like devices. It provides application software for messaging, Web browsing and general office use, connectivity with PCs and servers, a graphical user interface optimised for the phones and handheld devices, and sets of development kits. It is heavily based on the functionality of Psion's personal organisers, extended to include the wireless communication capability.

EPOC is a fully operating system, booted from ROM, and capable of handling scheduling, memory, power, timers, files, various i/o control devices (keyboard, pointer, screen), and plug-in memory, etc. cards. Currently, it runs only on ARM3 processors, but others are planned.

The graphical user interface for EPOC is called EIKON. It has been specifically designed to fit the footprint of the target hardware – devices unsurprisingly not unlike Psion's Organisers. Of particular interest to eCommerce applications is the Web application, which is implemented as a separate user interface and components. Currently it supports HTML up to version 3.2, with support for frames, forms, GIF and animated GIF, JPEG, JAVA applets. Included in the Web engine is rendering engine which interprets data tokens for display by the user-interface-specific part of the application, thus allowing for device flexibility.

EPOC is richer in presentation functionality than WAP and, being directly compatible with HTML, can capitalise on existing Web resources and is not incompatible with other PC application software, such as the range of Microsoft products.

Symbian is heavily based around the hardware platforms of the consortium, in particular the Psion organiser. One of the signatories to a memorandum of understanding with the Symbian alliance is, however, Palm Computing, who are the dominant player in the note-book computer market with the Palm Pilot, which is different in hardware and software terms from Psion. The Pilot has many attractive features, not least its very low power consumption and it also has a wide range of in-built features. It is not clear how significant Palm really is, consider the alliance to be and how their future strategy for their proprietary operating system stands in relation to EPOC.

3.21   MOBILE 'POSITIONING' SERVICES

Mobile terminals can be seen simply as a portable extension of fixed terminals, but perhaps their most interesting area of application, and one whose potential is least understood today, is through their specific geographical location. If we know where they are in the world, then they can be offered services that are specific to that place. We mentioned booking accommodation, local shopping, etc. There are likely to be many others. If a telephone exchange or mobile information provider knows where a terminal is – and it usually has to have some idea, in order to provide the basic communication service – then it can implicitly tailor its service to the mobile's current location. How do we therefore expect this location to be achieved, and to what level of accuracy?

One approach is to integrate a *Global Positioning System (GPS)* within the terminal. GPS is based on the reception of signals from the US NAVSTAR constellation of 24 Low Earth Orbiting satellites, positioned in orbits such that at least four or five of them are in radio contact at any one time. The satellite positions at a certain start time and date and a set of algorithms describing the subsequent evolution of these orbits is held in each receiver. The receiver continuously monitors the signals transmitted by the satellites, calculates the relative delay between each of their signals and uses the algorithm and start data to calculate its position.

Some natural factors limit the accuracy of the positioning, multipath interference and rain among them. Originally the biggest error was deliberately introduced. NAVSTAR is a US defense system intended for use by US and allied forces. To prevent it being 'misused' by their enemies, the signals had a coded pseudo-random error introduced into them. Without access to the secret decoder, users of the basic system would encounter an error which could exceed 100 m for 5% of the time and 300 m 0.1% of the time. Very recently, this randomising process was switched off indefinitely. Even this margin is acceptable for many applications, but, by using systems which average the signal over a period of time, accuracy of within 30 m 95% of the time can be achieved. Under a licensing agreement with the DoD, permitted users can also have access to an enhancement which uses differential reception and provides accuracy to within 5 m.

Currently, GPS receivers are sold into a different market from other mobile terminals, principally as route-finders for motorists, amateur and professional sailors, mountaineers and the like, rather than for electronic communication applications. Prices for self-contained terminals range from just over £100 for basic units to around £400 for complex navigators that compute estimated time of arrival, include digital maps and route-planners, etc. Most of this cost must be in aspects not directly related to the GPS receiver – power supplies, keypad, case, display, etc. – and full integration into a mobile terminal would certainly reduce the cost of a GPS option.

Probably there is no need, however, to use GPS for the positioning system, for many applications. After all, mobile systems require basic positioning facilities in order to function. In the worst case, a mobile can usually be positioned with a cell-site or two, on the basis of its signal strength alone. If cell-sites were equipped with direction-finding capabilities, then this margin could be considerably reduced. Another alternative is to use terrestrial beacons that a mobile receiver could use in order to confirm its position.

It is believed that high-accuracy positioning systems for mobile terminals is a very active area of research in the industry today, but performances achieved are still the subject of some commercial sensitivity.

A final thought on positioning services: it is technically possible to locate many fixed-line telephone customers today by using a simple look-up of telephone number against premises number/street name against post-code against map reference, but this is not done because of commercial and regulatory restraints. At least one pizza delivery company has been unable to provide a delivery service matching destination to nearest kitchen pick-up point, for this reason.

3.22   LAST WORDS ON THE 'HOME TERMINAL'

We have seen that there are a number of contenders for the 'home terminal' market. We have also seen that the term 'home' has to be interpreted loosely, as mobile communication is increasingly becoming integrated into our daily lives. Despite the market claims of the various vendors, it is likely that there will not be single model of 'home computer', perhaps not even a dominant one. The most likely scenario could be the 'diverse multicomputer household' where the 'computer in the study' is the future equivalent of today's full function machine (obviously grown more powerful with the power of new technology), accompanied by a range of simpler, more specialised devices, some constructed according to consumer electronics principles. These will include the TV set-top boxes, fixed function electronic books, mobile devices, etc. One other area to watch is the possibility of an increase in alternative input devices. All the goods we purchase at food supermarkets and many other stores are bar-coded, but very few home systems make use of bar-code readers. If these were available, we could keep track of the contents of our store cupboards; other readers could be used to programme washing machines with the right programmes for specific clothes. There are also a large number of ways of embedding machine-readable text into paper: recipes, maintenance instructions and so on could be read by computers from conventional manuals. Another issue is that of the *home-bus*, the notion of a domestic wiring or wireless standard for the intercommunication of domestic devices and also their connection to wide area networks. Various schemes for this have been proposed and it is probably only a matter of time before such systems become reasonably widely installed. From the point of view of eBusiness, a number of applications for them come to the surface. In particular, one could imagine networking one's domestic appliances so that the contents of one's storage cupboards could be compared with a recipe from a TV channel and cooking instructions down-loaded from the same source into the cooker. Less speculatively, remote diagnosis could be carried out of one's domestic appliances. (We discuss this further in Part 4, *Service and Support*).

All of these possibilities raise new opportunities for electronic retailers. They also create issues of standardisation and the need for multiplatform delivery.

4

The Retail (eCommerce) Server

In many ways, the most critical component of the shopping experience is the vendor's platform, on which is mounted most of the eCommerce software. As Figure 4.1 shows, this platform looks two ways, towards the customer and into the business.

The eCommerce platform is conceptual, rather than real. As we shall see, it consists of parts of a multi-tiered computer architecture, mounted on a number of hardware and software platforms. Indeed, parts of the eCommerce model will sometimes be hosted on the user's client terminal and not just on a vendor's server. But the concept of an eCommerce platform is nevertheless useful for it allows us to think about the necessary elements that are needed to support eCommerce. In our discussion of electronic retailing principles we mentioned the DAVIC set of requirements (page 15). DAVIC have also mapped this onto a base set of *functions*, performed by the end-user, the service provider, the content provider and the network provider [4]. Their wish list, shown in Table 4.1, provides a top-end specification for the shopping experience, and sets quite demanding targets for eCommerce developers, using existing technology.

As we also said earlier, real on-line shopping examples have been based on what is possible, perhaps relatively easy, to implement, rather than the DAVIC list which may or may not become realised when digital interac-

Figure 4.1 Positioning the eCommerce platform

Table 4.1   The DAVIC wish-list

DAVIC base set of functions

End-user
U1 Move through the shopping environment
U2 Select items of interest
U3 Receive (i) pictures of items, (ii) text, (iii) audio, (iv) motion video, (v) still and animated graphics, that describe items
U4 Talk to a real sales person (audio only or audio video), who knows the context of the application (for future consideration)
U5 Control media clips, including repeat, pause, and abort
U6 Authorise payment/purchase of goods
U7 Enquire about and alter previous purchase (orders) including requesting exchange/return authorisation
U8 Being able to make a hard copy
U9 Reserve products/services
U10 Select payment method
Service provider
S1 Provide the shopping environment
S2 Request media clips to be sent to the user
S3 Send media clips to the user
S4 Process user's order items
S5 Keep an intermediate list of acquired
Content provider
C1 Provide media clips for products
C2 Provide information about price, availability, delivery times, special conditions
C3 Categorisation of material for electronic selection
C4 Determine layout of virtual store
C5 Assign products to virtual departments
Network provider
N1 Transport various data formats down to the user including: motion video, still pictures, audio, text and graphics
N2 Transport information from the content providers or service providers to the server, in order to have rapid updates on product information
N3 Allow for the dynamic addition/deletion of connections between the end user and additional servers (i.e. if the user 'clicks' on an item that has a video clip, then a video 'pipe' must be set-up to the user)

---

tive TV becomes a mass-market service. Nevertheless, it may be worthwhile retaining some memory of the list to compare with what actually has been achieved.

4.1   CLIENT–SERVER MODEL OF eRETAILING

As explained in the other chapters covering retailing, there is a range of

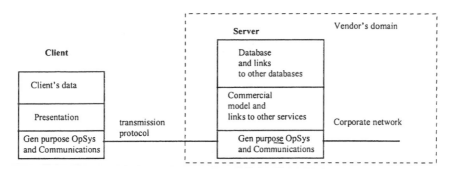

Figure 4.2   Client-server eCommerce

transmission techniques for delivering the eShop and an even larger number of customer terminals on which to receive it. However, the technology used by the vendor to serve the customers has much less variety. In all current models of electronic shopping, the common factor is a small to medium-sized computer operating as the server in a fairly simple, classic client–server configuration [26], as shown in Figure 4.2.

There is more variation in client hardware, as indicated in Part 1, *Retail Terminals*, but the general principles are very similar in all cases: the client (PC, TV set-top box, kiosk, etc.) contains local data and a general purpose operating system and communications hardware and software. In the case of a PC for example, these could be a Microsoft Windows operating system, a point-to-point Internet connection software and a hardware communications card. There is also a presentation layer, which allows the user to view the shopping scenario and interact with it. Again, in the PC case, this is embodied in the *Web Browser*.

Messages between the server and the client, and vice versa, are usually relayed via an independent Internet Service Provider, (ISP), (which has the responsibility for mediating between the point-to-point Internet protocol resident in the client and the full TCP/IP protocol) see page 135. Sometimes the vendor may provide the ISP function itself, but this is mere detail.

The actual shopping process is defined in the server's commercial model, which specifies the make-up of the screens that the client can view, the forms that they can fill in, the process of loading a shopping cart, the verification and basic security aspects, and so on. The server also contains the *catalogue* data, or has direct access to a separate database containing the information on the products and services for sale. As we said, this appears to be the universally accepted model at present, probably for a number of sensible reasons. The client machine is isolated from changes in the commercial model and users do not need to load new software every time the vendor decides to change the design or complexity of the shopping experience. If it were otherwise, then there could be compatibility and configuration problems with the client, perhaps inadequate memory or quirks in functionality. Working to the model as described allows for a market in which most of the cost and inconvenience of providing performance and upgrade will reside in the relatively few servers on the vendors' sites. In this way, hassle to the customer is reduced and barriers to accessing a specific vendor's offering are effectively removed.

Although this is the almost universally adopted design, there are variants and alternatives which require mention. For instance, the above example using a PC does mean that the client is relatively 'thick' – it contains a significant amount of software, including a full operating system. Despite this, in the past, the client's flexibility was limited: the browser was really only a document location and retrieval service, enhanced by the ability to send data to processes that ran on the server. In the last few years, there has been an increase in this client flexibility through the use of JAVA and ActiveX (see page 59) and other methods for downloading software from the server and running it on the client machine.

In one sense this 'fattens' the client – after all, more software now resides within it, but it also provides a potential for slimming as well: in many cases, the client does not require a large, general purpose operating system, at least while being used as a shopping terminal. Therefore, why not simply down-load a simple 'shopperating system' which combines the functionality of application and necessary operating system utilities, but with much reduced memory requirements? This could potentially reduce the cost of the client terminal and, in the case of mobile terminals, minimise weight and power consumption. An equally radical alternative would be to consider a customer terminal which was virtually nothing but a screen, with all the flexibility of presentation retained in the server.

This concept is not as fanciful as it might seem: television is today's embodiment of exactly that principle. With many digital channels available in stagger-cast or video-on-demand mode, there is the possibility of providing very thin client operation for home-shopping – and thereby using an extremely low-cost terminal, perhaps with only a simple telephony channel for order-taking on pre-set forms.

It has to be said that this *thin client* concept is a contentious one, with the major industry players taking sides which tend, naturally, to support their own product lines. Microsoft, for instance would tend to see the PC operating system as a growth area, rather than one likely to shrink; ORACLE, on the other hand have spent a considerable amount of time trying to promote the thin-client, database oriented model. We discuss this issue further, when we look at terminal equipment.

4.2 HISTORICAL DEVELOPMENT OF WEB-COMMERCE

Notwithstanding these theoretical discussions, today's reality is that on-line eCommerce is dominated by Web browser and Internet protocols. The Web was first developed as a tool to support scientific research, by providing a way for research workers, distributed about the world and connected to disparate computer systems, to access documents and communicate with each other reliably and efficiently, using the transmission protocols of TCP/IP as explained in Part 2, *eBusiness Systems Architecture*. The Web is an *application* (or, at least, a set of applications) that runs on top of the Internet. It is not absolutely necessary to run Web applications on top of the Internet and it is not inevitable that electronic retailing runs on either or both, although they both present, individually and together, a very good case. Indeed, today this is invariably the configuration chosen.

Most readers will be quite familiar with using the Web to gain access to remote servers across the Internet. Web services are provided via the standard *client–server* architecture described earlier: the user's PC, workstation or, perhaps, a digital set-top box, is equipped with a piece of software, a *browser*, which communicates with a remote *Web server*, using one of a set of application protocols, the most common of which are *http* for 'Web pages', *smtp* for eMail, *ftp* for accessing remote files. Note that the Web server is a piece of software, not an item of hardware (although, of course, it can be hosted on a computer reserved for that purpose).

Of greatest interest in eShopping applications is http, because that is the protocol used to access the Web pages on the remote server that collectively try to create the display catalogue for the goods on offer. Users access an eShop by sending a browser request, in the form of an http command. The command:

http://www.myshop.co.uk

begins by locating the IP address corresponding to the Web server for WWW.myshop.co.uk. It does this by first accessing a *Domain Name Server, (DNS),* which has been assigned by the system administration to the (sub-) network on which the client computer sits). An example of how this works, in the case of a home shopper connected to an ISP (Figure 4.3).

The DNS maintains a look-up table that sets up a correspondence between the (usually) memorable name ('www.myshop.co.uk') and the 'real' Internet address (the 'IP address') of the shopping server. If it does not itself contain this data, it can pass the request onto a higher-level DNS service.

Once the name has been bound to a real address, the client can send a request to the Web server. In this case, because it has not sent any additional commands in the http message, it is effectively asking the server to send its standard entry page. If the request is successful, that is exactly

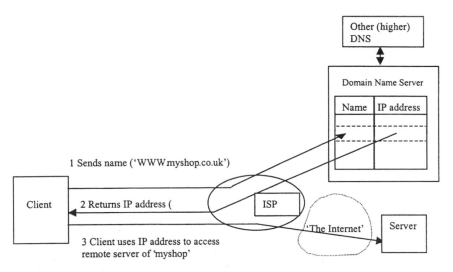

Figure 4.3 Operation of domain name server

what the server will do: send down a return message to the browser which, in turn, displays it on the user's screen. The message that has been sent, is a *Web page*, which in nearly all cases today, been composed using the *Hypertext Mark-up Language (HTML)*. In general, a *Web page* is a document, usually written in HTML, held within a file, where the file can be accessed over the Web.

HTML is a much-simplified version of the *Standard Generalised Mark-up Language, (SGML)* [27], a fairly complex electronic publishing language. This not only specifies the layout of documents but allows their detailed structure and, indeed, their context within an organisation, to be described in a manner that is unambiguous and which can be processed by machines. SMGL might at one time have been a potential candidate for the programming language to be used to describe complex inter-business transactions such we shall discuss in Part 2. However, it has probably lost out to a rather simpler candidate, the *Extensible Mark-up Language, XML*, which we discuss in more detail in Part 2, *Managing eBusiness Knowledge*.

Like most desktop publishing software, HTML surrounds the text intended to be viewed by the user, by additional *meta information* giving further instructions to the browser that displays it.

If you are not familiar with the concept, it is a good idea to look at some specific examples. An example of an HTML page is the default browser screen itself. Start up your browser. Then, to view the source code for what is visible to you, select 'source' from the 'view' option on the browser toolbar. For instance, doing so whilst viewing the introductory page of Internet Explorer, will display the following text:

<h2 style = "font:8pt/11pt verdana; color:black" id = "ietext">Internet Explorer </h2>.

This line is responsible for you seeing the text 'Internet Explorer' displayed as a 'type 2 heading' in black. Notice how this is done, including the delimiting *tags* <h2.../h2>, between which the operation on the text is to start and end.

Initially, HTML was mainly used on the Web as a medium for designing attractive layouts for text documents, and this is still one of its major roles today. It provides facilities for different strengths of heading, numbered and unnumbered lists, data tabulation and so on.

However, perhaps the most interesting feature within HTML, is the ability to include an http reference to another Web page – for instance, a page describing a product may contain a statement such as '*details* '. This underlined name is a *link*, clicking on which sends an http request to the server in control of that file to return the corresponding page. (The correspondence between the word 'details' and the actual address of the page, has been inserted by the programmer and can again be seen by inspecting the source code.)

Originally, links were indeed mainly represented on the Web page as an item of text – many still are – but Web page development environments now allow moderately unskilled programmers to include a button or an image, to represent the link to the viewer.

The development of browsers capable of handling *frames* was an early enhancement to the ability to display a single HTML page. Here the screen can be split into a number of areas that each can hold a separate file, sized and positioned according to a master *frame-set page*, which indeed acts as a frame in which the individual pages are inserted. This provides, for instance, a very convenient way to keep a control tool-bar on the screen continuously, whilst allowing the user the opportunity to see the results of a product catalogue search in the remainder of the screen (Figure 4.4).

Over-use of frames can cause problems and is discussed on page 370.

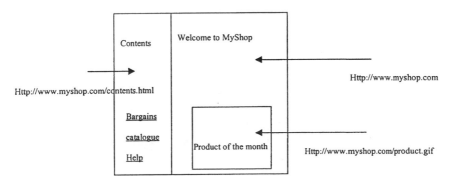

Figure 4.4 Use of frames to produce a composite page

4.3   WEB FORMS, CLIENT-SERVER INTERACTIONS AND DYNAMIC PAGES

So far, we have described the simplest of the facilities of the Web for retrieving information from a server. The basic instruction *http://www.myshop.co.uk* results in the loading down to the client of the Web HTML page stored at that address on the server as a static data file. When a client sends an http request to the server, the latter handles the request via an 'http demon', a piece of code generated to handle each individual request. The demon is called upon to run only one type of process: remote file transfer, nothing more. The pages are extracted from the disk, a few tops and tails are added and the data encoded for transmission using HTTP (Figure 4.5).

Sometimes this may not be the best or sufficient way to do things. Two particularly frequent operations come to mind: the selective retrieval of further details of information summarised on a Web page currently on view, and the returning of information from a client to a server, for example, a postal address, in order to complete a transaction. The obvious way of doing this is by including additional information in the http string from the client that can start up a process on the server and supply it with the information relevant to this particular event (Figure 4.6).

A client request to the Web server invokes an HTTP 'demon' in the normal way but, in addition, it passes the rest of the http string to the main operating system of the server, in a way that allows the operating system to start up the process named in the string and supply it with the data also contained in the string.

The specification for transferring this information and starting the secondary process was standardised some years ago, in a set of protocols

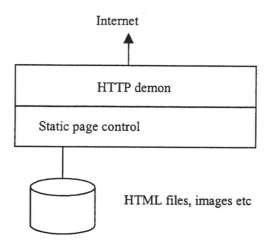

Figure 4.5   Handling a simple http page request

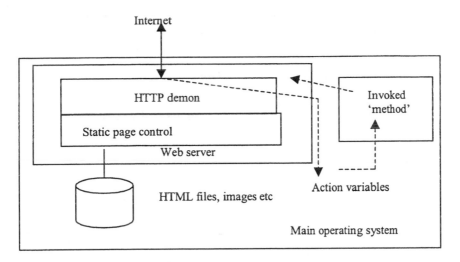

Figure 4.6 Dynamic page generation using invoked methods

known collectively as the *Common Gateway Interface (CGI)*. The process that was started up is known as the *method* and the parameters passed to it, that are specific for the current activity, are *action variables*. The CGI protocol defines how these variables are to be represented in the system and how therefore they can be picked up by the invoked method, irrespective of the source code language (usually, C, C++, Perl) of the latter.

Initially, CGI scripts were used for simple methods of collecting information from forms filled in by users and for inserting simple database responses within tables within created by the server, but the demand for more flexible solutions has grown. For example, consider the pages shown in Figures 4.7 and 4.8.

The customer has selected an item from an on-line catalogue by submitting a form to the server, which has returned a positive acknowledgement.

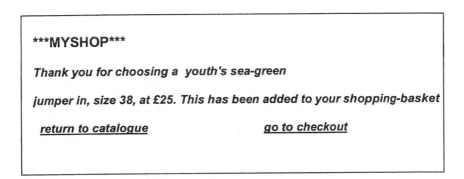

Figure 4.7 Successful addition to shopping basket

MY SHOP

Thank you for looking at our catalogue.

Unfortunately, we have no youth's jumpers in the colour

you requested, at the moment

Please <u>return to catalogue</u> and make another choice

Figure 4.8   Item out of stock

But suppose the jumper is not in stock? Then, the expected response would be something like the following.

The incoming request has resulted in a query to a database and, as a result of the query, two possible actions need to be handled: what to do when stock is/is not available. This is where the ability to run processes based on scripts comes into its own. The script not only provides the ability to return a 'goods available' or 'not available' page to the client, it can 'personalise' it to the specific request (' no youth's jumper…').

We could have created a set of static pages for each possible response, even for each possible product, and stored them in the static page store. Clearly, this would have been very time-consuming and presents a significant maintenance problem. We want instead to create *dynamic pages* which can be customised to meet every occurrence. Rather than having HTML page descriptions in the static database, one for each type of client request, the pages are generated dynamically, according to an algorithm encoded in the script. This can be done by using CGI scripts which invoke the required method which runs as a separate process on the server system (but outside the Web server itself). However, a more modern method such as *Active Server Pages* [28] from Microsoft is increasingly preferred. In this approach, there exists within the Web server itself, a *virtual machine* (defined in a set of processes) which loads the appropriate file from disk into main memory (Figure 4.9).

This file or *script* contains both the basic page data and the programme code for the method. We can think of this activity as that of holding a description of a blank template page, plus a method for finishing off the page in accordance with the current value of any action variables supplied by the client. The method is translated by software in the virtual machine within the server, usually an interpreter (although it could be a complete compile and run operation) and executed, having been supplied with the necessary method and action variables, in the http request.

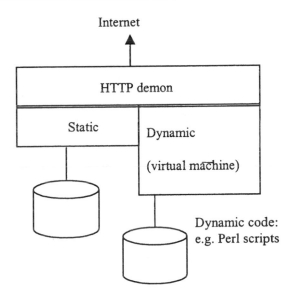

Figure 4.9   Dynamic page generation via virtual machine

Scripts can be written in any one of a number of languages, but Perl and JAVAscript are probably the most frequently used.

Apart from providing advanced programming features beyond that of simple CGI scripts, the active pages approach also has a performance advantage. As shown in Figure 4.6 the original CGI model involves the starting-up of a process that is outside the Web server. This takes up additional processing overhead and can, on a busy system, involve delay. It can even lead to an out-of-range memory error if the design is not properly specified. (An example of a security breach resulting from such a miss-operation is given in Part 3, *Security*. With active pages, the virtual machine which runs the process is part of the Web server and is therefore quicker and more reliably integrated.

4.4 SCALEABLE SOLUTIONS – DATABASE INTEGRATION

It is possible to create a Web site for a retail operation by writing a set of HTML pages that contain every item for sale, all its variations and price, within the pages themselves (Figure 4.10).

If the catalogue is not large and not likely to change substantially with time, then it is feasible to do it this way, altering the pages each time you want to add/delete/change a product, its availability or its price. In order to make sure that there are no embarrassing differences between any

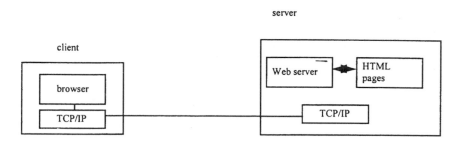

Figure 4.10 Static page catalogue

inventory, delivery schedule or stock-holding, and these Web pages, it is necessary either to take great care to update the pages regularly, whilst preserving a safe level of inventory. Alternatively one has to make it clearly understandable that customers will have to check availability, perhaps via a telephone call or eMail. This can be rather a turn-off. A better alternative is to link the Web service to the stock-holding database [29] (Figure 4.11).

In Figure 4.11 we have introduced two new elements, *application services* and a *stock database* or *catalogue*. Some people believe that introduction of the on-line catalogue will fundamentally change the business model of retailing operations. They see it as being a customer-driven, interactive process which overturns the earlier model business being driven from consideration of stock levels. However, in technical terms, this catalogue function can still be provided by traditional, legacy, *database management systems (DBMS)* that have been around for a long time. They can be small systems hosted on IBM compatible PC's for use by individuals or small organisations without much programming skill: Microsoft's Access is one

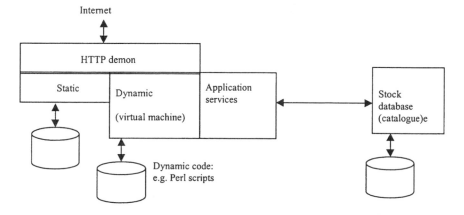

Figure 4.11 Retailing server with database catalogue and dynamic applications

such example, they can be medium-sized systems for use by skilled programmers and for use on larger computers, in which case they are commonly designed around the *SQL* query language, and running on ORACLE, Microsoft SQL Server or IBM DB2, or they can be very large, customised systems, perhaps running business-critical software, in which case, they may be quite old, legacy systems. Whatever they are, it is usually possible to integrate them into the Web environment, and this is done via the *application services*. These comprise self-contained pieces of software, often called 'middleware', that may run on the same physical computer as the Web server, or on a different machine. They must be designed to interface between the very different environments of the event-driven Web server and the transactional integrity of the database. This is more extensively discussed in Part 2, *e-Business Systems Architecture*. As its name implies, different application servers can handle other applications than that of databases, but the database requirement is one that is increasingly in demand.

4.5 OPEN DATABASE CONNECTIVITY AND COMPONENT ARCHITECTURES

We can briefly examine one popular database solution, (leaving detailed discussion to Part 2, *e-Business Systems Architecture*). Even before Web servers were much in evidence, database systems often acted as servers to application clients. Initially this was conducted using proprietary interfaces, which of course leads to lock-in to a specific vendor. To facilitate the use of PC-based applications, Microsoft produced middleware software that can run as an application service and offers an open interface. This software, *Open Database Connectivity* (ODBC) can allow PC applications to talk to a variety of relational databases, through standard SQL queries. ODBC also allows files within the DBMS to be referred to by name, rather than by their precise location in the database. Although it was originally designed for the Windows operating system, ODBC has now been recognised as a standard by the SQL Access Group and it has been ported to a number of mainframes as well as to Macintosh and UNIX environments and further developed to handle non-relational databases [26].

Note that the connection of a database to a Web server is one obvious application requirement, but there are others: in particular, the ability to complete a customer's order, by extracting credit-card details and carrying out a credit transfer is another that is frequently required. Harnessing these other applications is part of the construction of an end-to-end *n-tiered business architecture*, which generally requires a *component-based* solution. This will be explored further, in Part 2, *e-Business Systems Archi-*

tecture, when we look at the processes that are less directly visible to the end customer.

4.6 THE SIGNIFICANCE OF 'STATELESSNESS'

Returning to interaction between customer client and vendor server, we now cover one aspect which is not always understood: it is the browser, i.e. the software in the client, that remembers what was visited previously, whereas the server does not. As far as the Web server is concerned, each click is a new request and the previous ones are forgotten. The interaction between Web server and Web client is a *memoryless* or *stateless* process. This was a deliberate decision made when the Web protocols were first designed: it removes any problems of lock-up, etc. that might occur because of the rather unreliable nature of communication over the Internet, or overloading at the server. It also reduces the overhead on the server because the latter is not committed into a semi-permanent transaction with its clients and therefore does not need to hold status information on what has gone on before.

But it does have an unfortunate implication for eCommerce: each action by the client is seen by the server in isolation and actions cannot be aggregated into a complete process. To take a simple example (which also happens to be a DAVIC requirement): when we shop we often purchase a number of items, but we obviously prefer to settle-up with only one payment. In its simplest form, this is not something we can do on a Web operation: without adopting some more sophisticated approach, each selection would require an individual purchase form to be completed. What we would prefer is the analogue of the *shopping cart* that could hold information on the list of products we have chosen, even allowing us to remove items on second thoughts and then, finally settle up when we are ready.

The simplest way to carry this out is to make use of the *cookie facility* available on all current browsers. A cookie is simply a piece of data that is passed to a browser by a server and which can subsequently be retrieved by the server. In order to create a shopping basket, a server creates a space in a database and assigns to it a unique reference code which it will use to identify a customer. It then sends this number to the customer's terminal, where it is stored in the browser as a cookie. Each time the customer completes an entry in an order form, the cookie is transmitted back to the server, where it allows the customer-specific entry in the database to be updated.

Although the use of cookies is very convenient, it is not always possible: some users distrust them (probably without strongly valid reasons) as a mechanism for sucking information out of their machines without their control. Consequently, browsers are equipped with the option to turn off the cookie mechanism and some users make use of it.

An alternative, with reasonably modern browsers, is to make use of JAVA applets, which are actually rather more intrusive than cookies, being pieces of executing code rather than just data, but which tend to be more accepted by users. It is even possible to create a shopping basket that runs on the client, rather than on the server, only passing the purchase order to the latter, once the customer has finally made up their mind. For traditional PCs that are connected via a reliable wired network, this is probably not a good idea, as problems mid-way through the shopping are more likely to occur on the client than on the server, but it might be a good idea in a mobile environment.

4.7 SERVER INSTALLATIONS

The client/server model used in the above sections is, of course, a much-simplified schematic of the real installation required to meet a working e-commerce situation. Not every organisation can afford to run a Web server, let alone the time and effort to construct its own eShop from scratch. How can small and medium enterprises compete, on-line with the large companies, without having their own servers, staff to run them and creative Web page designers? To meet this need, a number of products are appearing on the market, which provide a range of facilities intended to allow the smaller business to have a Web presence. Companies such as HipHip and WebToolPro.com offer space on servers and tools to create eShops. They will provide domain name registration (http://www.myshop.co.uk, myshop.com, etc.) and also allow eMail forwarding, for customer queries. The tools usually also allow someone untrained in HTML easily to customise a series of Web pages, from a range of different page styles, into which can be cut company logos and illustrations. The business can create a catalogue and alter it as required, e.g. to show stock availability and prices. Users can browse the catalogue and put selected items into a shopping basket. Order forms allow customers to enter necessary details and security is typically provided via *secure sockets layer, (SSL)*. (See page 266.)

Of course, this is only an eShop-front, with minimal integration into the rest of the business's working processes, but it does allow a small business to go on-line almost immediately, for an annual sum of around £500. This would appear to be a very sensible approach for a small, business with limited stock variation and a product which is 'niche' but potentially with wide geographic market which could be fulfilled, e.g. by post.

Many larger organisations also prefer to hand over the running of the server to an outside company, either an ISP or a specialist outsourcing

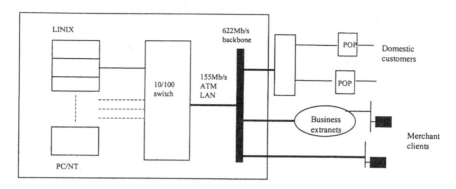

Figure 4.12   Internet eCommerce service provider installation

company. Some prefer to do it in-house. Whatever the route chosen, the installations generally follow a similar pattern.

Figure 4.12 shows a large installation such as would be provided by a company offering connection to the Internet plus the ability to host eCommerce applications, but is relevant to those of any size. One of the major uncertainties with an Internet site is in knowing how much traffic to expect. Thus, providers usually go for a scaleable solution which is easy to increment as traffic grows, simply by adding in additional servers and additional transmission and routing capacity. The Web servers themselves are on the left of Figure 4.12. They can either be UNIX (LINIX) machines, in which case they can handle several eShop installations, or be smaller, Windows-based machines running across Microsoft NT, in which case there is usually one per shop. Table 4. 2 gives a specification for an example of the latter.

Most eShopping users (on the right hand side of Figure 4.12) connect to the shopping service via dial-up modem connections to an Internet service provider's local *Point of Presence (POP)*. The POPS are connected across the service provider's network to the Web servers (on the left-hand side of Figure 4.12). In many cases, a company would have contracted with an Internet eCommerce application service provider for the latter to

Table 4.2   Web server specification

A typical specification for a windows-based Web server (e.g. Dell 'PowerEdge')

CPU: dual P3 533 MHz, both active and providing load balancing
RAM: 256 Mbytes
Hard disk: 9 Gbytes
Construction: 'robust' rather than 'ruggedised'. Equipped with 'uninterruptable power-supplies'
Approximate cost: £3–4 k

supply application services such as catalogue, credit validation, payment taking, in addition to the basic Web server. These application servers would also be provided on the service provider's enterprise network, which may be extensive. Alternatively, if the eMerchant wanted to handle most of these itself, then the service provider would be connected to the merchant across an intranet or extranet connection. This is discussed further in Part 2, *e-Business Systems Architecture*. At some place between the public 'face' that the Web server presents to the outside world and the internal, corporate databases and payment servers there will also be one or more *firewalls*, providing protection from accidental or deliberate actions that might damage the critical processes.

4.8 SERVER PERFORMANCE

Most performance measurements of Web services have concentrated on the speed limitations of the various links between client and server and until comparatively recently, little attention has been given to critical analysis of server performance and the effect this has on the user-experience. In fact, it is not possible to treat network speed and server performance as independent components that contribute to the overall performance: they are inter-linked and not simply additive. For instant, excessive queuing at the input to the server, results in the need to repeat commands across the network in order to keep the session running. Simple measures of server performance, measured at the server, are not necessarily indicative of the view seen by the client: in a study of one of the largest ISPs in the US, a research team from Hewlett-Packard Research Labs reported a number of such issues [30]. They carried out measurements on server performance against input demand (Figure 4.13).

As demand increased, the server's ability to process all the http requests gradually levelled off, as shown by the solid curve. Looked at this way, degradation in performance looks 'graceful', with no dramatic fall-off in performance. However, if the server does not return an acknowledgement of the http message, because of overload, then the request 'times out'. This leads to a repeat request being generated, which puts further demand on the server. As demand rises, a higher and higher percentage of this demand is, in fact, repeat requests. The net effect, as seen by the clients (who are, after all, the only ones who can really assess the service), is for a collapse in the number of successful sessions, as represented by the dotted curve. Service has actually degraded 'catastrophically'.

It is not trivially simple to avoid problems of this type: one needs to arrange to monitor at least some of the http requests to detect for repeat requests, and it may be necessary to include diagnostic software to look for instances of pathological behaviour. It may also be worth looking out

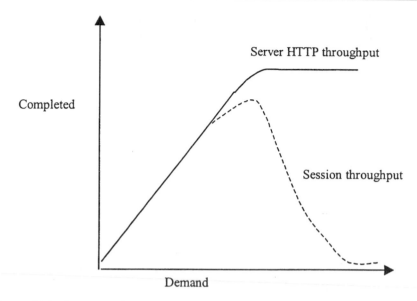

Figure 4.13   Server performance as a function of demand

for denial of service attacks (see Part 3, *Security*), in particular, large batches of very rapid 'refresh commands' coming from individual clients, which may be indicative of automated attacks. There is a complex issue as to who is responsible for loss earnings, under these conditions, the ISP or the client Web site owner.

4.9   SERVERS FOR MOBILE CLIENTS

As we discuss at several other places, there is a rapidly growing interest in mobile eBusiness applications. In many ways there is very little difference in server terms, between static and mobile access protocols and service aspects, but there are some significant variants.

Many mobile applications will involve simply the use of a personal computer device, in the form of a laptop. This can be plugged into a hardwired socket or operate over a radio link, for example using a modem connected to a mobile phone. There is no real difference between this and a telephone connection from a fixed point to a conventional ISP. Perhaps there will be the likelihood of a slower and less reliable connection and the desirability to use secure communication facilities, to avoid fraud, etc. but nothing fundamentally changes.

On the other hand, some mobile terminals, in particular those evolving from mobile telephone technology, may not be able to access directly a standard Web server. In Part 1, *Retail Terminals*, we discuss terminals based on *Wireless Application Protocol (WAP)* which is a case in point: the client–server communication language, the user interface scripting language and various security features all differ from stand Web principals and have to be hosted on clients and servers specifically designed to handle WAP. Usually there will be a gateway server that operates directly between the client and a Web server, in order to mediate between their protocols. See Part 1, *Retail Terminals*, for more details.

4.10 MULTIMEDIA – AUDIO AND MOVING VIDEO

The demand for seductive content has moved the Web from text-only, to static images, to moving images and sound. Because of the very restricted data rates generally available to domestic customers (see Part 1, *Retailing network technologies*) we have to warn against unrealistic expectations about this, in terms of its quality, but it is worth saying something about the currently available delivery mechanisms. This could just as easily have been included in the chapter on retail terminals, as the solutions all involve both client and server. In the case of the client, there will usually be the need for some plug-in or at least the possession of an up-to-date browser. We should note that there is a very real difference between the technology used for applications that only deliver multimedia in one direction, (for example, seductive images in a shopping catalogue, or movies-on-demand entertainment) from those that require two-way interaction, such as video telephony. In the latter case, we assume that any image processing and coding must be carried out equally at both ends, a symmetrical model. However, for one-way retrieval of stored images, we can relax this requirement. Moreover, in such cases there will usually be more receivers of such signals than there are producers, (many more TV receivers than transmitters, for example). These facts mean that coding can be asymmetric and coders can be expensive, though not real-time, whereas decoders should be cheap and real-time. This goes a long way to explaining the differences in the philosophy of coder design between the telephony/collaborative working community and those involved in entertainment and catalogue development. See also the discussion in Part 1, *Principles of eRetailing*.

Another aspect to consider is the distinction between media solutions that only allow the downloading into the client of a complete file, which is then played 'off-line', and methods that allow 'real-time' playing, where the server *streams* the content into a client which plays it back more or less as soon as it is delivered. (There will usually be a short processing delay.) Clearly, for continuous background to a shopping experience, a streaming solution is preferable, although off-line techniques can be used for special effects of short duration. When one considers that digital signals from audio compact disks, for example, are continuously streamed at $56\,kbit/s \times 20 = 560 \times 2 = 1Mbit/s$, one begins to realise that all audio and video coding standards must work to achieve quite significant reduction in data rate, in order to conduct good-quality signals over low-speed on-line links.

It is also necessary to separate out the multimedia 'product' from the basic technology: the tools and controls on offer, as opposed to how the signal is coded. A number of products have achieved significant market share, including *RealSystem, Shockwave, Quicktime, Microsoft Windows Media*. Some of them are rather proprietary in terms of the platform they run on, and each has its merits and demerits. Some are better for audio than video, or vice versa. In terms of how the signal is coded, there are again a number of proprietary solutions. One good, apparently unbiased, analysis is currently on-line [31].

The proprietary solutions may be around for some time yet for streaming applications, but one standard approach is beginning to predominate in the off-line audio case and where the connection between client and server can support quite high speeds. This is the *MPEG Layer 3 (MP3)* Standard, developed as part of a series of Standards by the *Moving Pictures Expert Group* [32] who, as the name suggests, are also concerned with video standards. One of the reasons for the popularity of the MP3 standard is the fact that it does not really provide much in the way of copyright protection!

In the wider band moving video case, there is a steady adoption of the *MPEG 2* multimedia standard, which covers a range of allowable transmission rates with broadcast TV quality at around 2 Mbit/s.

4.11   INTERACTIVE TV AND VIDEO ON DEMAND

With the development of transmission techniques that allow fractional Mbit/s or even higher rates to be delivered into domestic premises, there is increasing interest in providing high speed delivery of Internet services and/or digital TV by other means than simple broadcast. Although we have so far assumed that the client platform will be a personal or mobile computer-like device, there is clearly no reason why Internet and Web technology cannot also be used to provide eCommerce services on digital TV channels to conventional TV sets, either through digital broadcast or by video on demand.

As explained in that chapter, digital TV uses a set-top box, which is to all intents and purposes a simple PC, with a minimal operating system and separate image ('TV picture') and graphics memory planes. As an example of a simple eCommerce application, we could conceive of a hidden 'running commentary' flowing in synchronism with the video programming, which could be made visible on request (through the user's remote control) to provide captions to the programme. For example, a travel

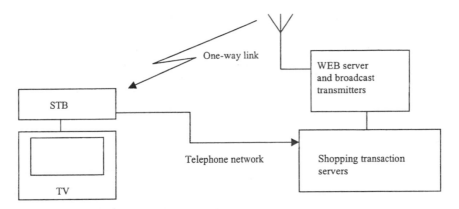

Figure 4.14   Simple interactive TV service

programme could contain an advertising commentary. Alternatively, we could imagine that a service provider broadcasts a recycling carousel of pre-selected Web pages, a Web shopping channel for instance. Users choose them from a menu page and then wait for the screen to be refreshed in a way similar to that used by current *videotext* services.

However, in neither case is there any real interaction between this client and the broadcast server: the latter simply cycles out all the information additional to the main sound and picture in a supplementary channel where users select simply by waiting for the information to arrive. The service is entirely *server-push* rather than *client-pull*.

(Obviously, there must be some interaction between customers and the remote shop, but this can be done by the set-top box making a call to the shopping servers [credit card details, etc.]) rather than through the Web server responsible for the broadcast (Figure 4.14).

Part of the Web transmission to the set-top box must be a unique key that uniquely identifies all the pages that can be viewed (and all the discrete items of data thereon) so that this can be used in the message between the set-top box and the shopping servers. Note that the broadcast can, in principle, include active pages with a code that can be executed on the set-top box client (Java applets and ActiveX, for example). These could allow order forms to be presented to the user and checked automatically, before automatically triggering the telephone call and sending the order data.

4.12   SERVERS FOR VIDEO-ON-DEMAND

We can consider the case of broadcast television to be an example of a very thin client operation – the TV set merely displays the picture exactly in accordance with the broadcast signal. Of course, we also have a very thin server configuration too: the server simply sends out a single signal to all receivers. This means broadcast is inherently an inflexible service. Despite superficial similarity, this is by no means the case with true *video-on-demand (VOD)*, which places considerable requirements on the server. This arises because VOD provides a one-to-one relationship between each active client and the server, and because it involves the storage and distribution of large volumes of video data.

In Part 1, *Retailing Network Technologies*, we explain the transmission principles for VOD and similar services relevant to eShopping. To recap: wide-band signals, of the order of a Mbit/s or even more and capable of providing high quality moving images, can be supplied to customers premises either by cable modems working on cable TV coaxial feeders into the premises or over conventional two wire telephone cables using *Digital Subscriber Loop (DSL)* technology. One option is then to use these systems to deliver data to customers using an IP-based service. In the case of telecoms delivery, using DSL, we may see architectures of the type in Figure 4.15.

Each customer has on the premises a modem capable of transmitting the high speed signal. These signals are *multiplexed* (combined onto a single transmission path) on a street-by-street or local telephone exchange basis, depending on demand and population density. The multiplexing may involve a low level of *concentration*, that is, there is a possibility that some signals will be squeezed out during busy periods (similar to what happens on an Ethernet LAN). At the *service node*, a combination of

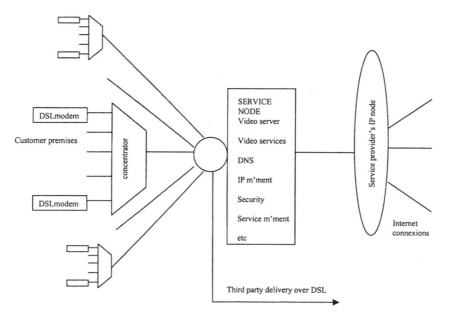

Figure 4.15 Video-on-demand service

servers provide control and access to a number of services provided by the ADSL: network service provider. These include wide-band video servers which can deliver a number of essentially uninterrupted, individual video streams to a number of customers, simultaneously, service management facilities for registering, de-registering, charging, etc. management of the IP streams and domain name management to allow for sessions between the client and any remote servers across the Web, security control, and so on.

The service node then connects to the Internet via the network service provider's backbone network. Until and unless network providers provide very cheap, wide-band connections across the Internet itself, the service node to Internet connection will experience the usual, high contention problems familiar today, unlike the concentrator to service node link, which will provide much higher speeds. This is why the architecture shown in Figure 4.15 is what it is: data (for example, video streams) that requires high-speed transmission, must be held locally to the client.

Thus, in a shopping scenario involving an on-line catalogue with moving video clips, the retailer would buy space on the video server in the service node. The moving video would be stored here, in a standard video format.

The retailer would then create, on its own server, a set of Web pages that defined the shopping catalogue, leaving 'holes' (in the form of frames or image anchors) for the video.

Users of the shop would still access the retail site at *http://www.myshop.-co.uk*, not the service provider's site, thus preserving the retailer's branding and day-to-day change control, but, each time a catalogue section was viewed, the appropriate video clip could be called up (automatically, or by spawning another Web browser). The architecture begins to become multi-tiered, with increased complexity but also flexibility: clients negotiate with more than one server, with the possibility of quite complex session control.

A further element of flexibility is also shown in Figure 4.15: although telecom service providers hope that they can provide all the service node facilities themselves, they are generally forced by telecom regulators to provide access by third-parties to their networks or to services such as DSL which run on them. Once services mature, there should be no problem in principle to a retailer connecting directly to the DSL connection on an equal basis with other providers. Thus retailers could position their own service nodes where they wanted and with their chosen functionality, being compelled only to connect into the IP management facilities used by the telecom provider on the link between concentrator and telecom service node. One issue is access control, (using authentication techniques such as RADIUS see page 139) to allow the customer to have wide-band access to the server. Perhaps a more likely scenario is that the

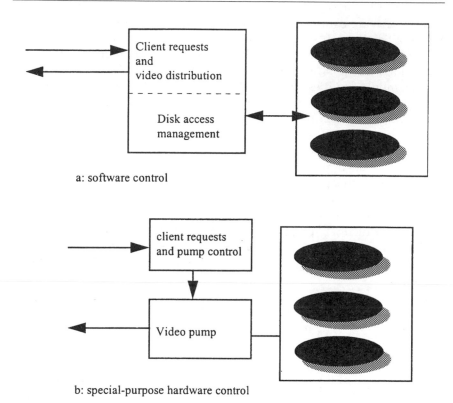

Figure 4.16   Two video-server architectures

option to connect at this level will be taken up by competing telecoms carriers, who will try to offer differentiated services to retailers (price, quality of service, low contention, etc.) just as ISP do today for their services. Yet another option is for ISPs to set up their own equipment within existing telephone exchanges. This has just begun to happen in the UK. One issue which is still at an early stage of research, is the best choice of location for video servers within a national or global network. Is it better to have a few large servers or a network of smaller ones and how should the data (the video clips) be split up between them? Indeed, if the technology of networked video servers is in its infancy, the service specifications can hardly be said to have been conceived!

4.13   SPECIFICATIONS OF VIDEO STREAMERS

As long as ADSL is used only as a means of providing fast access to the Internet, the servers that clients access will merely have to be able to process a supply of HTML files at a higher rate. However, if, as is likely, the demand for truly moving video grows, either because TV/movies-on-demand takes off or eCommerce sites want moving image catalogues, then server dynamics must change significantly. Any VOD service must allow individual users to treat the system rather like a high performance video cassette player: a continuous flow of programming must be available at any time, commencing anywhere within the programme, with pause, fast-forward, fast-back and scan facilities. To provide this using standard computer technology is rather demanding: the server must be capable of servicing a large number of simultaneous interrupts from clients and be able to handle the continuous streaming of high-speed data.

For nearly a decade, researchers have been investigating the options for adapting standard computer architectures, which are based on handling blocks of data, into systems which can support continuous streaming. There are two principal problems: latency and reliability, both of which are critically dependent on optimising the performance of disk drives and their interaction with caching facilities in the server. A large number of alternative variations have been proposed [33]. As shown in Figure 4.16 decision has to be made as to whether the management of data-retrieval from the disks is to be done using the server CPU or by means of special purpose hardware such as a *video pump* [34].

Earlier attempts at VOD tended to favour the special-purpose solution, but more recently, systems based on conventional servers have become possible with the improvement in speed that is now available from relatively inexpensive machines. Both approaches have their supporters: clearly, doing everything using software in the server will reduce costs but increase the load on the server; the reverse will be true where hardware is deployed. Hardware also generally provides a higher level of reliability.

Another way to reduce costs for video streamers (and for other high performance file servers) is to make use of conventional magnetic disk drives. To do this, requires a set of techniques that provide an unusual level of reliability and performance. The approach is called *redundant arrays of inexpensive disks (RAID)*. (Sometimes the word 'independent' is substituted for 'inexpensive'.) There are a number of variations on the theme of RAID and an attempt to provide an official taxonomy in 10 levels, has been derived by a RAID Advisory Board [26]. These levels are defined in terms of their error-correcting codes and by how the data is written *(striped)* across the disk sectors, including the possibility of writing it across more than one disk, in order to optimise some aspect of performance and reliability.

In the case of VOD, optimising the speed of writing is not generally an issue, as this can be done off-line, but achieving a continuously high reading speed is difficult, including the scenario where two or more people wish to access the same video file. In this last case, there is a severe

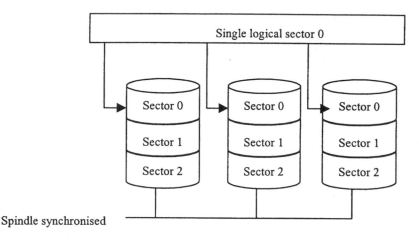

Figure 4.17 Spindle synchronisation of multiple disks

constraint placed on retrieval speeds if the file (typically a movie) is stored only on one disk. One way round this is to synchronise a number of disks and stripe the data across them, sector by sector (Figure 4.17).

The multiple disk array can be treated as providing a set of logical disk sectors each of which is n times larger than that of a single disk (where n is number of disks involved). Thus the transfer rate is increased n times also.

Where long files of movies, for example, have to be stored and are likely to be accessed at independent random times, the above method has problems with buffering. In this scenario, a better solution may be to interleave successive blocks of each movie across successive sectors on different disks (Figure 4.18).

This arrangement permits a number of alternatives for data retrieval, which can be designed for load optimisation [35].

4.14 FURTHER ISSUES OF eCOMMERCE SERVER ARCHITECTURE

In this chapter we have concentrated on describing the major components of a virtual shopping window, shopping basket and shelves, leaving aside the back-office functions such as order-handling, fulfilment and customer support, which collectively complete the complete eBusiness, rather than eCommerce, domain. These are dealt with in other chapters. But there are two further activities required, almost the most important, and which convert the shopping dalliance into a serious commitment: *order taking* and collection of *payment*. We could perhaps consider them to be simply adjuncts to the browsing of the store and the loading of the shopping cart, but there one very good reason why this would be unwise: security.

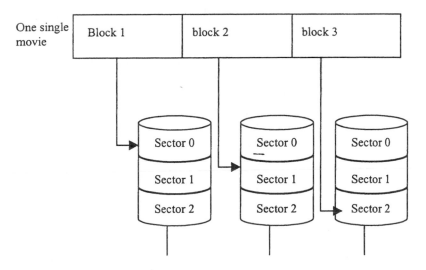

Figure 4.18   Data interleaving for movies-on-demand

Instead, order handling and payment should be considered as two distinctly different processes that occur concurrently in a phase that follows after the preliminaries of shopping (Figure 4.19).

Whilst it is desirable to provide a reliable, efficient and attractive service for the functions shown in state number one of Figure 4.19, none of them is, strictly speaking, fatally damaged as a result of failure or corruption. This is much less likely to be true than for the activities shown in state

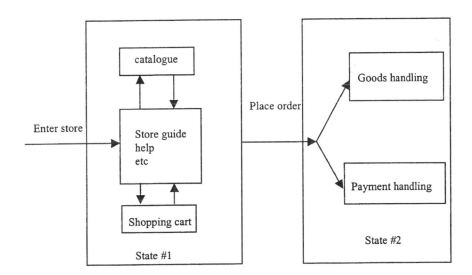

Figure 4.19   'Browsing' and 'Ordering' are seen as two separate states

two. Incorrect operation of goods order handling will often lead to very significant failure cost and customer satisfaction; problems with payment taking can be even more serious. Architecturally and in reality, designers adopt a separate way of dealing with them. Part 1, *e-Business Systems Architecture*, covers application processing, including goods order taking and subsequent processing. Payment handling can be briefly addressed here, with security details dealt with in Part 3, *Security*. The main message of payment handling is that the two-state model of Figure 4.19 is mirrored in the architecture – a physically and logically separated set of servers are used for each. Once the customer has committed to purchase items, then they should be switched through to a highly reliable, *transactional* (see page 154) set of processes that take payment and add the order to a queue. Typically, the payment server will offer the customer the opportunity to open a secure (i.e. encrypted) message-passing channel between them and it, where credit card or other payment information is collected. This process may even run on a different supplier's site, perhaps that of a payment agency. Details of the shopping cart requirements are passed from the Web server to the payment server (perhaps only the total cost), in some safe manner that cannot allow hacking into the latter. That at least should be the intention, as we explain in our discussions on security.

Part 2: Creating the eBusiness

Chapter 1: eBusiness Systems Architecture

In moving beyond the creation of a simple on-line store, a cultural shift from point solution, to end-to-end integration is required. This integration is complex, running across a heterogeneous range of hardware and software distributed-computing and communications platforms. We also need to consider how to integrate on-line applications that cover automated processes, human–machine interaction and computer-supported, co-operative working between humans. Increasingly, the standard communication services are provided by the Internet protocols of TCP/IP, supplemented by security processes that allow the construction of virtual private networks, intranets and extranets. The dominant computer architecture is *n*-tiered client–server, with object oriented components, exemplified by CORBA, DCOM and JavaRMI, that allow us to invoke processes without the need to know precisely where within the distributed environment they will run. Critically important data operations run within a distributed transaction-processing model.

Human beings, as part of distributed teams, achieve a level of empathy by using multimedia conferencing tools, although these are only partly successful, today. Activities across enterprises are supported by managed documentation, which contains heterogeneous content, under version control and synchronised across a number of platforms. Increasingly, this architecture, too, is influenced by Internet development, particularly regarding how much should be stored versus simply 'pointed to'. Automated workflow processes, which remove many manual re-keying and scheduling tasks, are outlined. The central role of wide-area directories is highlighted.

Chapter 2: Managing eBusiness Knowledge

Knowledge management is the conversion of raw data into meaningful information sets within the context of a business model and thus potentially capable of supporting planning and decision-making. One major problem is in the creation of a common understanding of information structure and semantics between organisations. Past EDI initiatives have gone some way to achieving this, particularly in some vertical sectors, but the task remains difficult. Internet developments such as the extensible mark-up language (XML), provide a mechanism for converting semantic descriptions into machine-processable format.

This paves the way for creating corporate knowledge portals which open up parts of the corporate knowledge repositories to customers and suppliers, thus working towards common understanding and purpose. Corporate information should not reside as isolated data sets within the organisation. Instead, the aim should be the creation of a unified data warehouse, consistent and consistently maintained, available in customised views to functional units. In order to be scaleable, automated data-cleansing must occur at a variety of levels within the warehouse, using a level of artificial intelligence.

Increased processing power and storage capability now allow for on-line analytical processing of data, to assist in strategic and market planning. Presentation in data-cubes and visually is discussed, and an overview and critical analysis of intelligent data mining and extraction are given.

1

eBusiness Systems Architecture

In technology terms, what separates an 'e-business' from an 'ordinary' one? Elsewhere we have remarked on the commercial models that emphasise globalisation, on-line trading and distributed enterprise. But when we come to look at the underpinning technologies, we see that many of today's organisations which, except possibly for publicity purposes, would not consider themselves to be eBusinesses, do in fact possess a highly developed and complex electronic infrastructure to support these facets. Automated business processes have been running for over 30 years. However, these processes, although perhaps designed by a single internal IS department, were often developed to meet the needs of specific individual functional units. There was little coordination between them and little thought given as to how they might inter-work. There was often even a complete cultural gulf between departments. Whilst technology can do little in the short-term to change culture, it can certainly challenge it. Therefore, to answer our question, if we were to find one single term to describe the architecture of a truly eBusiness, it would be *integration*.

Let us consider the historical background a little further. If one looks at the adoption of computer automation within an organisation it is often clear that computing operations preserved a complete split between *industrial process control* and commercial *data processing*. Often even their staff were placed in different career structures – 'engineering' and 'admin' or 'commercial'. Their computers were different also: engineering machines were real-time and interrupt driven, whereas commercial systems were data-oriented and carried out large batch processing jobs. In many companies, it was impossible to drive manufacture or supply in response to commercial conditions. Telecommunications, gas and electricity companies, for example, had no way of increasing or decreasing supply to business customers on the basis of new tariffing deals, because commercial and engineering systems did not interwork. It is now possible to rectify this.

Even within the engineering hierarchy we find technology divisions. One need only consider the profound gulf between 'telecommunications' and 'information technology', the first originally meeting the need for communication, the second for information processing. In the past, it was considered correct to treat these as two very separate activities, with their responsibilities in many organisations separately devolved to an IT manager and a communications manager, who might never speak to each other. Today, the distinction has become very blurred, with the arrival of eMail and the increasing need to operate computers within a data-networking framework. The situation is further complicated by the increasing interest in running voice services over data networks (in particular, voice carried via the *Internet Protocol*) with call-centre design being an important case in point.

The principal requirement for eBusiness applications is a low-cost, quality response, to inputs from a number of diverse, temporary channels. Integration of eMail, Web forms and voice calls, purchaser-supplier agreement over wide area distances and heterogeneous platforms, negotiations between people separated by distance: these, and integration of them all, are the main challenges in creating an eBusiness (Figure 1.1).

Systems design must be focused on strategies and solutions that enable this integration task, reliably, securely and in a way that can be built on by others without costly re-invention. Re-useable *components* that are platform and code independent (or at least tolerant) are required, with *object orientation* largely replacing other methods of *remote procedure invocation*. All of this must go on in the presence of many legacy systems, some mainframe oriented, although the dominant platform architecture will be that of *multi-tiered client-server*.

A number of solutions have been proposed and the debate on their merits is still very active. This chapter tries to give an overview of the principal activities involved and some insight into the issues still under discussion.

1.1 IS ARCHITECTURAL OVERVIEW

Originally, business computing was carried out as a point task, without any real concept of networked operation. There are two very simple architectural approaches to this: *mainframe and dumb terminal* or isolated *personal workstation*. Although the former may have been the preferred way of working for large companies and the latter the choice of the small business, in many ways they are equivalent. Their common characteristic is that all the business processes are run on a single platform

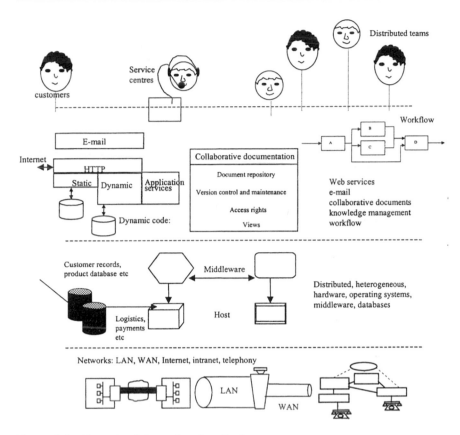

Figure 1.1   Complex integration for eBusiness

or *single tier* (Figure 1.2). Today, many systems have evolved to a *client-server (two-tiered)* approach, (Figure 1.2) where most of the business process runs on the server and the client is mainly concerned with presentation and only holds a limited amount of user-specific data. The next stage is a *three-tier* architecture with most of the process (in practice, a simple Web shopping procedure) running on one server and pulling down data from a database server, the *third tier.* After this, things get more complicated, with additional applications running in different tiers (Figure 1.2).

What is often not made clear in articles on the subject is that the tiers are logical, not physical. There is not one tier per server/computer or one computer/server per tier. One machine can run several business tiers and tiers can be distributed across several machines. This is known and admitted by system architects to be a cause of confusion [36]. However, the terms are widely used and it is necessary to be at least aware of them. For a good, worked example of designing a system for business evolution

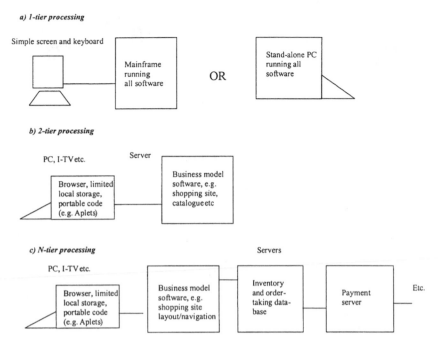

Figure 1.2   Evolution of tiered processing

from simple desktop to complex multistate trading, involving the progressive enlargement of the architecture, see the 'Duwamish Books case-study' by Robert Coleridge, at the time of writing available on-line [37].

   We should also note that multi-tiered architectures have arisen not necessarily because great thought was given to this choice of architecture; in truth, they are more the result of trying to make the best of what was there. One reason why eBusiness design is so complex is because we require several computers on a number of geographically separated sites to work together, despite, in many cases, their differing hardware and software. They may come from different vendors and may cover an age-span of more than twenty years. Today, if we were given the chance to design from scratch, we might prefer hardware designs based on clusters of relatively small, compatible computers and we would write programs using object orientation and portable code. However, many business systems still in service are based on centralised 'big-boxes', running all their processes locally and deploying relational or even flat databases and COBAL or proprietary coding. They may hold data that is difficult to integrate with the layered data models of more modern database structures. In compensation they may enjoy a very high level of security and transactional integrity. In many cases it will simply not be possible to throw away the old and introduce the new without jeopardising the business.

For this reason, solutions to building the extended electronic enterprise have had to be pragmatic. No one has yet seriously suggested that there should be an eBusiness operating system to replace existing ones. Nor would it be feasible to re-write all of the proprietary business-application software. Instead, the approach, as we shall see, is based on building distributed *middleware* that can mediate between disparate or distributed systems. In particular, Microsoft's DCOM, the CORBA open architecture and Enterprise Java solutions currently compete and co-operate in this area.

Because of the complexity of processes and the diversity of machines, these middleware solutions can provide severe speed bottlenecks, if care is not taken in the design.

1.2   CLIENT POWER VERSUS SERVER POWER

In much of the discussion that follows, we are going to concentrate on the server-side aspects of eBusiness integration. This is perhaps logical, as most of the processing power is required for handling operations inside the business. But there is another reason which is perhaps not so obvious. Indeed, the customer's client appears to be growing in power and becoming more integrated with the internal processes. Over the past few years, Internet computing has moved from simple pull-down of static Web pages to an architecture where mobile-code-enabled browsers on the customer's machine pull down Java applets or Active-X code from the server. These chunks of executable code then may interact with the Web server which now behaves like a gateway into the eBusiness, invoking processes within the business application machines (the *third-tier*). The applets may therefore require different access modes to different business application platforms. Hereby lies a problem. Systems that have to cope with this approach are complex to design and thus potentially dangerous for security of operation. This has led a number of commentators, e.g. [38] to postulate that there may be a move back to simplified client/Web server interactions and with the Web server interacting with the application tiers through an *object request broker* (see later) which can invoke longer, more complex and more assured applications. Essentially, the client is *dumbed* or *thinned-down* and the Web server fattened but linked with other servers via a well-conditioned interface. This places yet more emphasis on what happens on the other side of the Web server, within the eBusiness.

1.3 eBUSINESS SYSTEMS ARCHITECTURE

Figure 1.3, which we first encountered in Part 1, *The Retail (eCommerce) Server*, reminds us of the basic eCommerce view.

This is a model centred on retail customer transactions. It uses Internet connections handled via a Web server. This supplies the user's client with seductive and/or informative services based on Web pages, almost certainly in HTML or XML, perhaps containing small programmes (e.g. in Java) that run on the client. The pages may be of completely fixed format *(static)* but are more likely to be *dynamically* created and filled with customised data extracted from a catalogue. By use of other dynamic scripts, customer requests and personal details can be also be passed into the database, for order taking. Both activities are supported by a suite of software known as the *application services*. This very simple *three-tier archi-tecture* that splits out *presentation, (business) application* and *database*, is useful to position what follows, which is based on expanding significantly the application services and interactions with the database(s), into an *n-tier architecture*.

Let us begin by opening out the services into a layered model (Figure 1.4).

This architectural diagram is rather different from the previous Figure 1.3, in that it does not at this level separate front-office selling (HTTP to client) from back office data management (e.g. by ODBC to catalogue). Instead, it treats them equally, spreading their functionality vertically through a number of layers. In the parts that follow, we shall look at these layers in turn.

Notice also that we show a system management strand running verti-cally through Figure 1.4. This is perhaps contentious: architectural models such as IBM's *Application Framework for eBusiness* [39] tend to put system

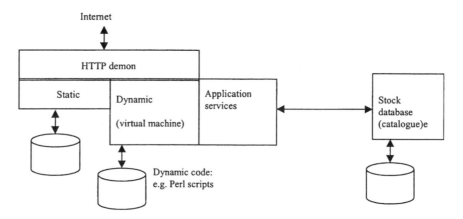

Figure 1.3 Basic eCommerce model

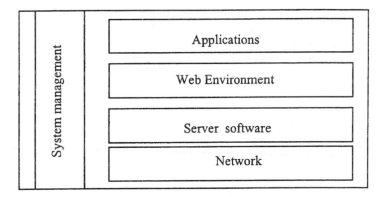

Figure 1.4   e-Business server architecture

management at the bottom layer of their models. Since one very large and important aspect of system management is security and, bearing in mind what we say in Part 3, *Security,* about security threats coming in at all levels, we prefer to make it explicit that these considerations must be treated at all levels.

1.4   NETWORK LAYER

It is rare, even in a single, unified business, to find all of the processes of order-taking, catalogue, supply-chain management, etc. being conducted on a single piece of hardware, within a single room. Almost implicit in our multi-tiered model is the assumption that a number of computers will be involved. These need to be connected *(networked)* together. If they are all in the same room, then, perhaps, we do not need to give too much thought to optimising speed and cost. But if they are separated across a campus, a country, or globally, then this becomes a big issue. This will usually be the case. Many companies exist on multiple sites and so the creation of truly integrated extended enterprises means that data and voice services must work over large distances. Almost any company that makes use of an ISP is going to have to install a connection between the ISP and its own computers. The way that these networked connections can be achieved effectively and a low cost has been, and continues to be, an area of rapid development. The easiest way to understand it is to follow the historical development.

1.5   THE EVOLUTION OF DATA NETWORKS

We saw in the case of customer-to-retail server links that domestic customers at least, usually had to make their connections via modems and the telephone network. The characteristics of this network have, over many years, been optimised to handle voice and are not necessarily so good for data. A 'telephone call' is set-up, end-to-end through all the intermediate exchanges for the duration of the conversation, with virtually no end-to-end delay and the path is guaranteed because the telecoms companies involved have centrally planned their routing and mutually agreed the connectivity beforehand. One consequence of this approach, and a very strong point of telecoms voice services, is *quality of service*. Calls that are connected are more or less guaranteed to proceed without fault or deterioration. If the network is congested, new calls are barred from entering the network, rather than being allowed to steal resources from existing calls. Furthermore, there is implicit in the nature of dialled calls, the concept of a *session*: this begins when the called party lifts the handset and ceases when the handset is replaced.

However, a basic voice channel has a rather low data rate, typically of the order of 64kbit/s and one is charged a time-based fee, irrespective of whether any information is sent or not. Circuits also tended to be one-to-one, rather than one-to-many or many-to-many. For many years, computer-to-computer connections made do with modified voice channels or bought *private circuits* which were hard-wired across the country to connect their sites directly. All the decisions about routing of traffic were made by the telecoms company, with the customers equipment being seen as 'dumb' in this respect.

The emergence of campus-size *local area networks (LANs)* in the 1970s gave rise to the first architecture which could be described as 'computer-based' (in contradistinction to 'telephony-based'). What makes a LAN architecture very different from a telephony one is the relatively short distances involved: a kilometre length of cable is large, in LAN terms. Now, the ability of a cable to carry high-speed data signals is a function of its basic construction and its length. Because LANs are short, they can support tens of megabits of data per second on relatively simple (thereby, cheap) copper cable. This allows for a very simple and 'dumb' network (Figure 1.5).

Terminals A, B and C are all connected to the same cable, Whenever they want to communicate, they simply squirt out packets of data with the send terminal name, the receive terminal name, some control data and the data itself. Every terminal 'hears' every packet and the signals are designed so that any data corruption caused by two terminals simultaneously transmitting can be recognised and the transmission repeated after a short, random delay. Notice that it is the terminal, not the network, which possess the intelligence. There is nothing equivalent to the central telephone exchange, and there is no need to restrict the data rate either (within the basic capacity of the cable).

(We should mention that there other, more structured and less 'statis-

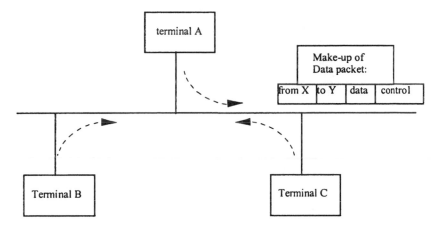

Figure 1.5

tical' LAN protocols which do not rely on random collisions being infrequent – IBM's *token ring* is a notable example – but the above method of sharing the cable as if it were an open radio 'aether', is dominant, under the appropriate name of *Ethernet*.)

One consequence of this approach is that there is no distinction between the transmission of control information (routing, connection-made, etc.) and message data, unlike the telephony case where the control information may even take a different path through the network from the data 'payload'.

Unlike telephone speech and fax data, which is generated at a fairly steady rate, computer data is typically very 'bursty' (Figure 1.6).

It is because of this variability that we can put several devices on the same LAN without their data colliding too much. As we said, the short range means that the cable can support the very high data rate, and the higher the rate of transmission, the less chance of collision.

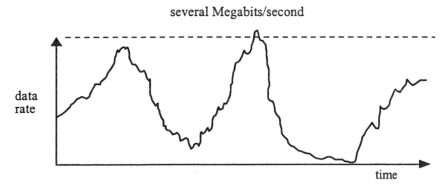

Figure 1.6   The 'bursty' nature of computer data

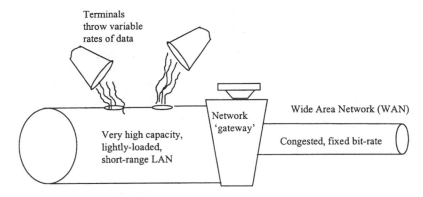

Figure 1.7   LAN-WAN connection

1.6   TAKING THE LAN TO THE WAN

This works well for short distances and networks which are not too busy – notice that even within one LAN, there is still the possibility of extended delay because of the need to repeat data bursts after a collision has occurred. But the position gets much worse when networks are joined together. The maximum data rate that a cable can handle is a function of its length. The longer the cable, the lower the rate. Consequently, we cannot pass high-speed bursts of data easily across long distance links. This condition is aggravated by the further condition we mentioned earlier, in our discussion of telecoms networks: the maximum data rate allowed onto telecoms cables is strictly controlled and priced. So, if you want to inject even the very occasional burst of high-speed data into a telecoms network, you have to pay for a high-speed circuit, whether or not you use it all the time. This can be very expensive. The alternative, is to severely control the rate of which data is injected into the wide area network from the LAN. This is done by providing some form of *gateway* between the local and wide areas. Computer networks must taper into the wide area (Figure 1.7).

The most common gateway control mechanism for this is to use a *router*. In order to understand more about how routers operate, we have to look at one of the key components of eBusiness: the *Internet*.

1.7   THE INTERNET

In the minds of many, eCommerce and the Internet are inseparable. Although this is too simple a view, there is no doubt that Internet-related technology, applications and standards dominate the implementation of eBusiness applications and in almost all cases, at least one part of process chain will be carried across networks operating to Internet protocols.

But first, we should recognise that there is no such 'thing' as the Internet. The real term is 'Internetworking', an activity rather than an entity, the activity of connecting together a bunch of networks and bits of kit, in order to provide some capability for end-to-end transmission. The desire to do this originally arose through the convergence of two distinctly different strands of thought regarding computer networks. One was a US Department of Defense initiative intended to develop solutions that would provide communications survivability in times of nuclear attack. The other was the commercial market which was seeking profitable solutions to a wider-area connection between local networks of a very heterogeneous nature. The Internet owes its positive and negative peculiarities to these twin desires.

Regarding the survivability issue: the US military were concerned that conventional telephony networks, which involved hierarchical connection via centralised telephone exchanges, were vulnerable to attack. It was therefore proposed that the data to be transmitted be broken up into small *packets*, (or *datagrams*), each packet be labelled with its serial number, source and destination and the packets fired off independently into the network. Thus, if one part of the network were to be destroyed, it would be possible to determine which packets were lost and quickly find a new route whereby to re-send them. The protocol for achieving this, the *Internet Protocol (IP)* concerns itself with 'best effort' delivery of the packets. Depending on the intervening conditions it may fail to deliver, it may deliver more than one copy of the packet, it may deliver packets in any order and they may contain errors.

To meet this survivable requirement within the commercial realities outlined above, IP breaks the data up into datagrams that are of variable length up to 64 Kbytes, (although they are usually about 1000-2000 bytes, in practice). These packets of data are frequently further broken up by the intervening networks into smaller 'fragments' to fit their local maximum size restrictions. These fragments are not recombined until they reach their destination and one of the complex tasks of the Internet protocols is in adding additional addressing and ordering information onto these fragments, so they can be delivered and reordered successfully.

Every host ('computer') and every router on the Internet has a unique address, called its IP address. This is a 32 bit address, which is attached to the datagrams to indicate the source and destination of the packet (Figure 1.8).

The address consists of two parts: one part identifies the network to which the host belongs and the other part is a unique number within that network. Networks vary in size and complexity, some are 'private', e.g. university campuses network, others are provided by regional or national authorities. All networks obtain their IP address from a centralised body,

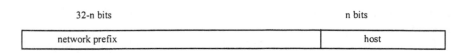

32-n bits	n bits
network prefix	host

Figure 1.8 Structure of an IP packet

the *Network Information Centre*. Networks and network addresses can also be split into 'subnetworks', which helps with system administration and allows the number range to be extended, rather in the manner that telephone numbers contain a local code, a regional code and a national code. Codes are usually expressed in octets, to make them easier to read.

Routers hold tables containing information on some (but not all) of the other networks on the Internet and also some of the addresses to subnets and hosts on its own network. There are a number of ways of creating these tables. For small networks, it is possible to set this information manually. For larger networks, there are a number of automated methods whereby routers acquire this information gradually from observing data passing through them. They can also periodically send out requests to their neighbours, asking them to return information on whom they are connected to. Now suppose a packet arrives at a router (Figure 1.9).

The router compares the packet's destination address with the table and may be able to send it directly to the local LAN where the destination host resides. If the packet is for a remote network, then the router will send it on to the next router recommended in the table. Note that the router does not necessarily send the packet by the most direct route; it may simply send it to a default router which has a bigger set of tables. In the example

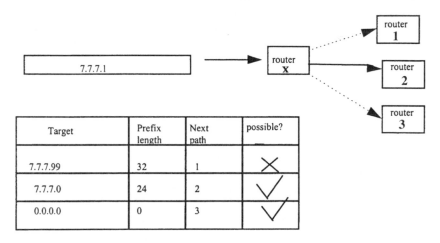

Target	Prefix length	Next path	possible?
7.7.7.99	32	1	✗
7.7.7.0	24	2	✓
0.0.0.0	0	3	✓

• The target and the packet must match for the entire prefix length
• The match with the longest prefix length is chosen

Figure 1.9 How Internet routing operates

shown, the packet destination address of 7.7.7.1 will not match with next path labelled 7.7.7.99, but does match to path labelled 7.7.7.0, that is, path 2. If path 2 did not exist and there were no complete match, then default path 0.0.0.0 would have been chosen.

Because of the resilience-to-attack requirement, it was decided that at each *node* in the network, (e.g. where a router connects between different organisations individual subnetworks), the routing to pass each packet onwards to the next node, would be calculated locally, on the basis of knowledge of what was happening nearby the node (for example, which route out of it was least busy and had not been destroyed). Packets would eventually get to their intended destination, provided a route existed at all, although they might arrive in any order and experience a significant amount of delay. It is important to realise that there was no intention to maintain a record of the 'best way', from end to end across the network. The idea of any semi-permanent best way, was felt to be contrary to what would happen when the bombs began to fall. No route would be guaranteed to survive and the packets would need to find their way through a network whose nodes were constantly collapsing and being rebuilt.

As well as the military imperative to create survivable networks, there were a number of commercial realities: computer networks grew up in a spirit of local service provision and within a fast-moving, competitive, non-standardised market. Consequently, it was inevitable that there would be a variety of communication protocols and little attempt to provide guaranteed quality of service outside the local domain. Nor does the data structure conform to identical layouts. There are a large number of ways that networks can vary, including how they are addressed, the maximum packet size, how errors are handled, how node congestion is dealt with, and so on.

That it is possible to achieve end-to-end transmission across this sort of complexity (and with bombs falling!), is a major achievement, but the Internet does it, although not without acquiring some less desirable properties. Packets can, for example, disappear down a dead-end of routers, either permanently or after a long delay. Packets can be broken up into fragments, by heterogeneous systems with different processing buffer lengths; duplicate packets can even be generated in some cases. For many requirements this is not satisfactory, what we require is some protocol to let us know where we stand and when to ask for information to be repeated.

1.8 LAYER 4 INTERNET PROTOCOL – TCP

Sitting above the IP protocol on the architectural layered model is the *Internet Transmission Control Protocol, (TCP)*, which is intended to handle issues of this sort. Of course, the computers at either end of this route do

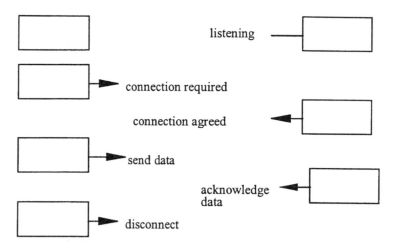

Figure 1.10   An Internet 'connection'

not wish to know any of the details as to how the packets wend their way over the network. All that they want is reliable transport to be established between them. In order to do this, the two computers establish a 'connection' (Figure 1.10).

The computer that will receive the data is first assumed to be listening in to the network for requests to set up a connection. One is sent from the other computer and it acknowledges it. Both computers are now setting up transmit and receive buffers, counters/timers and error checking. Then one computer transmits some data, hopefully the other host receives it and eventually transmits an 'acknowledge message'. Provided this comes within a specified period of time, the sender assumes that everything is OK and will transmit any further data in the same way.

All this activity is said to take place during a 'connection'. While the connection is in place, the two hosts continually check the validity of the data, and sending acknowledgements and retransmissions if there are problems. Interesting things can go wrong. For instance, data packets that have got lost within storage buffers used by a subnet may suddenly pop-up after the sender has given up on them and retransmitted the data, but all of this is taken care of by the TCP handshaking achieved through the acknowledge protocol.

TCP also dynamically controls the rate at which data can be pumped out of the sending host's buffer. It can assess the performance of the route by monitoring the progress of the handshaking and then adjust the flow of data in order to move it through at the maximum possible without overload.

1.9   THE INTERNET AND QUALITY OF SERVICE

Notice that TCP/IP is concerned to make sure that data is reliably transmitted from a sender to a receiver. If errors are detected, the protocol will require data to be re-transmitted until it is correctly received. This may result in many repeats, essentially without regard to the time taken. Consequently the quality of service specification for TCP/IP concerns itself with data quality, but not at all about the length of time taken to receive it. This is very different from the telephony specification, which is concerned to ensure a minimum end-to-end delay. Voice and streaming video services require this condition, the transmission of data for non-real-time processing does not.

Currently, this lack of real-time requirement is an accepted part of virtually all traffic over the public Internet. Because there is no central control of all the routes and routers involved, the transit time has been completely unspecified and there is no easy way to get round the problem, despite a number of ingenious approaches. It is certainly true that the current situation can be improved upon, to a degree: for example, it is possible to try to get the routers involved in a session to agree to try to keep a path open from end-to-end. Also under development are protocols which assign different 'quality of service' labels to different types of traffic (voice, moving video both having high priority, for instance). But it is not IP itself which is the problem; the real issue is in providing an end-to-end commercial accountability for the service, in a pluralist market-place. What is happening with private Internets (*intranets*) is quite significant: traditional network service providers (including telcos) are selling networks based on IP but with careful management of the traffic, so that delay bottlenecks are minimised. Often they do not use TCP to control the connection. Instead, they use more traditional switched, rather than routed, platforms, for example built on *Asynchronous Transfer Mode (ATM)* technology. All of this is a very active area of research and development, but it is beyond the scope of this text to cover this in detail [40,41].

1.10   VIRTUAL PRIVATE NETWORKS: INTRANETS AND EXTRANETS

Long before the concept of *virtual businesses* was part of the everyday conversation of management experts, businesses had become virtual in that they were often located on different sites but tried to operate as a coherent enterprise. The secret was, and is, effective communication. Radical improvements in this came about with the development of telephony, fax and point-to-point data circuits. Businesses have, for many years, used *virtual private networks (VPNs)*, for carrying their voice traffic.

These are services provided by telecoms companies where voice traffic is transported between selected intra-company sites through the national and international public telephone networks in a manner such that it appears to the company as if all its employees are on a single large PABX with simple numbering and billing broken out into whichever organisational structure is required. Traditionally, whereas voice and fax have benefited tremendously through standardised transmission principles and terminal equipment, data interchange has suffered from proprietary, incompatible systems. In addition, the bias shown by tele-communications voice network towards duration-based charging regimes, has until recently resulted in very expensive costs for shipping bursty data between geographically separated sites. As we indicated in the previous section, the Internet has dramatically changed these posi-tions, by offering a standard interchange protocol which can be put on top of many proprietary interfaces and, by calling on transmission bandwidth only when required, can significantly reduce telecommunication charges [42]. Many businesses are now moving over to Internet-based virtual private networks for their data transmission and, in some cases, for their voice traffic as well. These networks, where the prime purpose is to provide *intra*-corporate communication using Internet protocols, are known as *intranets*. The fundamental characteristic is shown in Figure 1.11.

There is no reason why this principle cannot be extended to *inter*-busi-ness communication, in the form of an *extranet* (Figure 1.12).

There are, as the figure shows, a number of differences though, between an intranet and an extranet. At the application level, it is clear that, in almost all cases, only some data or processes will be shared between the companies concerned. After all, the companies may well be sharing other data with each other's competitors. There needs to be a number of secur-ity measures put in place to make sure that only the appropriate data is

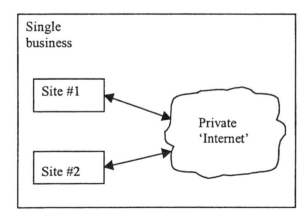

Figure 1.11   Basic intranet

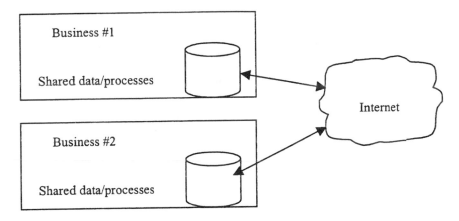

Figure 1.12 Basic extranet

shared. At the lower layers of shifting the data between sites, there may also be a difference, in that the companies involved may use the public Internet as a way of trading, rather than constructing a private network connection between their own intranets. This brings in additional issues of security and quality of service, particularly where the two businesses do not share a single ISP, whose roles we must now examine.

1.11 THE 'ISP'

Businesses do not in general administer the Internetworks that connect their businesses over the wide area. Instead, they purchase these facilities from any one of a large number of companies that specialise in their provision, the *ISP*. Some ISPs only offer a *point-of-presence (POP)*, a connection with unspecified performance to the global capability of the Internet; others, the so-called *Network Service Providers (NSP)*, also provide specified service contracts for conveying IP traffic over private, long-range networks they have constructed or leased. Many NSPs are organisations connected with traditional telecom network providers. (e.g. AT&T's *Worldnet*). A business seeking to maximise its networking performance may prefer, at a price, to run its multi-site intrabusiness processes over one such private IP network; a small business, or one which is cost-constrained may prefer to take pot-luck by using the fully public Internet connection. In practice, businesses will probably use both: the private NSP route for internal traffic and for connection to favoured partners, the public connection for other dealings. Figure 1.13 (repeated from Part 1, *The Retail (eCommerce) Server*), shows a typical ISP installation.

Beyond the point-of-presence, connection is made by the ISPs to the backbone of the Internet itself. This backbone comprises a number of very

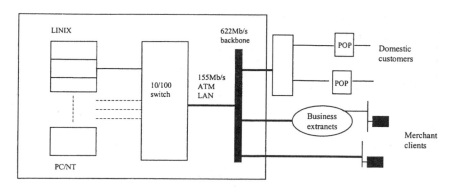

Figure 1.13 Typical ISP structure

large and high-speed private networks provided by major network companies, *Tier One Providers*, who agree and manage interconnection between themselves at Internet *Network Access Points (NAPs)*. To give some idea of scale: there are only six recognised NAPs in North America. Note that NAPs are administrative, rather than commercial, points in the Internet; businesses cannot buy access from them direct. The purchasing point is at the chosen ISP's POP; the ISP then provides or negotiates connection into the Internet via the Tier One Provider. Any business entering into commercial discussions with ISPSs should thus concern itself with agreeing service levels into and out of the ISP's POP and, if possible, try to assess the ISP's agreement with the Tier One company to which it is connected (Figure 1.14).

As Figure 1.14 stands, there appears to be little about the set-up that is virtual or private. In order to justify the terms, other elements need to be added: we need to provide some protection of the data against interception and corruption; we might want to hide the routing details across the public network from the two ends, so that it can be achieved over any arbitrary link without any need for them to reconfigure; equally, we may want to hide any of the interior details of the two ends from each other and from anyone on the public Internet; finally, we may want to set some end-to-end quality of service target. This last requirement requires co-

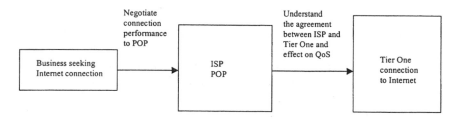

Figure 1.14 Negotiating a service level agreement with an ISP

operation with the ISP; the other requirements are met by the provision of *tunnels* and the use of secure gateways such as *firewalls* at the interface between the public and private networks.

1.12 INTERNET TUNNELS

In order to integrate the various distributed parts of the enterprise network, we would, ideally, like to be able to ignore the public Internet that joins them. Conceptually we use the strategy (Figure 1.15).

At the entrance and exit to the sites, we provide a logical gateway, which *encapsulates* the IP packets from one site to the other, in the sense that the Internet does not, perhaps can not, read their ultimate source and destination addresses, or even their contents. All the Internet does is route the packet from the source gateway to the destination gateway, because this is the only part of the packet it is able to read. At the destination the packet then has encapsulation removed by the gateway and migrates into the internal network, to reach its intended receiving host. In order for the gateway to do this with security-protected packets, it must have exchanged encryption keys with the other gateway. This can be done using protocols such as *IPSec* (see page 264), *PPTP* or *L2TP*. Also shown in Figure 1.15 is the access across the public Internet to a certification authority whose services are required in order to validate the encryption keys of the gateways.

One part of the gateway may operate as a firewall, as described on page 240. It can filter out dangerous or suspect packets and, if it includes an application gateway, maintain control on application requests from outside, also. This security gateway function is driven by a security policy server, which maintains a list of permitted and forbidden activities,

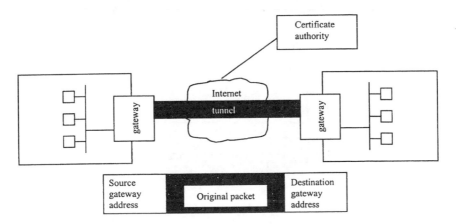

Figure 1.15

permitted or forbidden addresses that may or may not communicate between the interior and public networks, and, perhaps, log any such attempted violations or raise alarms when they occur.

1.13 DIRECTORY SERVICES

If different parts of an enterprise want to communicate with each other, they need to know each other's address. In a corporate telephone network, this is done by setting up a directory which associates a name to an extension number. Approximately the same process is employed for data routing across an intranet or extranet. However, this must also take into account the insecure nature of the public Internet route. The directory service will thus include the ability to store and retrieve the digital certificates that are needed to ensure the integrity of the VPN. It is possible to use a relational database to hold all this information, but this can become difficult and expensive to administer and there can be some performance problems.

A commonly preferred alternative is based on the *Lightweight Directory Access Protocol (LDAP)* [43], a reduced-complexity version of the well-proven X500 telecommunications directory standard [44]. In particular, it replaces the rather heavy OSI protocol stack with one based on TCP/IP. For more details, see [45] and [46].

LDAP is based on a hierarchical tree structure which effectively mirrors standard organisational structures and allows one to associate with any entry, typically a person or job function, a set of attributes such as contact details, digital certificates, etc. (Figure 1.16).

The tree-structure also means that position within an organisational hierarchy is clearly defined and, for example, superior and dependent units can be traced upwards or downwards. As a systems administration feature, LDAP also possesses distributed operation properties that allow

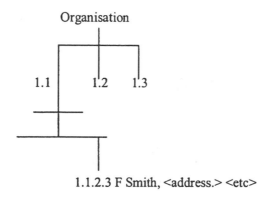

1.1.2.3 F Smith, <address.> <etc>

Figure 1.16 LDAP directory structure

for the full directory or sub-sets of it to be stored locally, e.g. to improve access performance, but to be automatically updated (*synchronised*) whenever the master directory is changed. Major vendors also claim that it is quite easy for systems administrators to fine-tune the caching and indexing of LDAP software to achieve very high performance. To give some idea of scale, Netscape claim their Directory Server can support 'over 5000 queries per second and 50 million user entries on a single server.'

Directory services are beginning to be seen as a major architectural component in end-to-end business process. We refer to this concept several times elsewhere in this book.

1.14   AUTHENTICATING REMOTE ACCESS

LDAP is an organisationally oriented approach, placing individuals and business units within an enterprise framework. A slightly different approach is necessary when individuals are 'outside' the corporate intranet, for example, customers accessing services from home or company employees working at home or on the move. One solution to this problem of remote access is commonly provided by a *Remote Authentication Dial-In User Service (RADIUS)* [42].

Within the intranet resides a RADIUS server, which holds data on each registered user's profile: passwords, privileges, usage statistics and so on. When users try to log-on to the network access point, the latter automatically queries the RADIUS server to check whether access can be permitted. Note that RADIUS's ability to hold usage statistics means that it can be an integral part of a *billing engine*, for charging out where access is not free.

The above sections only briefly touch on a number of security aspects. These are covered in more detail in Part 3.

1.15   SERVER SOFTWARE LAYER

Immediately above the network layer in the eBusiness architecture (Figure 1.16), resides the *server software*. This includes the basic operating system of the server and associated systems: UNIX/LINUX and Microsoft NT being by far the most common. The lowest part of the operating system is concerned with interacting with the network layer. As stated earlier, the latter is usually a TCP/IP implementation, but others are possible, of which Netware's IPX and Microsoft's NetBEUI predominate [47].

Most systems sold for eBusiness purposes will also include utilities other than a basic operating system. Usually marketed under the name of *application servers*, they will include a *Web server* to handle HTTP requests, an eMail server, security services typically based on *secure sockets layer (SSL)* (see page 266), usually some on-line conferencing facilities, directory and so on. Distributed management services, increasingly using HTML forms as the interface are also provided. Incidentally, although a hardware rather than a software issue, we should note that application servers need not be large computers: robustly designed machines based on Intel 486 processors or basic SUN SPARC chips will suffice. The application server is becoming not just a Web server plus a bit more, but a major controlling hub for external clients and for other servers running internal processes. Its connecting mechanism with these other processes is via *middleware*.

1.16   THE ROLE OF MIDDLEWARE

In the retail eCommerce case, at least in the case of providing customers with information and the ability to run shopping baskets and other basic tasks, the relationship between the customer's terminal and the eCommerce server is a relatively simple one. Most of the time it is simply concerned with pulling down Web pages in a client-server mode, perhaps also invoking simple CGI scripting or active server page generation on the server. There is little or no true sharing of processes in a distributed, peer-to-peer manner, reflecting the asymmetric relationship between customer and supplier.

In the inter-business case, things can be very different: business platforms switch from being servers to being clients and back again much more equally during a transaction. They do this between business processes within each individual business, as well as on a business-to-business process, and often businesses are in multiple trading relationships, even simultaneously. What is then required is a middleware solution that allows *remote procedure control (RPC)* or some similar approach to inter-operate between these platforms, perhaps in a true peer-to-peer manner, and one that can operate in a distributed and heterogeneous environment, principally across TCP/IP networks of computers, very often of different vintage and manufacture. Reference [36] divides such middleware solutions into a number of classes:

- SQL-based: this, which is one of the earliest approaches, is principally concerned with data-retrieval type operations. The development of ODBC (page 101) and other enhancements has given this a more procedure-oriented approach than the original database model.
- Remote procedure call: in its 'pure' form usually involves a neutral specification language for defining interfaces between *stub* or *skeleton* chunks of code on server and client.
- Object brokering: this extends the language-independent approach by

using heavily encapsulated (i.e. with internal workings concealed) objects.

- Message oriented and distributed transaction processing: are both concerned with guaranteed reliable implementation of inter-processor activities.

1.17   CONNECTING TO INTERNAL BUSINESS SYSTEMS

Perhaps the simplest example of connecting across the tiers of an IT system occurs when we consider operations within a single business which has introduced Web services in addition to its existing trading mechanisms. We saw one example of this with retail servers (page 101), in the case of ODCB, which allows Web servers to communicate with a variety of different manufacturers' databases. Part of any server software will almost inevitably be given over to suites of programs that carry out tasks like that. There are a number of middleware solutions proposed, of which the CORBA and enterprise JAVA environments are perhaps the most ambitious. We cover these in the sections that follow, but the general principles are the same, whatever method we use: interconnection can be achieved by constructing specialised connectors between the processes.

In constructing or even using these connecting programs one has to give attention not just to their correct operation but also to ensuring that they give satisfactory performance and are secure.

It is possible to do this directly by running *remote function or remote procedure calls (RPCs)*. These are messages from one computer to another that can invoke relatively restricted actions on the latter, directly through a firewall (Figure 1.17).

This approach, which is the core of architectures such as the *Distributed Computing Environment (DCE)* [48] allows client applications to start-up functions, usually written in C, on the remote computer. Simple RPC solutions, which involve ad hoc invocation of processes running on

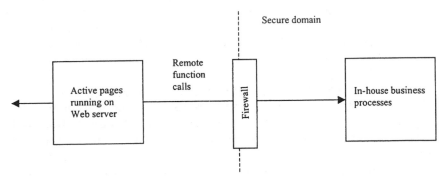

Figure 1.17   Remote function calls often traverse firewalls

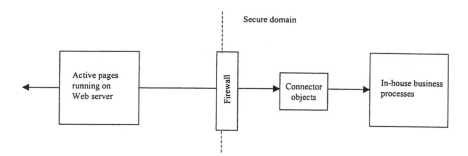

Figure 1.18   Object-based connectors minimise risk

other servers can be designed without adopting formal component or object oriented design. They are convenient for simpler tasks or where one has a good control over the environment both on the Web server and on the business one. However, if the IT system is technically intricate or the business exists in an environment where multiple, temporary alliances are created and torn down in a rapid and complex manner, then a more generic and structured approach to distributed processing is required. An RPC approach to problems of this nature can be complicated and therefore difficult to maintain and not necessarily very secure. With the growth in the use of object-oriented programming, the preferred course of action is now to interpose an object-based connector between the two systems (Figure 1.18).

Communication between the Web server and the connecting process may use, for example, the *DCOM* (reasonably) open standard developed by Microsoft. Companies such as *SAP,* (the vendor of one of the most popular resource planning suits of software, see Part 4, *Supply chain management*), provide toolkits that allow designers to create the necessary objects. For example, we frequently require to define a 'salesorder' object, which connects to the sales and distribution processes in SAP, a 'customer' object, from the financial accounting processes, and so on. The remote function calls can invoke these objects, but the actions that follow thereafter are entirely defined by the characteristics of the objects, which are developed and protected behind the firewall, rather than from the RFCs, which are exposed on the Web server. This approach to using well specified interfaces (almost invariably object-oriented), is known as a *component* architecture.

Incidentally, although it is claimed that good performance can be achieved via the component objects, it is also sometimes recommended [49] that it is better practice to reduce the number of accesses from Web server to business server as far as possible. For instance, it may be better to create a copy of the product catalogue locally beside the Web server, for use by browsing customers and only once they have made up their mind on a product, pass their queries through to the business server, to check availability, etc. This would make good sense, in terms of both performance and security.

We now look at some component architectures.

1.18   OBJECTS AND COMPONENTS

The object oriented approach to software design and coding was not originally developed for distributed processing but it turns out to have some advantages in that direction. Object oriented design (see, for example [50]) begins by defining the things we want to manipulate, their properties and the way they communicate with each other, in high level terms rather than in terms of data structures and the way that computers deal with the data. A 'thing' could be a customer, an account, etc.; obvious properties include names, addresses. Individual things can belong to a common group, a 'class' and share common properties, e.g. women and men belong to the people class and share one of its properties – an age. This is one reason why object oriented design is relevant to distributed processing: it allows us to retain the essential concepts, in isolation from incidental and transitory events. Customers are customers and they retain their particular properties whether their purchases are read by a point of sale terminal in a shop or made via a home computer. On the other hand, 'branch discount' is a property of the means of purchase, not the particular customer who purchases it. In a distributed environment, we often want to hide the geographical location of a service, because it is not material to the requirement and geography is a distraction. This technique of *abstraction*, where the essential properties of the processes are isolated from the nitty-gritty is clearly of value in the conceptual modelling of the business processes. But it is at least as valuable also in the implementation domain, where often we do not want to distinguish between local and remote processing or even whether a process runs on one machine or many.

As we said, objects are not just 'things' in the physical sense. Take the example of a 'value added tax gatherer' (TAXG). Applications programmers who want to use TAXG do not need to know how it works, but they do need to know some things about it (Figure 1.19).

TAXG is a 'black box' that is wholly defined by:

- The services it provides to applications.
- The way that applications should send messages to it.
- A set of details on how communication is handled that the applications do not need to know, but are there to inform the underlying distributed infrastructure.
- A unique identifier that will be used to locate it, even if it moves from place to place. (It is possible that an object that is calculating your tax

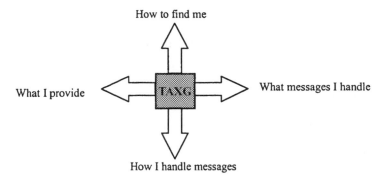

Figure 1.19 A CORBA tax-gatherer

can move from one computer to another. You still want to access that one, not one that is doing someone else's tax or doing the calculation on a different basis.)

In object oriented programming, objects are defined by *class* definitions. This is a way of defining a 'template' for the object which includes speci- fying its *properties* (for example, a bank account object may have balance, name and account number properties) plus some procedures or *methods* that allow, amongst other operations, for these properties to be manipu- lated. Each time a new *instance* of an object is created (for example a new person is issued with a bank account), this instance of a 'bank_accoun- t_object' is constructed and managed in computer memory as part of the compilation, loading and running of the object-oriented programming language (C++, ActiveX, Java, etc.), without us having to concern ourselves with the details.

Note that one aspect of class generation is the ability to incorporate or *inherit* the properties of other objects. (For example, classes 'male' and 'female' may inherit from the class 'employee'). This is claimed to be an aid to speed and accuracy of development. However, in the case of apply- ing object orientation to distributed systems, the inheritance property does not appear to be particularly important [51].

One facet of the object-oriented approach that is of significance to distributed systems is that it tends to make programming into a two- stage task. First, we define the structure of each object. Then we 'simply' write the rest of the program more or less as a series of communications between objects, using well-defined rules which fall out from the way the objects have been specified. It is often desirable to take this *encapsulation* approach to the extreme in that intercommunicating components need not even be aware of the programming language in which the other is written.

Object creation and management does pose certain new issues in the distributed case: the creation of a new instance of an object may be initiated by a client platform remote from the server on which the instance has to be created. Standardised procedures for the creation of these instances in *object factories* must be provided. Similarly, distributed facilities must be available for deleting these instances when no longer required. Usually hidden from view of the applications programmer, there also has to be a reliable way of transferring data between distributed objects.

In all, the move from object oriented programming to component design appears to involve a significant change in outlook. It is best not to ask what *components* are; it is better to describe what they do and how they appear, because it is possible to implement them in a variety of ways, not necessarily object oriented. First, their appearance: seen from other platforms in a distributed system, they are visible only through their interfaces. How the component is implemented on the host platform is of no concern to the rest of the system (Figure 1.20).

In the rather fanciful model shown in (Figure 1.20), the component interacts with the world via a completely isolating interface A client computer accessing the component does not need to know – and cannot tell – whether it is written in C, C++, Java or even implemented as a human ledger clerk. The advantage of this approach is simple: *pluggability*, the potential to connect processes together without having to worry at all about their internal workings.

We have already referred to another property of components: their *location independence*. Unless it is necessary to know where a component is located, it is an unnecessary distraction. Again, the basic principle is to hide things from the designer, in this case, the whereabouts of the component.

But if our components are going to be given freedom from the prying eyes of application developers, they must still be under the control of a number of management services, themselves invisible to the developer (Figure 1.21).

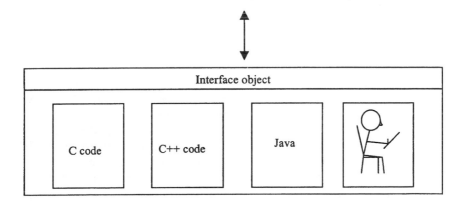

Figure 1.20 Component pluggability

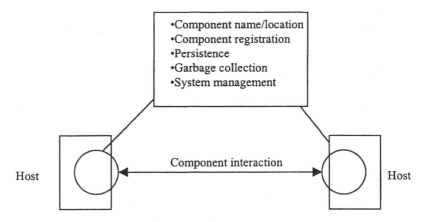

Figure 1.21 Managing the component architecture

The management system holds information on where the components are located and what they are called. It can therefore bind a name used by a developer, to an actual component. It also maintains a register of valid components and similarly a list of who is allowed to access them. Since objects will often be distributed, some overarching process must also be responsible for holding the current values of object states in a *persistent* store, which can provide recovery of data for hosts, for example, in the case of a transaction failure. (Transaction processing is described later.) Conversely, when clients are no longer using objects, there has to be some *garbage collection* whereby the memory and state records used to store redundant instances of these objects, can be freed up for new use. Finally, there will be a set of general management services to handle security and resource management. To varying degrees and with varying success, management systems try to handle fault tolerance, load balancing, etc.

Component middleware solutions have tended to have their origins either in the object oriented or distributed programming approaches. Pritchard [51] provides a useful distinction between *component architectures*, which concentrate on packaging of the code and in its cross-language operability, with *remoting architectures*, which are concerned with using distributed objects to run remote processes. In Table 1.1 we try to position a number of common middleware products.

DCOM, from Microsoft extends the idea of components within a single computer (typically, a PC running Windows) that was developed under the name *COM*, into the distributed environment, particularly a Windows client-server architecture. CORBA and Java RMI are both 'open systems'. CORBA is more mature and has the greater market share, but Java RMI is attracting significant interest.

Table 1   Predominant approach of middleware products

Component architecture	Remoting architecture
DCOM	CORBA
CORBA v3.0	DCE
Java RMI/CORBA	Java RMI

1.19 'COMMON OBJECT REQUEST BROKER ARCHITECTURE' – CORBA

CORBA makes extensive use of the programming concept of object oriented design, in particular, the defining of objects in terms of the four 'interfaces' to other objects and the infrastructure, as described in our tax-collector example.

These interfaces are specified in a specially designed *Interface Definition Language (IDL)*. IDL is not intended to be *the* language in which all programs for distributed computing are written; instead it is constructed so that the components of IDL can be mapped onto other languages, using automated procedures. This means that programmers can write applications in languages of their choice and, assisted by the IDL manuals and tools, import into their programs the necessary interface 'functions' (sections of code that carry out specific routines – 'print', 'read', 'write', etc.) from IDL, without needing to know the intricacies of IDL. Because the functions, in any of the languages, always convert into exactly the same set of IDL instructions, programmers can be assured that the program will run on the CORBA-compliant system, irrespective of the source language.

1.20 'OBJECT REQUEST BROKER' – THE ORB

Imagine the case of a request from a 'client' (that is, from a process executing anywhere in the distributed system) for an object. For example, the client may wish to invoke a TAXM object to deal with a tax calculation. In CORBA implementations, the client never communicates directly with the object; the request always passes via an *Object Request Broker (ORB)*, part of the CORBA software that acts as a simplifying buffer between the client and the object (Figure 1.22).

The purpose of the ORB is to hide from the application programmer the complexities of dealing with a distributed environment, one which moreover, may comprise a heterogeneous set of computers. The ORB finds the object and handles the detailed parameters of the request. Its task is, essentially, to provide *location* and *access transparency*. It does so, through

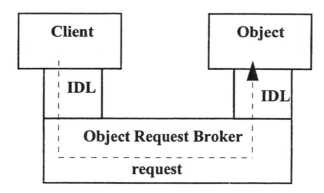

Figure 1.22 CORBA object request broker

IDL interfaces as shown. (Incidentally, it can be seen that the ORB fits our definition of a CORBA object.) Within a distributed system, there may be several ORBs, each mounted on a different set of computers. Clients send their requests to their local ORB. Suppose the wanted object is not in the area covered by the local ORB? No problem, the local ORB simply communicates with the object's local ORB (Figure 1.23).

A further advantage of this approach is that the different ORBs can reside on different vendors' hardware and software platforms and be written in completely different languages, to different internal designs. All that is required is that they are capable of providing the CORBA functions specified in their interfaces.

1.21 'OBJECT MANAGEMENT ARCHITECTURE' – OMA

CORBA itself is concerned with the interfaces between objects and how they are specified in IDL. This is the lowest level of a comprehensive *object management architecture*, ('OMA'), that comprises the vision of a non-profit-making co-operative group of vendors, known as the *Object Management Group* (OMG). Their idea is to create, by consensus, a

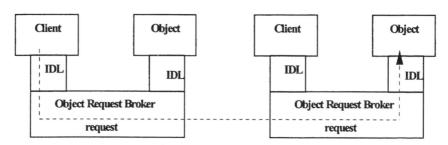

Figure 1.23 Distributed communication via other ORBs

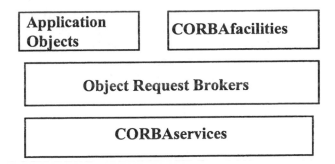

Figure 1.24   CORBA facilities and services

common architecture for distributed computing, built round the object oriented principles described above. They separate basic, lower level services for objects, the *CORBAservices* level, from the facilities provided to applications, the *CORBAfacilities* level (Figure 1.24).

CORBAservices comprise the more basic and generic needs common to most objects: things such as naming, querying, the grouping together of objects, security, and so on. A number of them are available in stable form from a several vendors.

The CORBAfacilities are rather less well developed. Four basic categories have been defined: user interface, information management, systems management, task management. These are clearly of use across all distributed business applications. 'Vertical' CORBAfacilities are also being generated for specific market sectors, such as Heathcare and Finance. The policy of the OMG is to adopt the existing models of information and processes already created by these sectors and integrate them into OMA, for example through creating new IDL interfaces.

1.22 CORBA NAMING AND TRADING SERVICES

Highly significant components of the CORBAservices are the 'naming' and 'trading' services. Naming is the simpler. CORBA has a set of system directories that set up a correspondence between the name of an object and its current location, just as a cellular telephone network can direct a call, described by a phone number, to a specific aerial serving the cell in which the called party currently is. (You simply use the name; the CORBA service maintains the path to the object.)

But that assumes you know who you want to call (and, also, possibly, for what purpose and on what terms). Suppose you want to find out who can sell you a particular type of car, at what price and how soon. You need to look at a Yellow Pages directory that lists companies under the services you offer, and obtain a contact number. That is part of what a Trader service will do. It will allow you to identify sources of objects you require. These may, in the early days, be simple services such as an application to securely 'watermark' your data against copyright theft, but, in principle, it should be able to find you a list of virtual organisations that can remotely run human resource management software for your organisation, for example.

The Trader not only gives you the names of the suitable objects; it also gives you details of the services they offer – speed, cost, security policy and so on.

At this point it has to be said that some of the Trader properties have still to be defined by the OMG and most of the benefits of Trader services can only be postulated. Nevertheless, the basic concepts are widely accepted and CORBA is becoming widely used as a platform for advanced services.

1.23   CORBA FACILITIES AND eBUSINESS

The Object Management Group has seen the significance of eBusiness to their areas of interest and have become highly active in trying to define CORBA functionality and OMA features in this arena. As a standards group, their approach is, naturally, thoughtful rather than precipitate, and therefore may suffer from the danger of being overtaken by events. However, they currently appear to be having some influence in directing the next steps in eBusiness architecture as well as seeing CORBA implementations on most of the popular platforms. It is therefore worthwhile considering some of the aspects of *CORBA facilities* (Figure 1.25) that are particularly applicable to eBusiness.

Figure 1.25 encapsulates a CORBA/OMG view of the processes that go together to make up eBusinesses, at least in so far as they require distributed computing to support them. Some, such as perhaps, payment, catalogue and Intellectual Property (IPR), are self evident, as are the generic

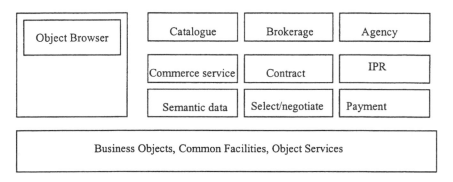

Figure 1.25   CORBA eBusiness principal facilities

business objects, facilities and services, shown at the bottom of the figure Some others may require some explanation:

- *Semantic data*: today's instantiations of on-line trading processes have been developed in an ad hoc manner, with each developer working with different core entities and with their own naming conventions. This means that it is not easy to join one person's solution to someone else's and even makes it difficult to get a common understanding on the requirements.
- *Negotiation facilities*: negotiations are necessarily multi-party (two-party at least). Thus these require to be defined in a consistent manner to allow the parties to extend their trading components towards each other's and to allow general values (for example, a price) to be bound to a particular value (US Dollars, Pounds Sterling, etc.) in a manner consistent between all participants.
- *Contract services*: likewise, these services which define the sub-actions within a contract or can refer out to generic contractual arrangements, also need consistent specification.
- *Brokerage and agency*: sitting above these facilities is a higher level grouping of facilities concerned with general market infrastructure, of which trading and brokerage services are particularly interesting. These provide the ability to distribute advertisements for services required and to recruit a remote service that meets the profile. The transacting process model for such activities, if they are to be carried out automatically, is clearly quite complicated and again standardisation will be beneficial.

The OMG and other organisations are working hard to flesh-out these ideas, many of which are still conceptual.

1.24 WEB APPLICATION ENVIRONMENT LAYER

The discussion above has strayed into the fourth layer in Figure 1.26, which is concerned with the development of applications in the eBusiness which really distinguish it as 'on-line'. In earlier chapters we referred to the various software elements necessary for the basic eCommerce retail applications: HTTP, HTML, active pages, active X, Java and so on. These are also required for intra- and inter-business applications, but with a much greater need for more complex interaction between client and server, for instance: transaction processing, electronic data interchange of standardised invoices, call-off orders, etc. These applications also need to run in a very distributed mode in so-called *n-tier architectures* (i.e. involving several application servers) with a high degree of equipment heterogeneity.

As Figure 1.26 tries to illustrate, we would like the applications to run

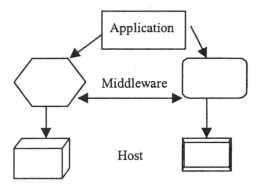

Figure 1.26 Heterogeneous operation

consistently in environments which employ a number of host operating systems, which may be running different middleware and using differing databases. It is possible to craft applications in a number of programming languages, to make them operate across CORBA middleware, for example, and similarly to write database query programs to access ORACLE SQL, IBM DB2, etc. databases.

1.25 ENTERPRISE JAVA

Recently, a rather novel approach, *Enterprise Java* [52], shifts the interoperability capability into a number of Java functions that run across the servers. Amongst other advantages, significant improvements in development time are claimed [53]. Java programmes work on the principle that they run on a JAVA *virtual machine* – a piece of software running on any make of computer or operating system, and which has been defined to behave like a standard idealisation of a computer, irrespective of what lies below it. One particular aspect of these machine independent programs is the *Java servlet*. Servlets run on most Web servers (just as their client equivalents, *applets*, run on most client browsers) and they run within the virtual machine environment (or '*sandbox*'). As shown in Figure 1.27, typical applications for a servlet might include a database access or a call to invoke a remote method (*RMI*).

A servlet is called and loaded into the memory of the virtual machine only once, at the beginning of the running program, and it can thereafter service several requests simultaneously, simply by creating a duplicate block in another part of memory for the new task. (This is a commonly used computing technique, with each new bit of memory-resident code being called a *thread* and the process called *multithreading*. Multithreading is efficient and quick because we do not need to load and run a new version of the program for each task.)

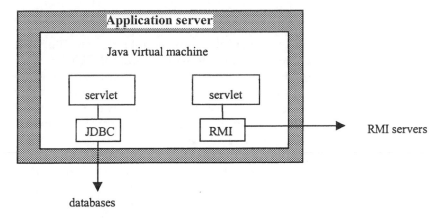

Figure 1.27 The Java virtual machine

Servlets carry out their specialised operations via interfaces such as *Java Database Connectivity, JDBC,* (for databases) and *Remote Method Invocation, RMI,* (for running remote processes, or *methods* as they are referred to in the object-oriented model that is used). RMI performs somewhat similar functions to CORBA; its object oriented approach allows objects on remote servers to be called up and interacted with, via *stubs* (client-side interface instructions for the remote object) and *skeletons* (server-side controls for invoking the object and passing parameters to it), similar to CORBA IDL stubs.

Java also possesses a *Java Native Interface, JNI,* which allows C and C++ code to be integrated into Java programs. This is a convenient way of integrating legacy programs into a Java environment, but only if they have C or C++ interfaces. Where this is not the case, one solution is to use CORBA as a middleware between Enterprise JAVA and the legacy application. This can be used, for example to integrate large COBOL programs and enterprise Java applications [52].

1.26 COMPONENT ARCHITECTURES: LIMITATIONS AND ALTERNATIVES

Component-based design is relatively new and still the subject of vigorous discussion, including a major debate regarding the relative merits of CORBA and DCOM. The relative advantages and disadvantages often result from the different approaches adopted by their inventors: DCOM is firmly from the Microsoft stable and, amongst other features, gives considerable power to Windows-based clients. To give but one example, DCOM garbage collection exercises the right to destroy objects when there no longer referenced by any clients. On the other hand, CORBA preserves the server's right to maintain these objects, thus providing better scalability but requiring specialised solutions to be written for each application. The facts are complex and cannot be given a fair airing here. Balanced and detailed analysis can be found in [51,54,55].

The new entrant, Enterprise Java, is clearly a very important solution to eCommerce requirements and, compared with most alternatives, it offers a conveniently 'simple' approach. It is also, at least in principle, a fully open solution: irrespective of the precise hardware and software environment on which it runs, the Java virtual machine presents the programmer with a consistent programming environment. Unfortunately, this can also be a problem: since code for the virtual machine has to be translated into 'real' code for the 'real' machine in use, there are inevitable performance penalties, notably speed. For this reason, developers in industries where transaction-intensive processes are required, for example retail banking, are generally reserving judgement as to whether enterprise Java will provide the appropriate solution. Another problem arises in the case of legacy code: as we said, it is possible through the Java Native Interface (JNI) to integrate with other languages, but difficulties can arise with object definitions, which must be coded in JAVA. This contrasts with CORBA, which being simply a specification, allows objects to be coded in any language provided an ORB library function exists.

It is early days to be able to comment on the security of the component architectures. One might worry about the power that DCOM gives to clients. It has also been claimed that CORBA security is still a rather grey area. Java security mechanisms should benefit from the sand-box approach, although it has to be noted that non-Java code, linked via JNI, will retain its own vulnerabilities. That said, following our dicta given in part 3.2 regarding the need to develop operationally simple security procedures, it is probably true to say that all of the component solutions we have discussed, will make it easier for programmers to assess the security of their code than the options previously available to them.

Finally, we should note that none of the architectures have been designed with truly real-time applications, such as continuously streaming video, in mind. This is not a pressing requirement, for the moment, but it would be foolish to rely on this being the case for much longer.

1.27 TRANSACTION PROCESSING

End-to-end eBusiness does not only involve more processes than simply providing an eShop; some of these processes also have more critical properties. We might be prepared to return to a shop which did not always show us all that it had in stock. Perhaps we might tolerate one which claimed to have items that it was out of. But we would be much less willing to revisit one which took our credit card and never returned it or charged other people's purchases to it. There are some tasks which need to be done completely and properly. The technical means for achieving this is through *transaction processing*.

We often use the term 'transaction' quite loosely, to mean any interaction between parties or systems. In this section, however, we use the word in a much more circumscribed sense, to describe interactions which have quite specific properties. To the most important of these is generally ascribed the acronym *ACID*, broken out as follows:

- *Atomicity*: transactions take place or they do not take place – there is no such thing as a partly completed transaction. If the transaction is not completely followed through, it has failed, and all data and states of the systems involved must fall back to their values as if the transaction had never begun.
- *Consistency*: the overall 'rules' of the process in which the transaction takes place, should at no point be broken – financial accounts, for example, should balance all the way before, after, and throughout the transaction.
- *Isolation*: transactions are carried out independently of each other, even if several are occurring at the same time in the same environment. Implicit in this is the requirement to forbid two transactions to operate on the same piece of data at the same time.
- *Durability*: once the transaction has been completed, its effects on data and states of the system should persist – the effects of a transaction should last beyond the lifetime of the transaction process.

Viewed in lest abstract terms, these conditions must seem obviously sensible: if money is transferred from one account to another, the end product must be a reduction in one account matched by an increase in another (less any charges for the service). Money should not 'leave' one account and fail to arrive in the other because of a network failure, for example. A auditing and balancing transaction that, mid-way through its operation, temporarily converts positive balance in my account to a negative one, should not allow an overdraft calculation transaction to see this altered data and charge me interest on an overdraft. Finally, once a transfer to my account has occurred, I would not expect that transfer to disappear because the processing program had moved on to something else.

Transaction processing is quite difficult and it does not really share the cultural mode of Web developers: it is not about swift entrepreneurial success; rather it concerns itself largely with consistently preventing failure. Fortunately, transaction processing has a long history in mainframe systems, such as IBM's CICS and Web-compatible off-the-shelf solutions have become available.

Details of these systems are complex, see, for example [56], but it is worth exploring some of the principles involved.

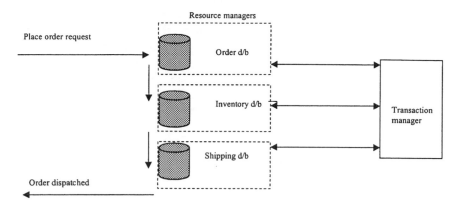

Figure 1.28 Typical approach to transaction processing

Central to the operation is the concept of a *transaction manager*, which interacts with a number of *resource managers* (Figure 1.28).

The role of the transaction manager is to act as a central coordinating point for the series of database operations involved in the transaction. It inter-operates with the resource managers which are protective *wrappers* placed around the various databases. When an instance of a transaction is instantiated (i.e. when a customer places an order), the transaction manager first creates a defined space in memory for this specific transaction. It then contacts the relevant resource managers to *enlist* them to create an instance of their data operations that is traceable back to this particular transaction.

To be more specific, consider how Enterprise JAVA JDBC methods allow us to handle a transaction made up of a number of database operations. For example, suppose we want to take an order for some goods, provided a credit check has confirmed that the customer has funds.

- Under simple JDBC operation, each time we carry out a database operation, this single operation is considered to be a transaction. A line of code equivalent to 'put customer's address into shipping database', would lead immediately to this happening. The process responsible for dispatching goods, which runs asynchronously would therefore include this item in the list to be sent out – which we do not want to happen automatically.
- So, we have to turn off the automatic alteration of the database data. We do this using a simple instruction which disables the *'autocommit'* function in the JDBC commands. This is done by a line of code which includes a statement: '.setAutoCommit(false)'. Essentially the change to the database is held in suspense. We probably also want to do a similar thing with a billing invoice to our customer. All the database commands within a defined block of code can be held in suspense in this way.

- Now we get a message back from the credit database saying that X's credit is good. We then reactivate the requests to the databases involved.
- But we are not quite done; many other things can go wrong: a data entry might be faulty or temporarily inaccessible, for example. First, all of the databases must return a message to say that they can effectively change the data satisfactorily. Only once they have done so, is a 'true' flag given to a '.commit' command, which signals to all the databases that it is safe to go ahead with the data change. The command also thereafter releases any lockouts they had enforced against other processes that wanted to access the same data. This process is known as *two-phase commitment* and is a fundamental component of transaction processing.
- But suppose something goes wrong: the credit is bad or a database has fallen over: in this case, the flag is 'false' and the program issues a '.rollback' command, which aborts the transaction, leaves the data in all the databases unchanged and removes any locks on the data. Things return exactly to what they were before the transaction was attempted.

1.28   COLLABORATIVE WORKING

So far in this chapter we have looked at the automated processes involved in integrating many of the eBusiness activities, but we have not said much about the processes that depend on the involvement of people within the organisation. Certainly, computing processes are being 'joined up' to avoid the need for any more human intervention than is necessary. This is not just to save on wages. Every time we require an operator to re-key data into a system, we introduce a significant failure cost owing to errors. (For instance, it has been reported that between 20% and 40% of all spreadsheets contain errors [57].) However, we cannot dispense with people altogether. In fact, human beings, in the right place, with the right skills, are the most valuable resource of most companies. Critical negotiations are still conducted between people, not machines, and there is no serious prospect of this changing in the foreseeable future, even allowing for rapid development in artificial trading agents (page 354) for non-critical tasks. Therefore, particularly for intrabusiness and inter-business activities, we require to create person-to-person environments that build trust and continue to support interpersonal operations throughout their life-cycle.

A common model of collaborative work emphasises the need to consider it in terms of three distinct, but inter-related activities (Figure 1.29) – processes, teams and individual activities.

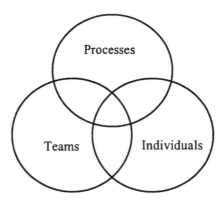

Figure 1.29 Three aspects of collaborative working

Processes, such a order-taking and fulfilment, or design of a product, etc. flow through an organisation, but only as a result of collaborative actions by teams of people who may belong to the same functional unit or to different units, even located at different places. These people need to be thought about as taking up particular roles, e.g. purchasing officer, design authority, etc. and also as individuals. The distinction is important: a purchasing officer has a set of accountabilities and rights, for instance the right to sign-off a purchase order. Organisationally, any purchasing officer 'will do', if a signature is required. However, as an individual, that person brings specific talent and knowledge to a particular purchasing domain – computers, travel and subsistence, etc. and that person needs to be consulted if a good judgement is to be made.

Relating this specifically to the operation of a computer-supported collaborative working infrastructure, we see that we need to preserve these distinctions and interrelationships. On page 313 we consider one example of this: the case of handling eMail from customers. Here we require software that can route mail to any one of a group of people, rather than to individuals, in the same way that an automatic call-diverter routes in-bound telephone calls to any one of a number of call-centre agents. The reason for doing so is that it will be the collective responsibility of the eMail reception centre to give swift and complete response to a customer query, rather than wait until one specific individual becomes available. The eMail is directed to the 'centre' not to an individual (unless, as is possible with such systems, exception handling does mandate mail direction to a specific person).

On the other hand, collaborative development of a product that involves interaction by specific members of a distributed design team requires them to be put into contact in a way that promotes free flow of information as well as encouraging empathy between these specific people. Conferencing systems are one such mechanism.

1.29 COLLABORATIVE WORKING TOOLS

If we are to achieve anything like the full potential of globalisation, we cannot expect all human interaction to occur face-to-face. Instead, we need to use tools that allow collaborative working at a distance. *Computer Supported Collaborative Working (CSCW)* is one of the collective terms used to describe activities which are usually not only supported by computers, but also by communications infrastructure. This is amplified in Figure 1.30 which lists the most commonly used tools, as a function of location and time.

If people can meet together at the same place, at the same time, then there is little need for technology (except, perhaps, to assist in recording the minutes). If they all live in the same place, then it is easy to provide access to common documents or by leaving memo messages on each other's desks. The more interesting technology options occur when they are separated by distance.

Even if they are separated by distance, they may be available at the same time. In this case, they can talk on the telephone or use video conferencing. A further real-time option is the ability for dispersed teams to share common data, on a *whiteboard*. This facility allows text and graphics from a common file-server to be displayed at each site. It is possible to manipulate the displayed information, for example, using the mouse to 'draw' a circle round a part of a diagram and displaying this circle on all terminals. Document text can also be edited and the results of agreed edits then used to update the file on the server.

1.30 VIDEO-CONFERENCING

The growth in the use of video-conferencing has been steady, rather than spectacular. It is not clear why this has been the case. Nevertheless, it is beginning to attract a significant number of enthusiasts and these will

	Same time	Different time
Same place	Conventional, 'physical' meetings	Memos and other documents
Different place	Audio conferences Video conferences Shared whiteboards	E-mail Voicemail Document servers

Figure 1.30 Taxonomy of collaborative working

continue to increase, as transmission capability, price-cutting and familiarity develop.

We can broadly divide video conferencing along three axes (Figure 1.31).

In terms of environment, the decision rests with either using a specialised studio, which can usually hold a larger number of people, has good lighting and sound conditions, may require booking and probably requires a faster network, or using a desk-top system, which is much more impromptu and ad hoc, in terms of numbers that can take part, (three is about the maximum at one end), quality of picture and sound, etc. Desk-top conferencing can be effective between small, non-collocated design teams, for example, particularly when combined with shared document and *whiteboard* facilities. (The ability to 'draw' with the mouse on images shared across the network.) Systems like these are available, as standard software, from most of the major vendors. The specialised studios, which often include document projection facilities, are better for the more formal occasions, but of course, the opportunity to call a spontaneous meeting is lost and the conference does not become a 'normal' way of working.

In practice, the choice of which to use may be as much dictated by network availability as by anything else. Studio conferencing services are usually based on 2Mbit/s telecommunication links, which can be expensive. Desktop conferencing, with reasonable quality moving images, can be achieved over 64 kbit/s ISDN (see page 28), although 128 kbit/s or higher is often preferred. These conditions apply to opera-

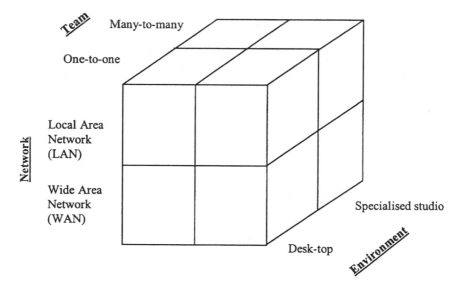

Figure 1.31 The configurations of video-conferencing

tion in the Wide Area. If operation on a LAN is possible, for example between people on a science park, campus or other set of locations where high speed data can be transported, then other options, such as simple Internet protocols can be used. Despite being sold for local or wide area use, Internet-based video conferencing over WAN usually gives appalling performance on most circuits, simply because of high network delay and low throughput. It may be better to use a series of 'stills'.

A further complication arises if the conferences are to involve more than two sites. In this case, there is a need to provide a *multipoint conferencing bridge*, which combines the bit-streams from the sites so that the parties involved can see the other participants at more than one site. These bridges can be located on company sites, but telecoms companies also sell bridging services within their public networks. Although usually better than Internet-based solutions, the quality of telephony-based bridging, particularly end-to-end delay, may not always be good enough to promote trouble-free interaction.

1.31 AUDIO-CONFERENCING

Although not as new or fashionable as video-conferencing, audio-conferencing is a rather underused, but highly effective, way of conducting distributed meetings. It is provided by a number of telecoms companies as a standard service, and one which is very easy to use. The participants either dial in to a number provided by the telco, or are called by a central operator, at a set time. Thereafter, for the duration of the conference they are connected together via an audio *conference bridge*, so they can all hear each other. Some telcos provide recording and/or transcription services, as well, which might be valuable particularly where contractual agreements are being discussed. Participants can use their normal telephone handsets or can use hands-free loudspeaking telephones whereby several people in one room can take part.

Although there is an absence of body-language and general visual empathy, audio does have the advantage that speech quality is generally good, no lighting or other studio facilities are required and participants only require a low-cost telephone. (It is also possible to have mobile-phone participants, although meeting structure might not be some controlled and there may be security and politeness problems of talking in public places.)

Costs for such meetings are relatively modest and the increasing quality of *IP telephony* which is very low costing, might make this an even more attractive proposition. IP telephony, which is of course packet-based, is easy to host on a PC and depends almost entirely on the loading of the transmission paths involved, for its quality. Usually it works well on a LAN and is increasingly available across corporate virtual private networks, which may be extended to other members of an enterprise. We could even imagine that voice interaction at a distance became very widespread and quite natural, rather than structured around semi-formal meetings: IP 'connections' would be always 'on', but perhaps muted unless deliberately triggered. We would thereby treat distant colleagues as if they were in the same room.

1.32 ADVANTAGES AND LIMITATIONS OF CONFERENCING SERVICES

There is a great body of theory (although none completely conclusive) concerning the effectiveness of these real-time, virtual meetings. It may now be apparent that the current performance of audio and video conferencing services is not yet up to the standard of a 'real' meeting. Nevertheless, they do offer a reasonable, low-cost solution for many purposes. It is probably true to say that they reinforce collaboration rather than create trust, ab initio. As such, they are very satisfactory for use by distributed project teams, who can be considered to have at least a minimum set of shared goals and purpose. Audio is quite sufficient for routine progress meetings of this type; video may be required when a bit more empathy has to be generated or when design information has to be shared.

In the eBusiness context, we ought to be aware that in using these virtual environments that are intended to build trust and support empathy between the partners in the virtual enterprise, there can be a danger that arises from their tendency to create an informal atmosphere. Open discussions and friendly feelings are essential, but they are not always structured and it is not always easy to recollect what exactly was agreed. There will exist a need to make proper records of the meetings, usually in an electronic text. Contractual negotiation is one case in point. Another case of this is in call-centre operations where integration of the call data (time, number dialled, etc.) can automatically be integrated with notes on the conversation typed in by the call-centre agent.

This, of course, captures only the minute-taker's perspective on what took place. The only absolutely accurate way to capture the event is by audio (perhaps even video?) recording. Although some financial institutions do this where high-value dealing is at stake, it is not something that should be entered into lightly. Maintaining a recording facility to handle the entire traffic from a large call-centre, involves a considerable amount of work and careful maintenance. The likely technology would be something like a bank of basic compact-cassette recorders, which are not really known for their reliability or quality, but are easy to administer and convenient in that it is easy to spool-back to the required conversation. Digital recorders are more reliable, but expensive. Recording on disks is more expensive yet, and not often used. Where network-based audio conferencing is used (see below) it is possible to arrange for the network provider to carry out recording and transcription, for a fee.

In the case of business meetings, it might be possible to record the meeting and store the tapes in case of future dispute. Where 'physical' meetings are involved, the recording conditions need careful monitoring – the tape will only be called for when there is a serious dispute and speech rendered unintelligible because of poor conditions will only give rise to further suspicion and ill-will. Perhaps surprisingly, it is easier to record a virtual meeting: the recorder and the parties involved both share the same message channel; if the listeners can understand the speech, then so can a tapped-off recording. Again, the recording machines need to be maintained and, in all the cases discussed, one has to have a strategy for disclosing the existence of the recording and being prepared for the natural inhibition that may arise.

Finally, it is almost certainly not worth considering the use of auto-mated speech-to-text conversion, as a way of creating a record of the meeting. Verbatim transcripts of spontaneous discussions are extremely poor. (What we actually say, is very different from what we intend to say, and often quite unintelligible without recording voice nuance, gesture and body-language.) Any automated recognition system will be worse. In any case, it is doubtful that voice recognition equipment yet exists that can offer a reliable conversion to text under realistic meeting conditions, with unconstrained speech.

For all these reasons, the predominant medium for negotiating critical agreements has been text. Modern technology now permits this text to be created more quickly, which is a convenience but not really a qualitative shift. What it also does, and which probably does have a qualitative effect on the way we work, is to allow this text to be transmitted and retrieved much faster, more conveniently and error-free. As we see in the next section, we can even expect nearly simultaneous, multiple authoring. With the existence of documents which can be read and perhaps worked on at different times and at different sites, comes an information explosion which needs to be managed and which places the document at the centre of a collaborative process.

1.33   GROUPWARE

We have cautioned against a too informal approach to collaborative work-ing and emphasised the need for formal, written records. Written docu-ments are of course of ancient origin; the very first were characterised by the fact that they had one authorship, one issue date, one master and resided in one library. The development of the printing press removed the single-copy restriction, but only at point of origin. Convenient, low-cost copying at point of receipt did not arrive until as recently as the 1960s, with the development of the Xerox process. For some years after, documents remained restricted in their availability and movement. They were mainly private, to selected individuals on specific sites. It is only very recently that all these restrictions have been lifted, thanks to technology, and organisations are still coming to terms with the implications.

Chief amongst these is the emergence of a document-centric view of collaborative processes. This treats documentation as no longer being static records, fixed at only one specific time or place, but rather as evolving, dynamic mechanisms for disseminating historical, current and emerging *corporate knowledge* in a timely and consistent way throughout the organisation. By this approach, it is claimed, people can, in a single operation, file all their reports, spreadsheets, forms, eMails, etc. in a place where they can be not only seen by others, but also revised, annotated, completed or incorporated into other documentation. We could, for example, imagine product information being collated and worked on collaboratively, across design documentation, sales reports, faults and customer complaints, so that an integrated view is obtained over the whole sales, manufacture, supply and support functions within an organisation.

As shown in Figure 1.32, it is not sufficient simply to provide a central *repository* for all the corporate documentation. It is also necessary to perform a number of control functions over this. There must, first of all, be some *version control* mechanism that looks after the version numbering, assigning dates and maintaining historical back-up of previous versions. This is a key requirement if documents are to be continuously changing, especially if more than one author is involved. This immediately brings into focus the need for *access control and rights*. Who is going to be allowed

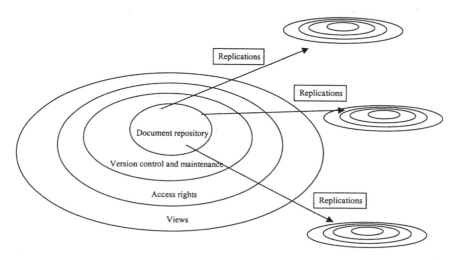

Figure 1.32 Conceptual view of groupware document system

to see a document; who is going to be allowed to modify it? Indeed, what do we mean by 'who'? We could mean a named person, or someone only identified by the role they play or the team they belong to. Increasingly, with Internet access, we need to consider the location of the user as an issue for access control. (Do we restrict access to known IP addresses? Do we provide password protected mobile access?) We also need to give some attention to what the user is going to see, when they accesses the repository. Documents will exist in many forms: spread-sheets, eMails, diagrams, reports, etc. and some will exist as modular sections which can be put together in many ways, to form a training guide, a user or repair manual, a sales booklet. We need to provide a series of *views* of the contents of the repository that meets the needs of the user and also allows for searching and browsing. Finally, we have to recognise that it is no longer sensible to think of there being a single, centralised, physical repository for all data, within many organisations. Instead, distributed satellite databases and proxy servers will hold local *replications* of parts or all of the data. They will need to be *synchronised*.

Groupware products that achieve robust solutions to these requirements have been available for some time, with IBM's (previously, Lotus) *Notes* being the dominant example [58]. Notes allows all the features described above and includes a security system based on public key encryption for authentication, access control and digital signatures. Two-way authentication is involved where clients and servers both have to prove their identity before dialogue can begin. Documents are uniquely identified by the database on which they reside and by a number within this database. They are also *time-stamped* with information on the last changes. There are complex mechanisms that ensure *synchronisation* of data is eventually achieved across all servers.

Notes were developed before the full significance of the Internet was yet realised, but it has since been enhanced with the development of *Domino*, to fit into the Internet environment. It is still seen, however, by some Internet aficionados as a data-base oriented approach, perhaps even over-designed [59].

One of the architectural issues arising from the Internet model, is the extent to which replication of data is really required. An extreme view is that is not necessary at all: there is no need to track within the groupware software, the version control of documents; instead, all that is required is to maintain a list of URLs where the versions can be found. Groupware solutions that have emerged in recent years do not all support this strong view, but do tend to have a lighter-weight approach to version control than that of the traditionalists. For a more detailed review, see [60].

Finally, notice that we have not said anything about finding relevant information within this potentially huge collection of corporate data. This in itself is a big issue and is the subject of Part 2, *Managing eBusiness Knowledge*.

1.34   INTEGRATED eMAIL AND GROUPWARE SOLUTIONS

Distributed work-groups may make extensive use of eMail, as a text-based, but rather informal way of communicating. It is, of course, possible to think of an eMail as essentially a fast way of sending a letter, or of sending a parcel when a document is 'attached' to it. Adopting that approach can lead to the same problems as one has with physical mail. What do we do with it, when we get it? How is it to be filed and how is it to be fitted into the other processes that run the business? Should eMail be treated a stand-alone entity, something simply that lands on someone's desk as a pragmatic commentary, which can be ignored, thrown away, or lost? This can lead to high administration and failure costs. In general, it is probably true to say that it is a bad idea to use eMail as the usual mechanism for an automated process, because of its loose structure. (Use a form, instead.) But sometimes it is necessary to deal informally and then what we want is some way to integrate eMail's informal and person-centred approach with the rest of the 'impersonal' business processes, such as billing, materials handling and so on.

To do this, we need to consider the eMail as part of a business process: 'customer responses to marketing mail-shot', 'customer complaint handling', 'order querying', 'exception-handling permissions to vary an engineering design', and so on. This means that we have to do at least three things: firstly, the in-bound eMail needs to be routed to the process-handling channel rather than to an individual's general mailbox. We discuss mechanisms whereby this can be achieved, in Part 4, *Service and Support*, where we look at integrated customer centres. Secondly, the eMail has to be associated with any appropriate other documentation. (For example, attachments must be linked to the eMail they were delivered with.) Thirdly, the eMail requirement must trigger the business process in a manner consistent with other parts of the process's operation, e.g. human agents must be automatically prompted to respond to an eMail within the designated time. In technology and interface terms, this means that eMail systems begin to share a great deal of the properties of Web browsers, file-handling interfaces and data-base oriented groupware.

The integration of global and local applications is also an area of interest, when constructing enterprise groupware that includes not just partners and suppliers, but also one's customers. *Corporate portals* which integrate external and internal data are one way of including customers in the continuing improvement of products and services, as well as reducing the cost of help-desk functions. Moderated *user-groups* which discuss problems and solutions by means of mailing into a common pool and mailing out selectively according to recipient interest profiles can be constructed around *discussion servers* which manage the profile segmentation and the mailing lists. See [59] for further details.

1.35 WORKFLOW

Turning now to an almost completely 'process' view of distributed working, we can consider any task that involves a number of people or organisational units, to be described in terms of a number of sub-tasks, which operate in parallel or series, similar to the way they are described in standard activity diagrams, for example, a *PERT chart* (Figure 1.33).

The sub-tasks must be completed, and considered to be complete by the appropriate *authority*, before their output can be passed on as input to the next in the series. Any messages or other outputs for the next stage must be defined in a set of *rules*.

Automation of these processes is achieved by using what is called *workflow* software. Let us consider the example of a purchase order request. An engineer wishes to purchase some further materiel for use on a construction task. They may send out an order request which goes to B, the team leader who has to authorise that the request is necessary, and to C, the financial authority who will confirm that it lies within the spending limits. The order may also, at the same time, go direct to the purchasing unit, D, to reserve a space in a queue. If all goes to plan, then the purchasing unit will receive confirmatory agreement from the two authorities and can go ahead with the purchase. Perhaps more to the point, the humans in the purchasing unit do not need to expend any energy on the purchase order, perhaps are not even aware of its existence, until it has been authorised.

The messages between the various parties to the transaction are usually electronic *forms* which have been created by the designer of the system using a workflow tool-kit. As the name implies, these forms then 'flow' through the system, which generates the next set of forms only when it has received a completely authorised set of forms from the preceding sub-tasks. Implicit in this process is the requirement for a *scheduler* function which monitors the completeness of the sub-tasks and controls the flow (Figure 1.34).

So far, what we have described is the state of affairs that occurs when everything runs to plan. The final principal feature of workflow is the

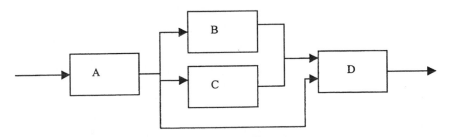

Figure 1.33 Decomposing a task into sub-tasks

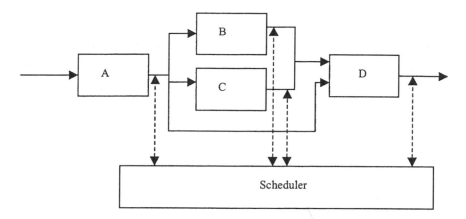

Figure 1.34 The scheduler function

need to include an *exception handler* to deal with the not uncommon occurrence of something that does not (Figure 1.35).

Typical examples include the case of an incomplete form, an order which exceeds a cash limit, or a query by an authority concerning the legitimacy of the order. For example, our engineer may wish to place an order for more materiel, but the team leader may want to question whether this is really necessary. They can therefore 'bounce' the form back with an annotation saying, 'Please justify'. This message can appear in a number of formats: a form on a Web application requiring to be filled in by the recipient, or an automatically generated eMail which requires a response to the groupware system (not to the human project leader). The exception handler and the scheduler are linked. The exception becomes a subtask in its own right and the scheduler, knowing that the sub-task is not complete, therefore suspends forward progress to the next sub-task until the engineer has responded and the team leader signed off this response as acceptable.

Traditionally, process-oriented software was written in-house and generally was developed piece-meal, with rules and scheduling implicitly, rather than explicitly defined, and covering only part of the end-to-end tasks of a business. There was still often a manual handover between these processes, often with re-keying and other failure-prone activities. In recent years, companies such as *Staffware* [61] have developed packages and tools that allow the creation of workflow systems with much less need to become involved in detailed programming and thereby freeing up effort to concentrate on modelling the process. Architecturally, workflow products sit naturally on client-server architectures, with the forms being delivered to desktop client machines and the scheduling and rule-based policies running on a server hub. Modern extensions to this now include Web-based interfaces and

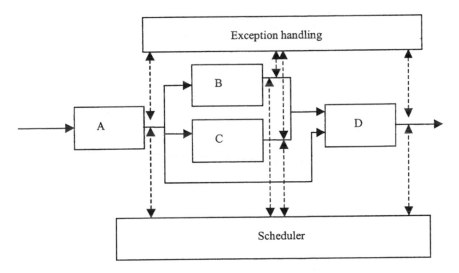

Figure 1.35   Exception handling

object/component methodologies of the types discussed earlier in this chapter. Moving to distributed processing, integration through DCOM, CORBA and Enterprise JAVA, means, for instance, that the workflow code can interact with already complex processes and job queues, developed earlier in-house or using other bought-in process and database products. Indeed, software sold under the title of 'workflow' is increasingly being positioned as integration software intended to link together partial, proprietary solutions to more narrowly specified tasks. Existing in-house *Enterprise Resource Planning (ERP)* which we mention in Part 4, *Supply Chain Management,* also needs to considered, when embarking on a groupware introduction. ERP has tended to be more of an integrated database solution rather than a process-driven engine for running an enterprise, but the distinctions are beginning to disappear.

1.36   THE ENTERPRISE DIRECTORY

Note that a critical component of the workflow system is an *enterprise directory* structure that can be scaled across the enterprise. We have seen that there is a need to hold details on people's access rights and passwords; the provision of call-centre, eMail and collaborative working applications require consistent addressing and numbering plans; elsewhere we describe how scheduling and other people-centric processes are critical elements of supply-chain management. In database terms, individual people act as unique *principal keys* to other information which may be spread out across diverse data repositories. A reasonable case can be made for designing virtual organisations around the concept of a single logical database or directory that holds information on everyone – customer, employee, contractor – that is involved in trading. All rights, privileges, skills, preferences payments, etc. should be linked to this data and it should be used as the prime source to bind variables in all business processes – financial approval, targeted marketing, payments, receipts, goods orders, etc. On page 138 we mentioned the *Lightweight Directory Applications Protocol (LDAP)* which is emerging as the de-facto standard for applications of this type. One of its significant advantages is that it is based on X500 directory principles, and with X500 being the basis for many LAN directory services, this makes integration between LAN and WAN much easier. LDAP has been adopted as the eMail directory for Netscape and Microsoft Web browsers. The significance of this, in terms of bought-in groupware solutions, is that they should be open enough to be able to take information from this database and use it to bind name variables to their processes, rather than have their own, closed directory structures. A test of a specialised product might be whether it supports an external directory, such as LDAP or whether it cannot be opened up to non-native processes. It is also important that it can be scaled to operate, not just on a LAN, but across a wide area which might require re-defining as partners to the enterprise come and go.

2

Managing eBusiness Knowledge

It is a truism that we are drowning in a surfeit of data. Some large organisations hold terabytes of information about their own processes and on
the customers. This was even true before businesses adopted a widely
networked approach to data collection and before they began trading
electronically with their customers and suppliers and it is becoming a
bigger and bigger issue. The buzz-phrase of today is *knowledge management*, the conversion of this raw data to meaningful information set within
the context of the business model and thus potentially capable of supporting planning and decision-making. Knowledge is not seen as passive;
rather it is considered to be a filtered, focused, attributable and intelligible
input to real business processes, without which they will not perform
effectively. Much has been written on the subject, but as yet, a lot more
remains to be proved. Rather than advance more theory, we shall concentrate on a number of implementation issues concerned with handling the
information explosion.

2.1 EXCHANGING BUSINESS DATA – EDI AND BEYOND

Where individuals or organisations wish to communicate with each other,
they need to share a common language and sets of protocols, in order to
avoid confusion and mistakes. For several decades it has been realised
that the advent of computers and telecommunications data links has
opened the way to automating business transactions as a way of speeding
them up and reducing their error, but only if this language and protocol
exist. The methods that achieve this are known collectively as *Electronic
Data Interchange* or *EDI*. EDI is really a complete business process rather

than just technology. Over the years, a number of standard approaches, in terms of agreed process definition, data formats and communication networks have been developed to a state that was relatively mature long before the Internet and the Web became all-pervasive. EDI has been successful, but not overwhelmingly so: one source claims that as few as 10,000 companies in the US currently use EDI [62]. The problem has been one of proprietary protocols and the need to create inter-business exchange agreements. Modern advances can reduce some of the difficulties, though not all. Today's challenge is to convert the lessons learned on 'old-fashioned' EDI into their equivalents in the Internet environment. How much to keep and how much to throw away is the big question.

Suppose organisation A wishes to use EDI to conduct business with organisation B. For example, A may wish to order goods from B through a call-off contract. As shown in Figure 2.1 a number of considerations and activities at different levels are required in order to achieve this.

Some parts are obvious. At the most basic level, we need to set up a reliable method of transmitting data from one computer to another. In the past, this was often achieved by connecting to a private network, offered by a third party *Value-added Service provider (VAS)*, usually a telecommunications or computer-networking company (for example, AT&T or IBM). Today, and increasingly in the future, these private networks will be replaced by public Internet services based on TCP/IP and its application protocols. The need to discuss transmission standards, either directly between the companies involved, or via the VAS provider, is disappearing fast.

We also need to encapsulate ('wrap') our data around with additional information to allow one database to correctly operate on data from another. At the simplest level, this is a formatting issue. For example, some databases may use ASCII characters, some binary. Much of the

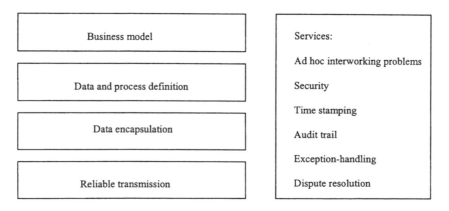

Figure 2.1   Structure of an EDI service

lucrative business of VAS providers in the past has been simple code conversion from antique coding schemes to more modern ones.

But what of the data? What exactly does it signify? Company X knows what needs it wants to pass over; so does company Y, and they both know exactly what to do with it and how to fit it into their business processes. They are both convinced that theirs is the only correct data and process model for the task in hand. The problem is, in more cases than not, their data definitions differ and so do their processes, since they have been designed in isolation, to meet their own perceived requirements. Look at Figure 2.2, which shows a simple example of the product assemblies of two companies, X and Y.

Business Y retails products to the consumer market. These products are constructed from sub-assembly units mounted in a case. Two examples are shown: Model A1 is the basic unit which comprises a case holding four 'standard' slide-in sub-assemblies. Model AB1 is a deluxe product, with one of the slide-in sub-assemblies (B21) built to a 'better' standard. Originally, Y constructed all the sub-assemblies of both of these products in-house. Consequentially, Y's financial database only distinguishes between the units as complete assemblies, that is, as 'products' to sell.

One day, however, Y decides to buy-in the sub-assemblies. The database of one sub-assembly company, X, quite naturally distinguishes its items at the sub-assembly level, as shown. X can describe Y's two different models entirely in terms of its own sub-assemblies, except for the case c1000 which encloses them, and which may be bought from a different

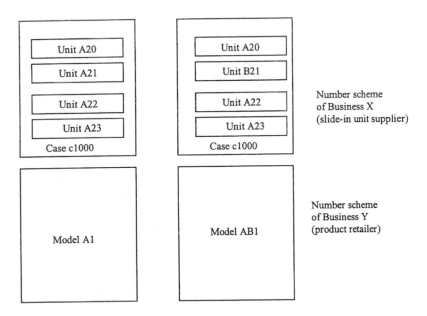

Figure 2.2 Data descriptions of product assemblies

supplier, or made in-house by Y. Additionally, Y may still want to retain the option to make some sub-assemblies in house, or have the opportunity to buy-in sub-assemblies from other suppliers. Obviously, although Y sells things at the product level rather than at sub-assembly level, Y must now create some way of separately identifying sub-assemblies as part of its manufacturing process (typically through its organisation of assembly drawings in its drawing office), but not necessarily on its financial database. Y has quite a problem: products do not map one-to-one onto subassemblies, at least onto the codes used by sub-assembly suppliers. There are further complications: X will use a number of different components to construct its subassemblies, some of those it may in turn buy in from its suppliers. X has to keep these components identified as distinct items, in terms of purchasing, but if the functionality is the same, then it does not need to distinguish completed sub-assemblies on the basis of the brand of components they contain. But 'functionality' is difficult to define: X may have been supplying sub-assemblies to company Z, who fits them into roomy cases, where a slight dimensional difference in sub-assembly caused by different components is not an issue, but Y's cases may be smaller. Thus there are cases where Y requires X to adopt a finer granularity of specification, than X had ever deemed necessary for its internal use. Confusing? That is what the example is trying to be, because it represents genuinely confusing data nomenclature issues that arise every day, whenever businesses decide to integrate their trading.

Defining what data to pass, what this data 'means' and what one is to do with it, is more complex than at first sight it appears. As Figure 2.2 demonstrates, we need to drive the whole thing from a business view – an understanding of what a product really 'is' – and then describe the data and processes that define and manipulate the string of information that is used to describe a product and all the computer operations associated with it.

Notice that, in the example discussed, we have considered a situation revolving round a 'product'. There are other possible viewpoints, (for example a process viewpoint for clearing items through customs), but the basic principles remain the same: define the business view first, then the data structures and procedures, before cutting code. The example we gave may have been confusing, but it was, even so, a simplification. It is considerably simpler than real life. To illustrate the point, we now examine a further example. We continue to look at things from a product viewpoint. In choosing an example of a product, what could be more appropriate than 'a book'?

2.2   PRODUCT DEFINITION: WHAT IS 'A BOOK'?

The *European Group for Electronic Commerce in the Book and Serials Sectors*

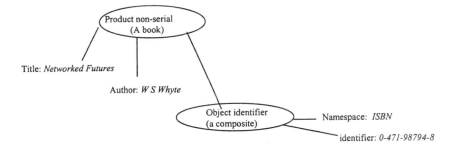

Figure 2.3   Example of EPICS description

(EDItEUR) [63],is a principal European organisation concerned with the co-ordination, development, promotion and implementation of EDI in the books and serials (e.g. journals and magazines) sectors. In January 2000, they produced version 3.02 of their *EPICS Data Dictionary*, which essentially represents their definition of products of type 'book'. Without the Introduction, which is still to be revised, it runs to seventy-one A4 pages! A book is obviously not as simple as it might at first seem. To understand fully why this is so, it is probably necessary to read through the whole document, but one can get some understanding by looking briefly at one of two facets and examples.

The EPICS document describes, at a number of levels, data items that are related to published products. At the highest level, EPICS defines (so far) five *objects*: 'Products – non-serial' (e.g. 'books'), 'Works', 'Product Components', 'Series', and 'Journals'. Below this level are *composites*, which in turn are made up of individual data *elements*. Thus, for example, a book (which is an example of 'product-non-serial'), called 'Networked Futures' by WS Whyte, can be partly described as shown in Figure 2.3.

This is represented in the EPICS schema (approximately, for we have slightly simplified things) as follows in Table 2.1.

The letters and numbers (e.g. C010) refer to their identifying codes in the data directory. As stated, this is an approximate representation: for instance, items such as title and authorship are not usually given directly as values as shown in Table 2.1. Normally they would be formatted in terms of data elements, that is, non-composite components. A *title element* (code C180) will contain a *title element qualifier*, and a value.

C180 title element:

  0608 qualifier = 03 (implying that the following will be the title text)

  0605 element = Networked Futures

The data directory is comprehensive: it defines ways of representing publisher's details, sourcing details, languages, geographical regions

Table 2.1

Example of object	Example of field within object	Examples of composites	Examples of field within composite
'Product non-serial' (in this case, a book)			
	C030 'title', value = networked futures C040 'authorship' value = Whyte WS		
		C010 'object identifier'	
			0001 'namespace', value = 02 (ISBN) 0003 'namespace identifier' = 0-471-98794-8

involved; it even covers the possibility of describing the prizes, if any, won by the product.

Comprehensive though it may be, EPICS is not the only solution to the complete description of the term 'book'. The reasons why it is not are various and perhaps at least as political as rational: does EPICS have the 'right' to specify a book? What happens if another organisation has a different view? For example, EPICS is European; the American Association of Publishers (AAP) separately began the creation of a 'standard' description, called *ONIX*. This is a somewhat simpler description and deals entirely in American terms. One instance of differences between EPICS and ONIX is that the latter deals entirely in Imperial measurements – sizes in inches, not millimetres, weights in ounces not grams. A further difference, which is political, is the fact that the AAP is not a Standards body, and there are some who believe that it is not up to them to try to mandate a Standard! (Be that as it may, the encouraging thing is that EPICS and ONIX appear to be working together to create a 'Universal' ONIX standard, which harmonises both schemes.)

Even solving the definition of a 'book' is not really sufficient to describe completely and unambiguously the process of trading in the book sector. There are other issues involved, for instance, *rights management*. Here, too, work is in hand to develop a model. This includes an interesting approach to the problem in terms of the triangular arrangements between *stuff* (the content), the *deal* and the *person(s)* involved (Figure 2.4).

The relationships between these entities are then described by verbal relationships: 'people write/sell/buy/hire stuff', and so on, and fuller definitions of the entities and relationships are under development.

All of this may be very boring for the keen entrepreneur of eCommerce:

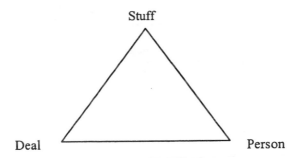

Figure 2.4   Publishing rights management

none of this is new and surely it simply reflects the previously failed attempts to roll out EDI solutions as a universal trading mechanism? To believe this would be wrong. True, some of the failures of EDI have been to do with transmission costs, protocol complexities and a number of other low layer issues. But, on the other hand, in cases where EDI has been successful, this has been largely due to getting the business model right. This issue will remain whatever the new technology invoked. Without giving really serious consideration to the business model and the proper definitions of data objects and process methods, advocates of Web-based EDI are in serious danger of creating dead-ends and significant failure costs.

A further cautionary tale: in our description of the structure of an EDI service (Figure 2.1), we have also shown a set of 'services' – ad hoc problem solving, security, date stamping, audit trail and so on. In the past, with VAS-based EDI, these functions were provided by the VAS provider. Inter-business transactions are based on limited, rather than perfect, trust. In cases of dispute, there was always recourse to the third party's audit trail, for example in verifying the time and date when a transaction was transmitted. 'Modern' EDI hopes to disintermediate the third party, to reduce costs and generally introduce more flexibility. This it undoubtedly can, but practitioners must remember to include mechanisms to replace these other added value services, rather than just addressing the technical issues of data-handling.

2.3   EDI IN THE WEB ENVIRONMENT

Despite the warnings given above, there is little doubt but that Web-based EDI trading will rapidly become pervasive throughout the trading community. As we have said, TCP/IP and Internet services will provide a much easier transmission environment for interbusiness transactions. We have seen how component-based middleware can allow us to connect clients and servers into multi-tiered architectures of arbitrarily complex

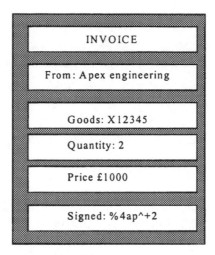

Figure 2.5   A simple invoice form

processes. Interoperability of databases across the Web is now a reality. What of data entities and the business models that drive them? What do we need to do in order to make it much easier to achieve inter-business interoperability?

In Figure 2.5 we give a very stylised and simplified invoice that might pass between two companies. In order for it to be processed automatically, certain specifications and constraints need to be created and imposed. Firstly, exactly as with the book example, we need to define the various elements which make up this transaction. We need to reserve a set of names corresponding to the various parts of the form: 'invoice', 'from' (which is probably an object called 'supplier'), etc. These need to go into an on-line library of allowed terms and terminology. We probably have to define and delimit the overall set of elements that are required to make-up a correctly formed invoice – in the example given, you might quibble that there is ambiguity as to whether the price refers to the unit cost or the cost of two items. We also need to set limits on the character sets used, the number of characters permitted, perhaps special symbols such as £ or $, and so on. One more thing: the invoice is signed with a digital signature, which probably contains the issuing company's authentication details as well as a hash function (see Part 3, *Security*) that makes sure that the data has not been interfered with. As well as specifying all the constraints previously mentioned there has to be some way of describing how this signature is to be derived and checked. In summary, our EDI process must:

- Where the transaction involves human beings, make sure that it can be presented on all likely presentation platforms.

- In some space shared by all parties involved, define the terminology of the transactions.
- Set limits to the values that each of the terms can have.
- Probably, check the compliance of these, on-line, in real-time and raise an exception handling routine to resolve any problems.
- Define and execute procedures to complete other aspects of the transaction.

So, we need to transmit data between the parties involved in the transaction, along with standard templates that control its integrity, its presentation, and which can activate any processes required.

2.4 XML

The usual way for sending Web data between on-line hosts is currently via HTML. The trouble with HTML is that it is principally concerned with how the information looks, rather than what it 'means'. We can use the terms 'title', 'author', 'date of publication', 'ISNB code', etc. but these have no special meaning in HTML. Therefore, they cannot be easily picked out automatically within a larger body of text; nor can the string that they represent be checked for correctness. We cannot, for example, easily check to see that the ISBN value that is given, (e.g. 0-471-98794-8) consists of a set of number-only fields, separated by hyphens.

This deficiency in HTML has been largely rectified in the development of the *Extensible Mark-up Language (XML)*. XML still provides a mechanism for layout and appearance of text, just as HTML does, but it is much stronger in its ability to give some form of 'meaning' to selected terms within the text, and this selection is largely at discretion of the writer. Take the example of our discussion on how to describe a specific book (*Networked Futures*): we want to specify that it has a title, an author, a publisher, an ISBN identifier, a date and a description. Then, a simplified XML description is shown in Table 2.2.

The <items> in brackets are *tags*. In the examples shown, they each represent a *type* of data. Information relevant to a particular type of tag is contained between the <item> and </item> pair. (Users of HTML will be quite familiar with this notation.) In the example we can see that there is a global item called <book>, within which are *nested* a number of other tags: <title>, <author>, etc. We have only shown one layer of nesting, but XML allows us to extend this to any degree. For example, we could imagine that <title> was made up of two parts:

Table 2.2 Simplified XML description of Networked Futures

```
<book>
  <title> Networked Futures </title>
  <author> W.S. Whyte </author>
  <publisher> John Wiley and Sons Ltd</publisher>
  < ISBN> 0-471-98794-8 </ISBN>
  <date> 1999 </date>
  <description> Technologies and issues of the Superhighway </description>
</book>
```

<title>

    <main_title> Networked Futures </main_title>

    <sub_title> Trends for communications systems development </sub_title>

</title>

and so on. The benefit for automated processing should be clear: because the tags are defined and enclose data strings, it is much easier for a piece of software to 'understand' that 'Networked Futures' is a title, not an author or a piece of a free-text description, provided, of course, it is aware of the way that we define a book and the terms associated with that definition. How is this last condition achieved? We need to make public our definition of a 'book'. We can do this in XML in a number of ways, one of which is by heading our example above with a reference to a *Document Type Definition (DTD)*. This is another document which contains the generic template for a 'book' and, optionally, additional rules that make it easier to check the data validity. The DTD can be located through its URL. The assumption is that trading parties, or even international bodies such as trade organisations, could put their DTDs and other XML documents on-line, so that trading could take place without ambiguity or error.

There is actually a lot more to XML than we have just discussed. In particular, there is a great deal of activity concerned with defining how XML should represent the presentation of data within electronic documents, and this is obviously of interest to Web designers in eCommerce. However, what we have discussed is probably the most significant impact that XML will make to advancing electronic business: its potential to allow the creation of standard business models for automated transactions. So far, this is only potential; the driving force has to come from the model into the technology, and the models have to be created by agreement. How this will happen is yet by no means clear: will it come from industry associations, technology-based collaborations such as the World Wide Web Consortium, or from dominant software vendors? Will it be achieved in vertical sectors (for example, EPICS and ONIX collaborating on a mapping between their model and XML), or by cross-sector, generic trading definitions ('invoice', 'call-off', etc.)? Indeed, what may happen is that some market-leading enterprises will follow the early EDI model and develop their own proprietary trading models. Although this in some way might be a pity, it will not have such a debilitating effect as did the same approach twenty years ago. Proprietary EDI created lock-in at all layers in the process, from the byte-level upwards. It should be easier to create a migration strategy from proprietary XML definitions to a common standard. But it will not be trivial. Enterprises may need to tread warily in this area, for some time yet.

2.5   XML AND DISTRIBUTED PROCESSING

Irrespective of the pace at which full vertical standardisation of XML DTDs occurs (whether or not, for example, that everyone agrees on the standard way to describe a product), there is undoubtedly a very positive future ahead for the technology in generally handling commercial transactions, particularly when combined with programming languages such as Java.

On-line order forms are a simple case in point: the simplest way to program an on-line order form is to create an HTML form, which allows the customer to type in details and post them off to the Web server. Unfortunately, as we have noted elsewhere, there is a high degree of error generated when data is keyed into a system, perhaps as high as 20% on a multi-line form. There is no easy and inexpensive way of correcting this and it will lead to customer dissatisfaction and loss of business.

Instead, increasingly the solution has been to send the form to the customer's terminal with an embedded form-checking piece of executable code (e.g. a Java applet), which runs on the client machine and checks the data that is inputted against a template. XML provides an excellent way of coding up this template. It is in a language that is relatively easy to understand; it can be generated separately from the Java code and made available to the code and the coder when required; it can be developed around the business process, not the client form-checking application; it can be synchronised to changes in the business process, ensuring integrity. Moreover, it is a way of publishing widely on-line the trading rules of the organisation. These XML trading processes can be readily accessed by remote intelligent agents, to allow fully automated tendering, collaboration, etc. as described earlier. Even where different organisations have used different XML descriptions of their trading rules, the structured and clearly defined discipline of coding the process, may, in many cases make it possible to build mappings between them, thus allowing trade to take place.

2.6 THE CORPORATE KNOWLEDGE PORTAL

Supposing we have been able to define, capture and analyse a sizeable body of accurate 'corporate knowledge'. What do we do with it? Specifically, who should have access to it, perhaps even with interaction capability to annotate or change it? The traditional model for access to corporate data was rather restrictive: finance departments had access to finance data, marketing departments to marketing data and so on. Part of the reason was that the databases were separate and only configured to deal with the processes of the functional unit that set them; part of it was the difficulty in providing convenient and secure access to them, from distant sites. The multidimensional presentation of data and the transmission capability of corporate intranets have changed all that. In theory, anyone with access rights to data should be able to view it, perhaps even structured to meet their business model rather than that of the originator. Sometimes, these access rights could be extended outside the company itself, to partners, suppliers or even customers. Some companies, cautiously or otherwise, are now experimenting with this knowledge-release approach, in the form of the *corporate portal*. Essentially the idea is to drive many of the knowledge-centred processes within the company which directly or indirectly interact with people via a set of Web interfaces, each with a set of access rights to some of a single, consistent corpus of enterprise data. If this is possible, then some clear gains can be had, including:

- *Both the public and private facets of the company are driven by the same information model.* Customer and intracorporate views are not therefore inconsistent and, where they differ in detail and knowledge, there is a clear explanation as to why this is so. We discuss further this role of the portal in Part 4 *Electronic Marketing*.
- *Customer data and internal process data are the same.* When a customer chases up the non-delivery of an order, the help-desk agent and the dispatch clerk that they query, will be working to the same audit trail. All will work to the same headline pricing and stock ordering details (although there may of course be 'special terms' data joined with full referential integrity to this and only visible to the company representatives.)
- *Marketing and design departments will both have access to customer comments on the company's products.* Product flaws, suggestions for new functionality, comparisons with other products can all be worked on together, sharing common knowledge and perception.

- *Functional units and executive planning teams will all share a common semantic view regarding the manageable elements of the organisation*: a sales or geographical 'region' will be consistently defined and discussions about performance and targeting can be conducted efficiently and without misunderstanding.

Notice how the customer has become part of the knowledge team. One critical component of building an eBusiness will be the ability to capitalise on customer on-line data. The Web gives much greater power to collect on-line behaviour patterns that demonstrate likes and dislikes for products. Consideration must be given as to how best to integrate this information into any corporate data store. Obviously, one should try to avoid basing this mainly on eMail input, as this is difficult to convert to automated input. On-line questionnaires are a possibility and one can also get a lot of information by designing forms that can be completed by call-centre agents and transferred into an accessible database.

2.7   DATA WAREHOUSING AND DATA MINING

Good organisations collect data in order to run their operations, but also for future analysis to guide their strategy. The logical (not physical) integration of this data into a single manageable information space, especially where this is intended to promote managerial decision-making, is known as *data warehousing*. A data warehouse is a conceptual approach, rather than a physical thing, which aims to integrate across a number of real databases, which may be mounted on a heterogeneous collection of hardware and software platforms, usually distributed. We discussed the general case of technical integration of such systems in the previous chapter, when we referred to component architectures and the need to interpret between data bases, from low-level data formats to high level data semantics. What we shall mainly discuss here are the application-based technologies that lie on top of these.

In designing a data warehouse a decision has to be made as to whether it should be *data-centric* or *application-centric*. In the former case, the data model for the warehouse is designed (as far as possible) to be independent of individual applications. Rather, it tries to be comprehensive enough to cover existing and likely future applications. The customisation to meet the needs of any specific application or user is provided by the query front-end and presentation software (Figure 2.6).

In Figure 2.6, it is the bottom layer of data that is prime, with the application-centred views being generated from it. This holistic approach is frequently not feasible for organisations that have grown their data handling in a decentralised manner. More often, the solution will be an application-centric one, (or several application-centric solutions). The

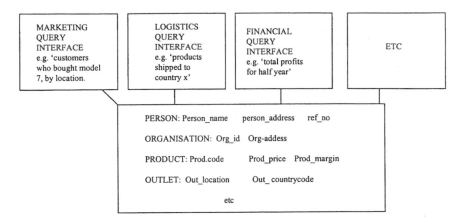

Figure 2.6 Structure of 'data-centric' warehouse

warehousing solution then is based on integrating a number of databases that all serve one functional unit, marketing, for example. As shown in Figure 2.7, the data often exists as *tables* in relational databases. For historical reasons, these are often of different vintage and design. The application warehouse tackles the more modest problem of integrating this data into something which can be queried in a homogeneous way, without redesigning the basic data model.

Nomenclature is rather imprecisely defined, but application warehouses tend to be larger and more ambitious examples of *data marts*, which are formed from new or from existing data bases and are intended to support a relatively compact business process such as customer management. A very real danger occurs when data marts overlap in the data which they contain. Maintenance and integrity becomes a serious

	DELIVERY CHANNEL					
	COMPETITORS					
	COUNTRY PERFORMANCE					
	SALES CHANNEL					
	PRODUCT DATA					
	#1	#2	....		#n	
profits						
sales						
mkt share						
delivery						
quality						
etc						

Figure 2.7 Example of application warehouse tables to support marketing department

problem and much time can be taken up in resolving disputes between finance and marketing departments, for example, if each accesses a different set of sales figures and margin calculation algorithms. Nevertheless, many organisations take this route because of the difficulties and delays involved in getting management approval for a global solution [64].

Although the goal of warehouse design may be to create a generic solution that is independent of the individual application, the main characteristic of such a system must be that it is business-centric rather than simply a huge repository of data. Specifically, it should be aligned around a process model, rather than a traditional data model [65]. One reason why this is important is that of performance. Queries on a traditional, one-subject database used to support a single operation, can relatively easily penetrate the data in a manner which optimises performance – the data modeller knows which are the dominant entities and attributes and can design the database accordingly. However, where the warehouse has to serve several 'masters', their queries will hit into the database from directions which could not necessarily be foreseen. Consequently, database custom and practice may need to be modified. For example, *normalisation*, the design principle that entries should not appear more often than necessary, frequently has to be abandoned in the interest of speed [66]. Also, although the central, generic warehouse concept is worth preserving, it may be necessary to create subsets of the data on local servers tuned to the needs of specific business units. Warehouse software will in this case include mechanisms for synchronising these local databases with the master system, with regular updates and marking of the data with details of when it was last updated.

2.8   ENTERPRISE-WIDE DATA MANAGEMENT

Although, in the best of possible worlds, the design of enterprise information systems should begin with a top-down determination of what we want to manage followed by the integration of the nuts and bolts that will deliver this, we almost never start from that favourable vantage point: legacy and multiple initiatives within the organisations concerned, have usually made that impossible. Consequently, quite a significant amount of work must be done, simply in joining together what already exists.

Typically, an organisation's data resources will reside in a number of places, in a number of formats (Figure 2.8).

The position shown in Figure 2.8 represents the realities of many organisations. They are often the battlegrounds between the integrated, but perhaps unresponsive or bureaucratic, control exerted by the company's IS department, and the more recent fragmented initiatives undertaken by individual functional units who have more rapidly created their own data marts. Into this already confused scene, has been thrown the corporate

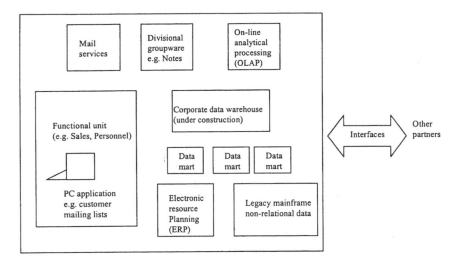

Figure 2.8 Typical mix of data resources

Intranet, which is intended to give everyone in the organisation direct on-line access to corporate knowledge.

In situations like these, the expectations of these individuals are sometimes very much let down when told that the only way files can be accessed from one system by another is by physically transporting data tapes or disks. That is not all, often the data then needs to be run though off-line converter programs which take some time to develop, and can only provide very restricted windows onto the data, because of data-model differences. Even if physical transportation can be replaced by the often expensive, and still rather slow, bulk transfer of files electronically, (Figure 2.9), using methods such as Internet *File Transfer Protocol (FTP)*, application integration is not easy to obtain across different platforms.

Moreover, data is not static. It changes and these changes must be synchronised across all applications that make up a business process. File transfer between processes therefore would need not to happen just once, but every time the relevant data changed. This quickly becomes unmanageable. Instead, data should be maintained only on its host platform and only relevant parts of it fetched by dependent processes residing on other systems, as and when required. In order to achieve this, one approach is to use *component* principles, as described in our earlier architectural discussion (Figure 2.10).

The components, which interface between the two systems, must be capable of maintaining a reliable communication across the interface, perhaps with full transaction processing integrity, commitment and roll-back. In practice, the management of this may be done by a separate suite

Figure 2.9   Application processing on copied data

of software run on a separate server. A fundamental requirement is that the data structure of one database must be mapped into that of the other: one typical example of this is the conversion of unit values – pounds weight to kilograms, US Dollars to French Francs, etc. – that might occur in the case of multinational enterprise building. Another common example involves the data tables of relational databases: some may use one column to contain the address and the post-code, whereas others, which have adopted post-code marketing, may have split address and code into two separate columns.

Clearly, there is advantage in creating a solution (component-based or otherwise) which provides programmers with a set of tools that easily allow the construction of rules for carrying out such conversions. A number of commercial solutions now exist. See, for example Humming-bird's Genio Suite [67].

Of course, there are issues of performance involved, when we run applications which may not only require complex data queries but even more complex and time-consuming intermediating processes to be run on the data. One solution is to provide a hierarchical set of processes, with simple ones invoked for simple tasks and delegating as much of the work as possible to the existing database engines which are optimised for their own data. For more complex data interchange, the interchange may be conducted by the management program, running on a separate server. Also, where data bases reside on the same server, or are directly linked via

Figure 2.10   Application processing of remote data using components

Table 2.3   The 'Hummingbird' suite

Hummingbird Genio bi-directional connectivity
Delimited text files
Fixed text files
AS/400 flat files
Standard flat files
MS SQL Server 6.x, 7.0
MS Access (native, ODCB)
Oracle 7.x (native, ODCB)
Oracle 8.0 (native, ODCB)
Informix 7.x (ODCB)
Generic ODCB
Sybase 11.x (native)
Sybase SQL Anywhere (ODCB)
Teradata (ODCB)
IBM DB2 AS/400 (ODBC)
IBM DB2 CS (ODCB)
IBM DB2 MVS (ODBC)

a communications gateway, it is sometimes possible to allow them to communicate directly, thus reducing the need to 'trombone' the data stream from source to management server to destination.

At the time of writing, systems are just emerging which, for the first time, promise the possibility of making it feasible to integrate across a wide number of database platforms. For example, the Hummingbird Genio Suite referred to previously, claims to achieve this for a wide range of systems (Table 2.3).

2.9   DATA DEFINITION, QUERIES AND META DATA

One of the several conceptual challenges that the World Wide Web sets to the database community is in the matter of *query* logic. By far the dominant paradigm in today's database architecture is that of the *relational database* with its well-known tabular model, entities and relationships. The adoption of this framework naturally gave rise to a specific approach to conducting queries on the data, in particular, the development of the *Simple Query Language (SQL)*. SQL is dependent on the data being, at least conceptually, in rows and columns, on one or more tables. Queries are formed with reference to them: 'find all customers in India ordering more than $1 million of stock', can easily be represented in this manner. But much on-line Web data is not tabulated. Queries are usually string-based, searching for 'India', for example, perhaps enhanced by Boolean operators: thus, 'India' AND 'computer'. This is a simple, dirty and quite effective way of gathering a comprehensive load of information, but also retrieving quite a lot that is not particularly relevant.

In trying to integrate 'pure' databases with 'dirty' Web data, we have to demand compromise from both sides. This is already happening. In the case of databases, there is an increasing realisation that *indexing* of the contents is important: summaries of the contents are created automatically and become searchable with Boolean strings. On the Web side, the emergence of relatively standardised meta data to describe the contents of Web pages is increasingly practised. We discuss these aspects further, when we look at knowledge extraction. Earlier in this chapter, we discussed XML as a means of mapping between the trading documents of interacting organisations. In particular, we made the point that different views are often held by organisations about the necessary fields on invoices, order forms, etc. This is not peculiar to Web trading; it is also a problem whenever databases designed by different teams are brought together. In general, this data issue is more complex than the trading case. The meta data held in databases is often described as serving two purposes: *administrative* and *business*. Administrative meta data is concerned with how the data is to be managed: when it has been/is to be, updated, whether it has been *cleansed*, its source, and so on. Business meta data describes what the data is: its relationship to some business process, the rules by which it is derived from other items ('profit' = 'income' − 'cost') and so on. In creating a data interchange system, it is necessary to include both kinds of meta data information, as well as conversion rules between the meta data for different databases. Tools need to be provided to allow non-experts to construct these mappings.

2.10 DATA CONTAMINATION AND CLEANSING

There have always been problems with data integrity, but this will be an even greater problem in the eBusiness, for at least three reasons: firstly, the sheer volume of data will increase, therefore, the potential for error will rise proportionally. Secondly, the joining together of existing organisations into temporary extended enterprises will lead to incompatibilities between systems and, perhaps an unwillingness to accept responsibility to put things right, especially to support what is perceived to be a transient relationship. The thirdly, the loss of control of data inputting. Instead of having a relatively few, well-trained and supervised data input clerks, the Web environment envisages a significant percentage of data being inputted by customers and suppliers with little knowledge or willingness to take care of a system which is not their own. There will also be an increasing amount of data coming from relatively inaccurate bar-code and other readers. In general, data will come from sources with much less inbuilt quality checking. The problem to be solved is how to correct this data after it has been collected, or at least, how to minimise any damage that can be done.

Earlier in this chapter we have described the three conceptual stages in the creation of eBusiness databases and warehouses. Data integrity can be improved by processing at three main points in the cycle (Figure 2.11).

The first place where we can verify the data and exclude errors is within the specific application, for example in an order form. Clearly, it is very important to ensure that as many checks as possible are carried out, before the data is accepted. Standard practices of creating data checksums should be enforced and perhaps re-enforced. (Checksums are an early form of digital signature, usually applied to a product or payment mechanism: imagine a case where a company's products are assigned numerical codes in simple integer serial order. A mathematical operation is performed on this code and results in the generation of additional digits which are appended to the original code to create a full code for the product. A *check-digit* operation is performed by any system which receives a product code. This reverses the mathematical operation and checks to see if the check digits and the basic code are compatible. For instance, a very simple approach might be to append a '1' to all odd product codes and a '0' to all even ones. In both cases, summing up the digits would result in an even number, unless data had been corrupted on the way. This is a very rudimentary and not very accurate or efficient solution, but demonstrates the principle.)

As far as possible, the bounds on data values should be checked at the earliest opportunity. Web forms could be checked on the client, before they are posted to a server, by using a Java applet sent with the form. This could make sure that the zip code field had been completed, check date of birth to see if it fell within feasible limits, and so on.

However, it is not always possible to carry out these checks at this point. The clients may not be able to accept applets or they choose not to accept

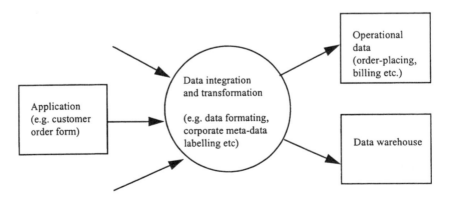

Figure 2.11 Three stages in populating a data warehouse

them. The application may be too old to be re-written effectively with integrity checks included. It may be that it is not possible to carry out all the checks at the input stage for a number of other reasons. In any case, some errors will get through.

There is also a problem with local versus corporate *data semantics*. The meaning of a data term within one application may not correspond exactly with the meaning of a term with similar name in another: a common example would be the term 'location'. To one application or to one company within an enterprise, this might mean a post-code district, to another, the specific delivery bay of a warehouse. There is a need to reconcile these different viewpoints, which are not errors but can still screw up the business processes.

This reconciliation problem is generally given a much more positive name: *data integration and transformation*. Here we require a medley of software that can first map the low-level data structure – e.g. relational and flat databases – into a common format for processing, and then analyse it to see if it contains anomalies. A very simple example might be in discovering that a company apparently always located at one address in one application was always located at another in another application. Even more basic issues such as databases which truncate input at different lengths need to be picked up. Clearly, these types of problem could not have been detected at the single application stage. The integration and transformation stage therefore adds value to the individual application data streams by checking them out against the corporate data models that are used by the data warehouse and the operational databases, flagging inconsistencies and then adding the meta data descriptors before the data is allowed to go into the warehouse or drive the operational process.

2.11   DATA INTEGRITY WITHIN THE WAREHOUSE

Beyond this point you might believe that the data that goes into the warehouse is now as near to perfect as it can be made. This is probably true at the time it is written into the database, but surprising though it may seem, this may not be a sufficient condition. So far, we have only discussed what are essentially problems of multiple input points, i.e. an essentially spatial problem. Databases also suffer from temporal problems: a database can remain unchanged, but the circumstances of how it is populated may not. First of all, there is the issue of *referential integrity*: if, for example, data item Y inherits values, properties, methods, etc. from data item X, then the justification for continuing the existence of Y within the data model, depends on the continued existence of X. A subsidiary company can only exist as such, if the parent company continues to exist, to give one simple example. That is not to say that the subsidiary ceases to trade, it may continue to do so, even under the same name, but it is important that the database has some way of distinguishing the change in circumstance. For relationships like this, it is often a good idea to extend the data structure to include *time stamping* information, which may be updated periodically, which defines the beginning and end dates for which the information is considered to be valid. It is possible to run intelligent software over the data warehouse with a view to testing issues like this, for example, examining all entries on parent and subsidiary companies to check for significant changes in behaviour (e.g. change in an invoice address, implying a reorganisation or dissolution).

2.12   THE DATA WAREHOUSE AND ON-LINE ANALYTIC PROCESSING

The very act of designing a data warehouse may in itself be advantageous: it brings much more clearly into focus the true nature of the vital business processes of the organisation. But warehousing is intended to do more than that. With a fully populated and reasonably accurate warehouse, businesses would hope to be able to extract information that would help them with running and/or redesigning business processes in a better way.

There is at least one fundament difference between a database built to drive a business process, such as order handling and a *decision support* data base from which is extracted information intended to drive a strategy, for example assessing regional sales performance and planning a marketing campaign: operational databases are usually read or written one data item at a time, for example, placing an order for a specific product, for a specific customer. This is one reason why it has become standard industrial practice to use relational databases whose individual records can be accessed and updated in this way. Databases for decision support, on the other hand, essentially aggregate and average information across a sample of data records. ('Provide the average sales per customer in North Region, South Region, East Region', etc.) Because of good practice data maintenance, such as *normalisation*, retrieval of this information may involve reading across a number of different tables. Even if this is not the case, simply accessing each individual record that matches the criteria, can be a very slow process, where large databases are concerned. It may not be possible, by this method, to provide *On-line Analytical Processing (OLAP)*, the name given to decision support services that are to-hand, rather than take hours or days to supply.

The solution to this problem is, in principle, simple and can be built on top of existing data bases. All that is required is to decide in advance what are the categories (or 'dimensions') of data that you want to manage –

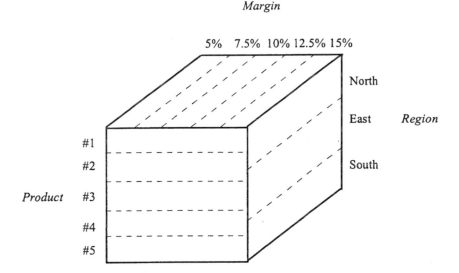

Margin

5%   7.5% 10% 12.5% 15%

North

East   *Region*

South

#1
#2
Product   #3
#4
#5

Figure 2.12   The 'data cube'

sales, regions, customer demographic groups, etc. – and then precompute the aggregates for each of these dimensions, periodically updating as new transactions occur.

We used the term *dimension* for each of the management items. This is the usual term and has given rise to another concept widely used in OLAP – the *Datacube*. Suppose, for simplicity, we consider just three dimensions: product, region, profit margin. Then we can represent our data aggregates as being placed within a 'cube' (Figure 2.12).

The axes of the cube are columns or fields in the operational data model and a value, the profit margin for sales of product 3 in East Region, is located at the corresponding set of three coordinates (a *3-tuple*) in this space.

There is no mathematical reason why we need to stick at three dimensions. All the rules for processing data will work for 'hypercubes' of arbitrarily many dimensions, but it makes it much easier for human visualisation and for processing performance if we do restrict the number of dimensions to a low value. Of course, we can calculate cubes whose axes are different: customer age, socio-economic group, postcode, quarters within a year, and so on, and it is usual to define the management data as a series of these cubes which feed into the multidimensional *decision support system* (Figure 2.13).

There are some limits to numbers of cubes and the total number of dimensions that an OLAP can be expected to handle, in terms both of processing time and storage. To get the best out of these systems, thought

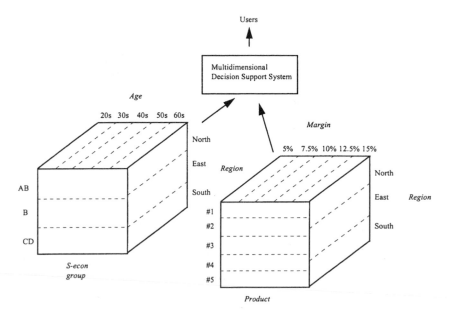

Figure 2.13 Data cubes and multidimensional decision support.

must first be given to selecting the dimensions that most compactly represent the key variables of the organisation. Warehousing and DSS vendors sometimes argue that it is best to choose these largely from customer-centric data, as these are most likely to reveal trends which can be used to enhance profitability. See for example [68].

OLAP database architecture is also an issue. Although the concept is a logical cube, the database that underpins it can be designed in many ways. The most radical is to mirror the logical cube by a cube of memory, or, really a set of arrays which together form a reduced version of the cube, omitting empty data points. Other approaches modify or augment the traditional relation database. For more details on this, plus an argument that sparse arrays are likely to be adopted for small/medium solutions, see [69].

2.13 DATA VISUALISATION

Given our various databases constructed in a way that allows us to use their contents to help our business planning, we still need some way to interact with them in order to obtain this guidance. There are many ways we can do this, and in doing so, various ways we can partition any 'intelligent' behaviour between the human user and the machine. The earliest systems simply supported human judgement by allowing queries and aggregations to be formed according to the user's intuition regarding what was significant, or by presenting the data in a way that made it easier to interpret, for example, by graphing it or producing pie-charts. This is still a very powerful aid to finding out what is or is not significant in data. Simple packages that can, for example, link a Web site activity monitor to an Excell spreadsheet, can be extremely powerful in identifying who are the repeat visitors to your site. *Data visualisation*, which capitalises on the human brain's ability to detect significance in two or three dimensional patterns is an extension of this. The success of data visualisation has been, in recent years, very much as the result of increased computing power. Faster machines have meant that it is possible to view dynamically altered renderings of the data: changing and compressing scales, rotating three-dimensional plots of data, better colour resolution and so on. It has also meant that tool development has been freed from many performance constraints. As well as making it feasible to construct user interfaces that are more friendly, it also means that a number of visualisation operations can more readily *pipelined*, so that, a scaling operation could easily be connected to a 3D plotting routine, for instance, and then viewed in a different way. Strategists can use these tools to give real-time demonstrations of scenario-playing to board-room decision-makers (who, incidentally, seldom themselves wish to 'play' with such systems, but do require convincing business-case evidence [70]).

2.14   AUTOMATED KNOWLEDGE DISCOVERY

Increasingly, vendors are offering solutions where some of the recognition of significant patterns is carried out by the computer. Rather than simply responding to sets of queries generated by users and based on the user's hypotheses, more advanced systems can themselves carry out *exploratory analysis* of the data, creating hypotheses themselves and clustering the data into meaningful, or at least, useful, patterns. To these patterns it will usually ascribe some statistical level of accuracy. For instance, a retailer might want to find out 'relevant' patterns of customer characteristics within a given postcode area. They will then request the intelligent software to 'find patterns related to postal area X'. The system may respond by computing the statistical co-behaviour of a number of parameters for that area and return a rule which states, 'within area X, there is an 83% chance that people over 45, who have bought life assurance from you, have also inquired about health insurance'. Of course, the parameters 'post-code', 'age', 'life assurance' (or simple derivatives of them) must be entities or attributes in the underlying data model. There is also the possibility that the software may generate too many statistical 'rules' of this type, if simply programmed on these lines. It is more rewarding to ask it to come up with rules which describe atypical behaviour, in our example, perhaps informing us of post code areas where the inquiries by 45 year plus people have been unusually high.

Notice that the rules mentioned above are essentially ways of compressing a large number of data sets into a much more compact and efficient processing method, and one which is, moreover, easier for a human being to understand. Rule-based reasoning also has the advantage that it can be expressed formally, in logic terms, leading to ease of programming. Like all systems which form general views of specific data, it is however, an approximation, and may need refinement over time. (It may even have to be abandoned, if circumstances change.) Sophisticated OLAP schemes carry out repeated checks on rules of this type. Some schemes even 'experiment' with rules, modifying their parameters by the addition of a small random element (*simulated annealing*) or by combining parts of rules in a manner analogous to sexual reproduction (*genetic algorithms*).

Rules are 'explicit': provided we understand the nomenclature of the logical language in which a rule is written, we can in principle at least (in practice the rule might be quite complex), understand its derivation. There are other powerful techniques which can be used to direct our strategy, which do not allow us so readily to understand why they work. This is not magic, although sometimes almost sold as such. One example is the *neural network* which consists of a very large number of simple, interconnected units that can perform logical functions. The ability of these networks to carry out their tasks comes about because of the high degree of connectivity between them, thus giving rise to another name for this class of techniques: *connectionism*. Typically, a single element in a neural network behaves as described in Figure 2.14.

Into the neural element are fed a number of the outputs of other neural elements, each via a weighting element which changes the strength of its effect. Each element may have a different weight applied to it. The neural element then adds up all these weighted inputs and produces an output which is not their simple sum. In fact, it has a shape similar to the 'S' shape shown in Figure 2.14. This shape, which approximates to the way that a biological neurone in a brain behaves, reacts very little to small inputs and tends to saturate with big ones. This means that in the connectionist

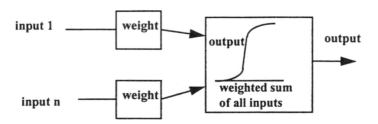

Figure 2.14   Single element in neural network

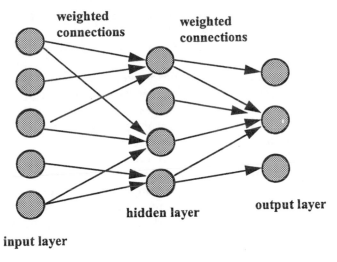

Figure 2.15 Typical connectivity of elements in neural network

processes that comprise the network, the very small is ignored and the very large is not allowed to dominate overmuch.

The nature of the connectivity is shown in Figure 2.15.

The neurones are stacked in three layers: an input layer which receives the data to be analysed, the output layer which gives the response, for example the recognition of one object amongst many, and a hidden layer.

In a typical OLAP example, we might wish to separate our customer base into those who are worth account managing, those who should simply be sent a mail-shot and those not worth contacting at all. We have a number of examples of data for each of these types of customers. We call this the 'training set'. The data table may be quite complex: post code, age, profession, etc. Each of these dimensions in the data table is arbitrarily assigned to one of the input elements of the network. Our example will have three output neurones, named as follows: 1 = 'manage', 2 = 'mail', 3 = 'ignore'. We begin by assigning random values to the weights of the network and feed the values of the first record from the training set into the input layer. It is extremely unlikely that this will result in only one output node having a large value. However, it is possible to have a simple algorithm that, when told that the input corresponds to customer type say '1', then varies the weights in such a way as to drive the first output to a higher value and the other two to smaller values. A second example of the same object class is applied to the input and the process of adjusting the weights carried out again. Eventually the weights approach a steady value, which, in some way is a collective 'memory' of type 1. We can store the weights as a recognition pattern for type 1 and then repeat the process with type 2 examples and then for type 3. In the end, we have patterns for each customer type. So, when we get data on a new customer, we can test this on the three patterns and would expect to get a high output value from the one which best predicts the customer's management preference.

In the marketing example cited, the network was trained on the basis of adjusting the weights according to which class the object belonged to. It is even possible, in some cases to carry out 'unsupervised training': the network spontaneously begins to separate out different clumps of data corresponding to the different classes of objects.

It sounds rather magical, but it is not. It relies upon there being separable properties inherent in the data from different object classes. These differences may be very difficult for human observers or classical statistical methods to discern, but they must exist. In fact, this is one of the problems with neural networks: they do detect differences between different classes of objects, but it is not always possible to see why. This means that one cannot in general use a neural network to extract 'features' that could be used by a simpler or faster classification techniques; it also means that it is difficult to predict how a network that was successful on one class of problems will perform on another. Neural networks were greeted enthusiastically after an initial period of scepticism; some (though not all) of that scepticism has returned in recent years.

2.15 'ABOUTNESS' IN KNOWLEDGE MANAGEMENT

Moving the interface with your customers from a call-centre to an eMail handling facility may reduce costs but it merely shifts the pain of management from one operation to another. There still has to be some way of separating out queries for information from customer complaints; there still has to be a way of noticing from this input that there are problems with specific products and alerting the designers to this fact. There are many other areas where the increased volume of information, both into the organisation and across it, need to be analysed and understood, preferably by a machine. Specialist companies in the field of knowledge management software are beginning to offer solutions which 'intelligently' analyse corporate data and target it at the correct area for resolution.

It has been said [71] that we can only decide whether information is important if we first know what it is about. The concept of *aboutness* may seem highly abstract and difficult to define in a formal way, but, increasingly, companies are constructing knowledge extraction tools that can produce operationally satisfactory explanations of aboutness. A reasonably accurate way to think about organising knowledge is to consider it to be submitted to a number of queries that can be phrased in natural language terms such as, 'Who is involved?', 'What is it about?', 'Why was it produced?', 'When did things happen?', 'Where?', 'How (as part of which process) was it created', 'How much (or what are the numbers involved)?' It is not particularly difficult to list a number of ways that the aboutness of an electronic data item can be described:

- By its title.
- By a piece of meta information whose definition has been published.
- By who wrote it.
- By when it was written.
- By its contents.
- By links to and from it by other items.
- By its appearance (coloured brochure, spreadsheet, etc.).

and there are others. It can be seen that documents do not just exist in isolation, they exist within a number of contexts that include the people that work with them and the business processes they support and they alter with time, rather than being fossils that do not change. Figure 2.16 sets out to explain this further.

In Figure 2.16, a document is represented in two ways. Firstly, by its 'content', which consists of the core, subject matter and any additional meta information added to it by human or machine interpreters. Secondly, by its history – which process it was involved in, who read it, included it in a personal profile as a typical example of things it wanted to hear about and so on.

It is obvious that some of this information is overtly, even consciously, applied to the data: most reports these days are written using standard 'styles' for different levels of headings, etc. most have contents sheets; spreadsheets and databases have column headings that relate in some

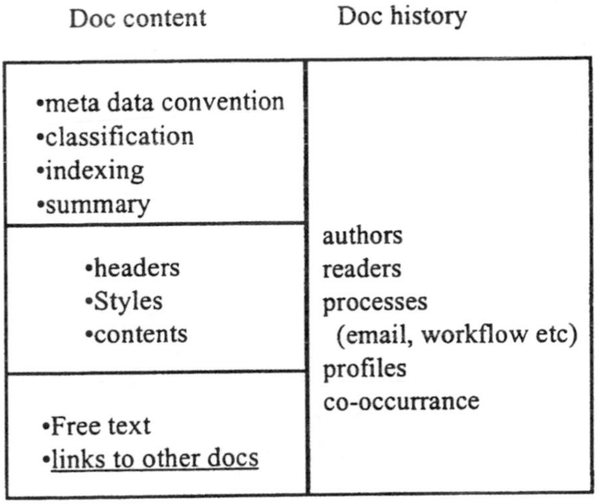

Doc content	Doc history
•meta data convention •classification •indexing •summary	
•headers •Styles •contents	authors readers processes (email, workflow etc) profiles co-occurrence
•Free text •links to other docs	

Figure 2.16 The richness of document structure

way to corporate data models and dictionaries; sometimes authors or interpreters apply *keywording* to a document, in an HTML or XML <meta> definition; sometimes database engines automatically index the free text body of the document, extracting words and word-frequencies which are intended to act as a machine-searchable summary of the document.

But much of the information about a document's use is covertly (although not in the deliberate sense) generated. If this is to be used as part of the aboutness evidence, then it too, must be written into a meta description and updated as often as necessary.

Thus, a basic task of knowledge management solutions is the integration of a number of sets of process-generated data by marking them up with index labels and source information, then observing the subsequent history of the data and continually updating this record, together with using *inference* engines, equipped with rudimentary intelligence, that try to achieve a simple pair of goals:

1 Supply all the right people with all the data important and relevant to their area of working (the *recall* requirement).
2 Without supplying anything that is not necessary (the *precision* requirement).

In general there is a reciprocal relationship between the precision and recall performance of an information retrieval process (Figure 2.17).

Note that this can make it rather difficult to compare knowledge retrieval systems. Some may err on the side of giving too much, others prefer to give less, with the chance of losing something vital. Again, the choice is not absolute; it will depend upon the current requirements of its current user population, given today's data set.

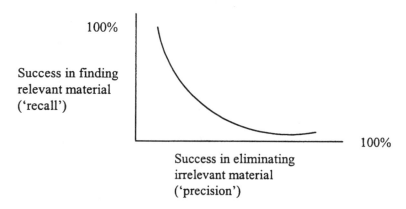

Figure 2.17 Inverse relationship between 'precision' and 'recall'

2.16 UNDERSTANDING TEXT

The starting point for most practical systems is in trying to extract subject matter from data sources. The process of extraction necessarily consists of taking a longer text and reducing it to a shorter one. This is done partly to improve processing performance in the stages that follow, but also because extraction in some sense gets at the 'essence' of the document. One of the simplest methods relies upon producing a list of individual words used in the text. Users who request information on that word can be supplied with all documents which contain that word in their lists. The process can be improved, in terms of recall, by *stemming*: although users request information on 'dog', the process has a database which extends the term into 'dogs', 'lapdogs', 'gun dog', etc. A simple refinement can be to index documents to find out how many times each word occurs as a fraction of the total length of the text. It is usual to exclude all the very common words such as 'the', 'and', 'is', etc. by putting them into a *stop-list* that is used by the indexing program. Thus, given a user request for texts with 'gas' (say) in them, the texts can be listed in descending order of frequency, thus promoting precision.

But there is more that can be done by pursuing this simple strategy of including word frequency counts: it is possible to describe a text in terms of how this whole list compares with the lists generated for other documents (Figure 2.18).

The document on the left hand side of the figure is one which a user has found to be of interest. Part of its word index is shown, in particular, the words 'lpg', 'octane', 'gas', 'auto', together with a measure of their frequency of occurrence. On the right hand side are the statistics for these words, averaged across the whole database.

Document of interest		Document X		Document Y		Database average	
lpg	000	lpg	033	lpg	000	lpg	001
octane	010	octane	000	octane	009	octane	003
gas	034	gas	045	gas	039	gas	012
auto	025	auto	001	auto	033	auto	020

Figure 2.18 Word counts

Now, without even knowing that the indexing had occurred, the user could say, 'Find me more documents on the same subject'. The search engine looks across the database, comparing the index of the interesting document with all the others. Consider two cases of such a comparison, with documents X and Y. Without resorting to complex maths, we can see that document Y is 'much more like' the document of interest than either the average or document X. In particular, we notice the frequent occurrence of 'gas' AND 'auto' in both Y and the interesting document. Probably this means that they share a common interest in 'gas' in terms of its American usage for 'petrol'. By using the joint probabilities of words, we can begin to define rudimentary 'meanings' for them individually: 'gas' = 'petrol' versus 'gas' = 'natural gas' because the former tends to associate with items such as 'autos' (= 'cars') rather than 'lpg' = 'natural gas'.

If we were to look more deeply at how we did this intuitively, we would find that we essentially were measuring for each document, how the ratio of (frequency in document)/(average frequency across all documents) for each term varies. A ratio significantly greater or less than one for that term, implies it is highly atypical. Documents which share many atypicalities are likely to be on the same subject. The ratio of frequency of term within a document to the average frequency, is known as the *Term Frequency Inverse Document Frequency (TF-IDF)* measure and commonly used in locating 'similar' documents.

In this example, for simplicity we have assumed that the user asks explicitly for similar documents and does so as a single query. In fact, knowledge management engines tend to remove the need for users to make this request. Many are instead designed to operate in a proactive, *push* mode: they gradually build up lists of preferred words as part of acquiring *user profiles*. These lists are modified by usage and by periodically asking users to express preferences. Having acquired a set of profiles for each user, the system can then, without prompting, send messages to users alerting them to the arrival of documents which match the profile.

2.17   NATURAL LANGUAGE PROCESSING (NLP)

The principles outlined so far, have been concerned with the behaviour of individual words or the statistical connection between words, but without any attempt to relate them grammatically. A slightly more ambitious approach is to make use of the fundamental property of language: that it possesses a *grammar. NLP* is the grand title that can be applied to rather simple processes that operate on free text, making use of elementary rules.

In general, we would expect the NLP engine to have two parts to it: a *lexicon*, which comprises a list of words, their 'meaning' in some sense, and their *part of speech* values (noun, verb, etc.), together with a *syntax*, a set of rules that operate on elements of the lexicon to generate 'grammatically correct' and 'meaningful' phrases and sentences. To give a very simple example, a lexicon could contain

'DOG':

PART OF SPEECH = NOUN (80%), member of ANIMAL, synonyms = HOUND,

MAN'S_BEST_FRIEND

PART OF SPEECH = VERB (15%), transitive, synonyms = PURSUE

PART OF SPEECH = ADJECTIVE (5%) etc., refers to ROSE

and a rule might be PLURAL $<x> = <x>$ + 's'. In practice, a real grammar would be more complex than this. (It might, for instance, be able to handle the fact that the plural of 'sheep' was still 'sheep', not 'sheeps', as posited by our example.) Nevertheless, even the simple example has some power. Note that three possibilities are given for the part of speech of dog: 'a good dog' (noun), 'to dog the footsteps of someone' (verb), 'a dog rose' (adjective) and probabilities of occurrence of these are given in brackets, based on measurements across a large number of texts whose parts of speech have been marked up by humans. Other statistics can also be gathered about words and classes of words, for example for parts of speech, and embodied in statistical rules. For instance, very seldom does the word 'to' come directly before a noun, but very frequently before a verb. So, since 'dog' reasonably frequently occurs as a verb (15%), it is very likely to be one in the phrase 'to dog someone's footsteps'.

Comparatively simple NLP grammars of this type have been used to analyse business data to allow highly accurate extraction of company names, location, etc. and the analysis of eMails to determine whether or not they are complaints. A rather different application is the *parsing* of queries put into a search engine which queries a database. Here the parser uses NPL techniques to correct any errors made by the person making the query. The parser uses a grammar which allows it to check on whether queries have the correct structure and vocabulary and may also hold semantic knowledge about the database that allows further interpretation. One example given by one vendor [72] is the parsing of '6 Pak 12oz Diet Cola', where it is claimed that semantic knowledge is required to separate bottle size from the packaging.

2.18 TEXT SUMMARISATION

Often the problem is not so much in finding a document, but in having time to scan through it to find out if it contains anything of value. What we

The tagging in the document unambiguously splits the rows in the table into two, using the <td> (table data) and <tr> row tags and it can also be seen that the names 'ABC' and 'DEF' fall under the heading of 'ORGA-NISATION NAME'. A relatively unintelligent programme, written probably in a language such as PERL, could be used to inspect Web pages for tables containing headers such as 'ORGANISATION', etc. and extract all entries that fell under these headings. Elaborations to this approach can be used as a way of converting a temporary partner's on-line data into a format that can be processed within another organisation's data architecture. It can also be used as a way of gathering marketing intelligence.

2.20 THE ROLE OF CONVENTION IN KNOWLEDGE EXTRACTION

Although, from what we have so far said, the task of automated understanding of data created by others without regard to explicit data conventions and meta information, may seem a daunting task, it is not always as difficult as it might seem. Human beings do bring special skills to understanding the aboutness of documents, but, when we do so, we usually also rely on a lot of more mundane features. The creation of documents, whether technical reports, sales brochures or newspapers, usually follows a number of conventions. Banner headlines cover the most important and general topics, with more specific issues appearing beneath them in smaller type. Stock-market reports use reportage virtually stereotyped in terminology and sentence structure. Web pages are differentiated on the basis of their functions: it is often not too hard to pick out a page listing products and their prices. One hope for automated extraction of information is founded on the belief that this conventional behaviour can be categorised or even learnt by machines. Some successes have been reported. In Part 4 *Electronic Marketing*, we describe how advertisements can be detected on Web pages and rejected. We also mention artificial trading agents, which normally would be expected to trade within a well-conditioned DTD-defined process, but might be able to carry out opportunistic bargain-hunting elsewhere.

2.21 KNOWLEDGE MANAGEMENT – SOME FINAL OBSERVATIONS

The increased availability of electronic communication inside and organisation and between it and its partners, suppliers and customers has certainly led to a flood of unstructured messages that have not submitted themselves to traditional rules of corporate data management and it is reasonable that attention should turn to methods of managing this flood. One message is clear and we must agree with the IBM comment that 'relational databases are not designed for the ever-growing demands of knowledge management'. But what is? The problem, let us remember, is not to store the data, but to get selected parts of it back to us, with good precision and recall and with a good idea of the bounds of its operation. We need to be able to integrate the various inputs to the organisation and the integration of the various systems that store this data. But this integration is not without its problems. It comes with the big price of trying to deal with heterogeneity at all levels, from the plug and socket to the deep semantics of the information model. Traditional relational databases at least have the advantage of consensus on the data model, usually some kind of quality control on the data, and a well-defined query interface, for example, *SQL*, which carries out information retrieval according to principles that are generally understood by the users. The alternatives, some form of meta information or a higher layer machine-understanding approach are not yet proven. We have seen a number of these in the sections above, and all of them have had their successes. However, we must be cautious; a large proportion of the advertising surrounding artificial intelligence solutions to this problem, is misleading, to put it kindly. Yes, there are working examples of intelligent software that can detect letters of complaint; there are even systems which provide reasonably accurate order-taking via speech recognition (see page 318). But these all operate within fairly constrained environments, on selected and usually rather narrow data-sets. The 'intelligence' behind successful systems is very rudimentary. It has to be said that there is a severe, almost complete, lack of theory or algorithm that can be described as demonstrating true *machine understanding* of complex data. The achievement of an artificial intelligence product should only be assessed by its continuing performance and value on your own data-set, over a period of time, not by claims concerning its IQ. It should be treated like a human analyst, indeed, its results compared directly with that of human competitors, and rewarded accordingly. In the short term, perhaps more modest approaches might yield more useful results.

Part 3: Trust, Security And Electronic Money

Chapter 1: Trust

Trade is based on *trust*, not security. People will deal with others, or not, on the basis of belief, not objective technicalities. We discuss a simple, four parameter, model of trust, and examine in general how technology can be used to address each parameter. Third-party qualification of Web sites is outlined and the content-labelling approaches of the Platform for Internet Content Selection (PICS) and the Platform for Privacy Preferences (P3P) explained.

Chapter 2: Security

There are a number of reasons why attacks are mounted against eBusinesses. These, and their likely consequences, are classified. The case for approaching security from a policy and process, rather than a technology point is argued. Some fundamental principles of security are discussed, including the adequacy of military security as a model for the commercial scenario. There are different aspects to security protection, including confidentiality, integrity, service availability and authorisation. Computers are, formally and practically, complex systems, and thus inherently weak from a security point of view. This is aggravated by the layered approach to system design. A number of attack methods are discussed, including eavesdropping and snooping, source routing and defragmentation attacks, the use of covert channels, problems with remote procedure and CGI calls, and out-of-the-box vulnerabilities. A variety of firewall variants can be introduced to protect against network attacks and to enable the construction of secure virtual private networks. Private and public key encryption and hashing systems are explained, together with their use in common digital certification and secure data transmission applications. On-line services such as Napster and gnutella have threatened copyright protection. Some relief may be had through use of digital watermarking but this is not foolproof. Providing access rights to legitimate users can be enabled by means of smart-cards and biometric systems. The latter are a special class of pattern recognition applications. Some basic theory is reviewed, particularly with regard to what makes a 'good' classifier and a range of biometric alternatives are assessed.

Chapter 3: Electronic Money

In the area of electronic trading, undoubtedly the highest profile application is in the transporting of credit and money, in electronic form. Apart from the traditional credit or charge card there is no single preferred approach. This is partly because of natural caution and conservatism among customers and financial providers, but also because of lack of standardisation and the varying sizes of transactions involved. Low value transactions require cheap solutions; a number of electronic cash methods exist, of which the WWW consortium's Per Fee Link Handling system is only one. For larger or more complex transactions, card providers have pressed for the adoption of the Secure Electronic Transactions (SET) approach. Electronic token holders may play an important part in eMoney transactions and the SIM smart chip within mobile telephones offers an intriguing challenge to traditional card services.

Many on-line retailers may prefer to leave payment service platforms to specialised suppliers and this is reviewed.

1

Trust

Trade is about trust, not security. People will deal with others, or not, on the basis of belief, not objective technicalities, and in the arena of trade they deal on the basis of trust. Trust is essentially subjective; it may be backed up by processes that objectively deserve that trust, but the subjective dominates. Trust will not necessarily be unconditional; every day of our trading lives we accept a degree of risk that the other party will not play entirely fair. We do so on the basis that, on balance, we shall benefit more than we suffer. Of course, the various measures that, collectively, are described under the term 'security' are intended to realise this trust, but they are enablers, not ends in themselves. This is probably wise: it may be better to rely on the 'gut-feel' of many customers than on the considered opinion of a single security officer; there is no better audit of the integrity of a system than barrage testing by external parties.

Moreover, there is another reason why we should not too readily use the term 'security' in the domain of eBusiness: it is too much tied up with *military* security. Certainly the basic techniques, the algorithms, are the same, but the outlook, the security processes, are often a long way apart. Military security is about external attack; almost invariably the parties engaged on legitimate exchange of information are assumed to trust each other. This is not the case in free trade. So, although in this book we shall talk a lot about security, it is important to remember that the real issue is trust.

1.1 WHAT IS TRUST? HOW IS IT ASSESSED?

There are of course a near infinite number of possible ways of defining trust and a similar number of ways of assessing it. A simple way of doing both has been described by the author and colleagues:

An act of trust involves placing yourself at hazard to another's actions, in the belief, at least partly without explicit computability of risk, that they will act to your benefit [73].

Trust is therefore something which is generally uncomputable, involves you staking something that will lead to loss if things go wrong, but doing so in the hope that they will not and you will thereby benefit. In making the decision to trust someone or some organisation, there are a number of criteria that need to be satisfied. In our model, we define four of these:

- *Need*: I trust you because you need to do the business fairly. Perhaps you need to make the sale; perhaps you need me to come back for more; perhaps I have fenced you round with guarantees and penalties so that you cannot safely cheat me, maybe using technology that preclude you from denying that it really was you that signed an agreement.
- *Identity*: I trade with you on the basis that I feel we share a common set of beliefs, a common vision for the future, that makes cooperation the natural way ahead. Companies with, for example, a common belief that they are leaders in the use of technology, will find it easier to take risks in their relationships. In retail trading, Branding is the equivalent, having the image that you can plausibly share a set of attitudes and solutions that meet the lifestyle aspirations of the buyer.
- *Competence*: I can trust you because your processes are good: your inventory control gives me confidence you can deliver when you say you can; if I tell you a secret, I can feel happy that you have the means to protect it.
- *Evidence*: What is more, I have knowledge of past dealings with you that you have not let me down.

Of these four, it is perhaps the 'evidence' criterion that is the Achilles' heel of the 'instant enterprise' or the Internet start-up; there will be no track record of fair dealing.

We therefore hypothesise that the lack of *evidence*, must be compensated for by a corresponding increase in the positive features of *need*, *identity* and *competence*. Trust technologies must therefore always be looked at in the light of these parameters. Perhaps we can compensate by offering online very visible demonstration of competence – we use a widely recognised encryption method for handling credit card details. Maybe we involve a third party, such as the trustE or Verisign organisations (see below) whose whole business credibility relies on the need to demonstrate fair dealing among their members. In the sections that follow, it may be helpful to remember this message: without trust, no trade.

1.2   WHAT TRUST DO CUSTOMERS PLACE IN SUPPLIERS?

When customers choose to deal with a particular supplier, there are certain expectations regarding the latter's behaviour that have influenced that decision. Of course, it is expected that the product or service purchased will be fit for purpose and not significantly over-priced. But there are other aspects. We expect financial controls to be secure and honest, so that no one can steal our account numbers, charge goods to us, or take money from our bank through a fault of the supplier. There is an expectation about the confidentiality of the transaction – customers generally do not feel that their purchase is something that should be broadcast to the world, at least without their consent, either leading to their discredit or to unwanted selling approaches.

All of these become bigger issues with the arrival of on-line trading. We pass out our credit card details to someone or some machine that may be located at the other side of the world. We have no visibility as to how they handle this information once they have used it for the intended purpose. For instance, it becomes extremely easy to collect information automatically on each customer transaction. Thus it is also easy to subject the sum-total of transactions to profiling software that can create a pen-picture of each individual's buying and payment habits, to be passed on to the marketing department or sold on to third parties. These organisations might thereafter approach us with outbound selling of loans to pay off the purchase, offer other products that they think meet our purchasing profiles, or even secretly use the data to construct credit assessments without the customer's agreement.

We may accept all of this and it can even be used to provide customers with a more satisfactory service, but it can be open to abuse and it can be seen as an infringement of our rights. Companies should consider how prospective customers view these issues and it may be necessary to put in place measures of reassurance.

Some Web sites declare an explicit policy on their handling of customer details; in order to give additional assurance that this policy is adhered to, they let themselves be audited by a third-party. For example, the eTrust consortium [74] describes itself as a 'global initiative for establishing consumer trust and confidence in electronic information exchange'. They promote a 'trustmark scheme' whereby Web sites are allowed to display one of three labels guaranteeing that:

- No data is collected on users; or
- Data is collected only for the site owner's use; or
- Data is collected and provided to specified third parties, but only with the user's knowledge and consent.

These assertions are backed up by self-assessment and professional, third party auditing.

Security of personal data and of financial aspects is another area where customers seek reassurance. This applies not just to naïve users of home services; quite a number of eBusiness experts have expressed their reservations about buying things over the Internet because they do not trust the way that companies treat their details. Probably the commonest complaint is that the Web companies may take great care to use well-qualified security products on the link between the customer and the Web site, but the back-office operations are totally insecure, with electronic and/or paper files being open for anyone to see. These anecdotal tales need to be sensibly rebutted by companies, as regards to their operations. Probably the only way to do this is to carry out a security audit of the end-to-end process, correct any deficiencies and then write-up the process in words that the company's senior executives, company legal department, the non-technical customer-handling staff and the customers themselves can understand. (No prizes for guessing who are the most difficult to get through to!) This could then be converted into a short statement, published on the site. There is probably no better way to be able to convince your customers to trust you, than for you to have already a well-founded confidence in the processes, yourself.

Another way to gain customer trust is to hand over parts of the operation to third party organisations which have the leading brand-name for providing trustworthy services. This is a sensible interpretation of a security maxim that secure systems not only need to be secure, but also provably so. A trusted third party fulfils this requirement at the subjective level. A relatively few organisations such as Verisign [75] provide heavily branded services of this type. Technically they offer mainly standard digital certification processes, which we discuss in the chapter on *Security*, but they offer these within a widely known and generally highly regarded operational framework. One is buying from them the assumption that they dare not abuse the trust you place in them, by letting down your customers. It is a probably a subjective decision by eCompanies regarding how much they are prepared to pay for the brand against doing it themselves.

1.3   QUALIFYING THE PRODUCT – PICS, P3P

In our discussion of the *need, identification, competence* and *evidence* components of the NICE model of trust, we mentioned the positive role that could be played by involving third parties that commit their own credibility in acting as guarantors of the relationship. It is not in their interest to support an organisation that defaults on its responsibilities. Therefore, it is reasoned, customers can trust companies that they endorse. In terms of the NICE model, if they are well-known, with track-record, they can substitute the risk of dealing with an unknown with that of past evidence of their own performance and competence. In some cases, particularly when the third party is a celebratory or a lifestyle company, we may even identify with the aims of the third party and want to buy what they promote.

One particularly good example of a way to develop identification between customer and vendor, via a third party guarantor, is a rating service. Even before commencing a shopping interaction with a vendor, customers sometimes want to make sure that they are not entering into something they might later regret. A specific case is in issues of taste, for example, is the definition of 'decency'. The Web is international and easy to publish on. This means it contains a considerable amount of material that could be considered offensive to many individuals and cultures. How can we be sure we are going to find material that is, by our definition, suitable for us or our children to view?

One approach to the problem is to adopt the policy of 'buyer beware', but to provide tools that make it difficult to receive inappropriate material by accident. The *Platform for Internet Content Selection (PICS)* initiative was set up in August 1995, by representatives from twenty-three companies and organisations gathered under the auspices of MIT's World Wide Web Consortium to discuss the need for content labelling. The idea behind PICS is to provide a way whereby the producers of content and third parties, such as magazines, consumer organisations and industry watch-dogs, can insert information onto a supply of World Wide Web content, using standard protocols that can be read and acted upon automatically by computers, to provide local screening out of offensive material (and, indeed, the selection of desirable content) without introducing global censorship.

PICS was concerned to establish only the conventions for label formats and distribution methods, not the label vocabulary nor who should police it. The emphasis is on freedom of choice. Although labels can be created in line with, for example, movie rating categories, ('PG,', etc.), it accepts that personal standards and attitude must be expected to vary, particularly in a global context. In any case, publishers can get round these classifications by publishing in countries where standards are different.

(It is not simply a matter of overall laxity: some social groups are very anti-violence but tolerant of explicit sexual material, in other groups the reverse is true.) Although there were a number of such rating schemes in existence, PICS attempts to standardise the approach so that it is vendor independent and the labelling can be done by any number of third parties. Indeed, any number of labelling organisations can mark up a page at any time.

On the left-hand side of Table 1.1, we give an example of a simple PICS label. If one were to visit the labelling service site, 'http://www.here.-

Table 1.1   Example of a PICS label

Label	Explanation
(PICS-1.0 'http://www.here.co.uk' labels on '1998.01.01' until '1999.01.01' for 'http://www.newtv.uk.co/film.html' by 'Bill Whyte' ratings (12 s 1 v 4))	First the label identifies the url of the labelling service and gives beginning and end dates within which the labelling is valid Then it gives the url of the page that is labelled and, optionally, who labelled it Finally, it gives the rating (language = 1, sexually explicit scenes = 2, violence = 4)

co.uk', one would expect to see an explanation of how this the rating was created, as shown on the right-hand side of the table.

As we said, the labelling service can do what it likes. It does not need to restrict itself to moral judgements. In fact, it does not need to rate pages from a moral point of view at all. It could choose to rate recipe pages on how expensive it thought the recipe was, or even the spiciness of the food. The user of the service would set the filter software on their machine to select on any rating basis that they required. The rating service they would choose would depend on the credibility to them of the service: vegetarians would be unlikely to use a food-rating service provided by the beef marketing association; Baptists might choose a film rating service offered by an organisation with religious connections, whereas humanists would probably not. It is even possible to include a time of day filter on the pages: a PICS filter on a server in a school could ensure that, for the duration of a self-learning period, pupils could only access material relevant to what they were supposed to be studying.

Regarding the management, distribution and control of labels, two questions might be asked, how do we make sure that the PICS label covers the material to which it is attached; how do we know that the label has been applied by the claimed labelling agency? One simple way is to access the information via a reputable service provider, who will take steps to ensure that the material on the server is not provided under false pretences. An even stronger guarantee can be got through two other features of PICS: one is the ability to apply a 'watermark' to the material, for example in the form of a hash function, (see page 261). This appends to the text an encrypted summary of the page, that cannot easily be interpreted or corrupted by a malicious process. In addition, there is also the ability to include a digital signature of the labelling authority, on the page.

There are a number of ways the PICS labels can be applied and filtered (Figure 1.1). For instance, the originator of the page can insert the label as

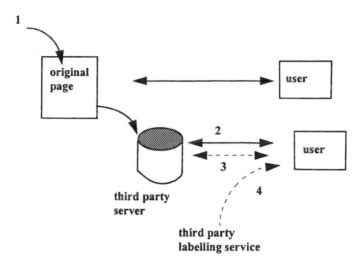

Figure 1.1   Applying PICS

an HTML or XML meta-tag. Then, the user's browser can read these and accept or refuse them according to profiles set-up in the browser.

Alternatively an intermediate server, possibly operated by an independent third party, can be asked by the user's browser to supply labels along with the document.

An extension of this is for the intermediate server to supply its users with pages it has pre-filtered, to save them the problem of doing so.

The client could make use of an independent labelling bureau, which only provided labels. Essentially this company is creating selective filtering, based on value judgements of assessors, critics, etc. These third party labels are 'seals of approval' and content producers might pay to have their products assessed by leading brands of labelling authorities. (Hypothetically, the seals now granted to certain manufacturers could be paralleled by on-line content labelled with, 'by Royal appointment', for example!)

It is even possible to imagine a democratic labelling service where users who access sites individually rate the material, from which an average set of data could be used to generate a reduced set.

1.4   PLATFORM FOR PRIVACY PREFERENCES (P3P)

PICS is a way of presenting data within an HTML or XML page, so that users can make decisions regarding its suitability. A logical extension to this is to provide more active (or interactive) behaviour between the user's browser and the service on offer, so that the personal choice can be used to trigger a process. This is the aim of the World Wide Web Consortium's *Platform for Privacy Preferences (P3P)* [76].

Users are considered to access services by using *agents*: an agent can be a standard Web browser, a special plug-in to a browser, or server-based agent located on a proxy server.

When a user accesses such a service, the agent is sent one or more P3P *proposals* – machine readable statements, contained in the HTML or XML reply or pointed to as a URL, which describe the privacy (or other labelling) practices of the service. The proposal will clearly identify and cover a specific *realm*: a URL or set of URLs.

The agent already contains the user's preferences and can compare them with those on offer. If the preference and the proposal are in agreement, then the agent sends a description of the agreement to the server, in the form of an *agreement ID*. On receipt of the agreementID, the server sends the requested Web page to the agent, which then displays it to the user.

Supposing, however, there is no initial agreement between the server and the agent. What then? P3P contains the ability to carry out further *negotiations*, and has the valuable property of allowing services to offer multiple proposals. In some circumstances, it is possible to route these negotiations via services which provide anonymity to the user. Clearly, this has implications for on-line contract negotiation, and converts the simple, trusting shopping transaction into one which recognises the gradual build-up of trust between the two parties, something more akin to the realities of inter-business purchasing.

One significant element of the P3P process is its ability to put in place a 'memory' of the previous transactions between the parties concerned, as part of a process of building trust and reducing the need to go through an elaborate re-negotiating protocol. Although some of this can be done at present by using *cookies*, the process does not lend itself easily to naive users wanting to be selective in their trust. P3P offers an alternative in the form of *pairwise id (PUID)* messages.

If a user agrees to accept a Web service on the basis of a service, which it receives indexed with a *proposal id (prop ID)*, it returns this prop ID together with its indexed PUID, to the URL specified in the agreement (Figure 1.2).

Apart from providing a two-way reference for the current transaction, the PUID and the prop ID can also be used in the future, to indicate the previous mutual agreement. Again, this can be conducted via an anonymising service, to protect the identity of the user.

One advantage of the P3P approach is that most, if not all of the above detail is hidden from the users and users will not need to make an explicit attempt to read the security policies of the sites that they visit.

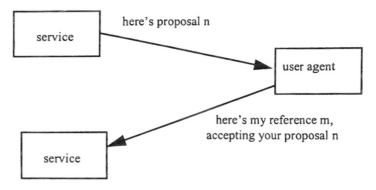

Figure 1.2 P3P pairwise messaging

1.5 DEMONSTRATING COMPETENCE

One's trust in a company is often based entirely on our assessment of their competence, measured by observing how they treat our current transaction. So it is worth remembering that an on-line store is only as good as it behaves: broken links, messages of the 'JaveScript error at line xxx' are interpreted by the potential customer, probably correctly, as a sign of how well resourced and managed is the on-line operation. The only real way to make sure that the quality of a company's brand is not tarnished is by testing, testing, testing, not just on the desk, but also in the field using naïve as well as expert users. Different browsers throw up different problems, as do different default colour and font settings. 'Back buttons' and other hot links that are easy to see on a developer's large monitor may be off the screen of a smaller monitor. Consider how your site will look when accessed at the end of a slow link as well as from the company's LAN. Code walk-throughs are standard practice in good software companies, but far too few do field testing of user interfaces. Probably it should be the responsibility of non-software-oriented people, perhaps a bunch of loyal customers can be recruited instead.

Finally, as research on the subject has confirmed [77], trust is more easily lost than re-gained. One bad experience with an on-line financial transaction may be all that potential customers need to avoid your site in the future. Thus, although we emphasise that it is trust that is the issue, it is now time to look at the single most powerful way that this trust can be gained: the issue of security.

2

Security

2.1 THE IMPORTANCE OF SECURITY

Increasingly, the pundits of eBusiness are raising the stakes on on-line security. So are the criminals. Frankly, the increased attention by the good guys is not before time. No one can reliably estimate with any reliability the cost of Internet crime, but it is known to be large and rising. Sadly, the number of competent good guys is relatively low. Out of interest, the author recently surveyed the teaching of computer security in universities and colleges around the world and was depressed to see how few of them taught anything that was of real practical use at undergraduate level. At postgraduate level, much of the teaching is in business and accounting schools, rather than in computer engineering departments. The position in the UK seems particularly bad, perhaps because we have a sneaking feeling that 'police interference' and 'snooping on staff' are not 'the thing to do'. Americans seem to take a more robust attitude to it: there is a patriotic strand which believes in defending Western democracy and a complementary 'freedom of speech' approach which vigorously practises 'white' hacking, thus advancing knowledge and theory. But this is getting outside the realms of technology!

2.2 THE NATURE OF THE THREAT

Why do people want to attack on-line sites? The obvious answer is 'greed', simply to steal money or fraudulently acquire goods. Clearly this is a major motive for some people, and some of these may belong to organised syndicates which have access to large resources and to very intelligent technicians and confidence tricksters. These may be however, a minority, albeit an important one. Probably the largest amount of on-line disruption and loss of business is caused by people who do not at least start out with the motive of personal gain. Some of it is simple curiosity aligned in some cases to social inadequacy – the traditional 'hacker' or 'anorak'. There is an above normal level of obsessiveness in some of these people, which means that they can and will devote a large amount of time and personal savings to mounting attacks on on-line systems. The privacy, anonymity and ease-of-access to the Internet, together with the widely held belief that it is a 'victimless crime', provide an ideal environment for this obsession. It has been said that some hackers feel they are doing customers and companies a favour, by exposing such weaknesses and it is up to companies to ensure that they can protect themselves against attack. Whatever the strength of their argument, the damage they can do to the businesses they attack, can be very serious. Even if they do not deliberately or accidentally destroy or disclose critical data, they can seriously damage a company's reputation.

Increasingly, we also see the rise of 'single-issue' protests, directed against 'government', 'big-business', etc. Sometimes the resulting activist behaviour can go well beyond any legitimate democratic rights to protest, but large numbers of people may still passively or actively support it. Some of these will have the capability to assist in these 'victimless crimes'. Others will look on indulgently at friends or children thus involved. The attackers in these cases are not motivated from sheer curiosity and they may feel entirely justified in doing as much damage as they can.

Returning to directly criminal activity, we note that the competitive business environment has always had its share of unscrupulous participants. It is impossible to put a reliable figure on the gains and losses accrued, as businesses will seek to minimise publicity on the activity, whilst security consultants will do the opposite, but it undoubtedly occurs to a greater degree than we hear about. On-line systems provide mass, global access to corporate IT systems, on a scale significantly greater than ever before. There is also a semi-clandestine network of hacking resources available on-line and through closed eMail groups. Through this network, one can gain access to a fund of information on weaknesses in specific systems, much of which is unknown to or forgotten by legitimate systems administrations. Since take-up of any practice, criminal as well as legal, is directly related to opportunity, we need to give increasing attention to the protection of sensitive data, either our own, or that of other parties we deal with.

These different reasons for attack manifest themselves in different sets of visible symptoms which have implications on the technologies used for the attack and for its defence (Table 2.1).

Recreational hackers do not really care which part of the system they get into; the challenge is the most important thing. They are often not even

Table 2.1

Type of attack	Target	Appearance
Recreational hacking	Indiscriminate	Varies, usually discreet, sometimes whimsical, may leave re-entry trapdoor, often copy-cat inspired by conventional hacker lore
Protest and personal-vendetta	Sabotage, denial of service, alteration of Web pages	Highly visible
Criminal	Financial transactions, credit card details, creaming-off of transactions	As hidden as possible, for as long as possible, especially in-house attacks
Industrial espionage	Corporate data	As hidden as possible. May never be discovered

particularly interested in whether they are spotted or not, although they may sometimes leave a secret trap door for later re-entry.

Although superficially similar to hacker attacks and, certainly, often carried out by the same type of attacker, protest attacks are, by their very nature, intended to inflict damage, particularly highly visible damage. The simplest way to achieve this, and one which many are prepared to consider the least 'illegal', is to modify a target's Web pages, perhaps to carry a counter-message against the company. An approach with perhaps more damaging consequences is to effectively barricade the site, using so-called *denial-of-service* attack, so that no one can access it and the company loses business.

There are other ways in which a business can incur material loss by on-line attacks, but they do begin to require a degree of ruthlessness more akin to out-and-out war. Personal-vendetta attacks might be included here: the disgruntled IS employee (or ex-employee) may have both the means and motive to carry it off.

Criminal and espionage attacks are, generally, much more unobtrusive, in order to achieve their aims, and they are both highly selective. Here, again, the insider possesses a great advantage.

Finally, a warning on the dangers of being obsessed with technology: most of us in the IS world suffer from two factors that make us poor security guards. We are pre-disposed to fair dealing, and we also tend to be rather blind to non-technical ways of achieving a goal, indeed, often rather dismissive of the simple approach to a problem. Expert criminals do not suffer from these defects. They cheat! They will 'forget' their PIN codes and are adept at persuading bank staff to install a new one for them from a PIN mailer behind the counter [78] in exactly the same way that computer help-desks have been defrauded since their services began. An equally simple scam was perpetrated on a diamond house in New York, some years ago. An agonised and panicky phone call was made to the company warning them that the security codes, used by couriers when they called to pick up packages of stones, had been compromised. With some show of reluctance, the caller released a new set, for emergency purposes only, over the line. Of course, the courier who quoted these numbers and received the diamonds was never seen again. Nor were the diamonds.

2.3 FUNDAMENTAL ASPECTS OF SECURITY

Because security specialists are only too aware of the truth of the last paragraph, they generally support an approach to security that operates at a number of levels, of which technology is only one (Table 2.2).

It is important to start looking at these issues from the top, not the bottom-up. It is well established that, in order to ensure that your security features deliver what they promise, they must be driven from a sensible security policy. Any such policy should start by looking at the business processes involved, beginning with an assessment of the skill, training and commitment of every one involved. In nearly all cases this will usually be depressingly low (and the opposite may be true of the attacking team) and any security implementation must take this into account.

Complementary to this assessment, there should also be one which tries to estimate the actual amount of material loss versus cost of protection. This is not just for the purpose of avoiding unnecessary cost in the purchase of equipment; it will also provide an estimate of disruption to the rest of the operation. In general, the more protection afforded by a

---

Table 2.2 Security on three levels

Security policy
What do we want to secure, how does this relate to our business processes, our staff and our customers, who is likely to attack us, why and for what gain, etc.

Security system
what are the procedures for authentication, access, key-management, audit, IT staff issues, etc.

Security technologies
(a) Implementations – networks, computer systems, key-management channels, etc.
(b) Algorithms – *DES*, *RSA*, etc.

security system, the greater the inconvenience to those legitimately involved. Consequently, if your security scheme is disproportionate to the perceived threat, then you can be sure that it will be compromised by its users who will treat it with contempt. There is a traditional argument for separating your critical operations from the less critical ones and only applying onerous security procedures to the former. This way it is easier to get buy-in to the procedures, as everyone involved can see the dangers and it also down-scales the size of the task. However, in computer operations there are dangers to this approach. It is not always easy to see the boundary between the critical and the non-critical. Later in this chapter we shall look at attacks on processes that run at the other side of a Web server by simply sending corrupted HTTP messages to the Web server: the attack leaps from the open Web access point into internal processes that are believed to be private and protected.

Finally, you also have to carry out a risk assessment as to the likelihood of attack, before and after you introduce any security scheme. This should be measured in terms of value to the *attacker* (not just to your company) and set this against the likelihood of the attacker possessing the necessary skills and access.

These security assessments are best left to security professionals, who are aware of human frailties as well as technology.

So, a secure system is not simply a set of passwords and encryption algorithms. It also comprises the day-to-day processes that administer the security: the role of the help-desk in dealing with people who 'forget' their passwords, the frequency and depth to which audits are carried out, the reality of the knowledge that attackers have about your system and the sloppy way with which your own employees administer it.

Although commercial security principles should not be considered to be a simple copy of those used in military situations, there are some general lessons that have been learned in the latter that are sensible to apply to the commercial case. Professor Fredrich Bauer, of the Munich Institute of Technology, has defined a set of maxims for cryptographic systems that can also be generally applied to commercial computer security [79]:

- *'The enemy knows the system'* (also known as Shannon's Maxim) – you must assume that the attacker can have access to the design of the system, (although not necessarily to the actual system under attack, or to specific security key-settings). Not just the principles but also often the fine details of most commercial systems of encryption and system security have been published and are widely known. Many of the best-know weaknesses and methods of attack will also be in the public domain. You can expect attackers to be at least as familiar with the algorithms and operating systems of the eBusiness servers, as most of your IS staff. Any algorithm must not only be strong in itself (difficult to break when primed with a specific key) but also supported by a process which makes sure that the attacker cannot get their hands on the key. Secure key management is one of the very big issues of security.

- *'Do not under-rate the adversary'* – do not be complacent about your systems. As we have mentioned, there is a large community of computer-fluent amateur hackers and professional criminals, some with a great deal of money and time on their hands. There is also a vast amount of guidance and system-probing routines available on the Internet. It would, for instance, be perfectly feasible for an activist organisation to run a *Chinese lottery attack* (see page 249) involving a hundred of so members using continuously running software in parallel, to probe your site twenty four hours a day, every day.

- *'Only a [security expert], if anybody, can judge the security of a system'* – it is not a job for amateurs. You must buy-in the technology and, for any reasonably complicated business process, recruit the services of a security professional. In fact, it is very difficult for anyone, expert or otherwise, to prove the strength of a security implementation. This is another reason why we have to rely on a limited range of commercially available algorithms and their implementations. Although they may turn out at a latter date to have weaknesses, at least they will have been subjected to extensive testing, which, in many cases is the best that we can do in proving their worth. Perhaps the most important asset you get from an accredited consultant is that they will know which systems in the market have a good security record. In general, it is better to buy system components such as encryption algorithms than write your own. Do not invent them!

- *'Security must include an assessment of infringements of security discipline'* – this is particularly true in commercial situations; the real weaknesses often lie in mis-operation, as for example in the case of credit card PINs mentioned earlier. The strongest algorithm is pointless if today's key is post-noted to the terminal. Perhaps in military circles this can be made into a hanging offence and thereby discouraged, but eCommerce vendors cannot do that to their customers. The fundamental principle of security measures for retail eCommerce must be a process which is simple enough to be used in a customer-supplier relationship, *without requiring training or unusual diligence on behalf of the customer.*

- *'Superficial complications can provide a false sense of security'* – this is particularly true in computer systems, where complexity can make it very difficult to audit a system's weaknesses. Although security demands experts, the experts responsible for it in any specific installation will have a vested interest in believing that what they have developed is uncrackable. Do not believe them, and insist on a system which can be audited by a third-party. Resist suggestions from staff who are not security experts that they can install additional features of their own invention.

We want to add an additional, depressing maxim to Professor Bauer's list:

- *Complicated systems are a security risk in themselves– and computers are inherently complicated systems.* We return to this point later, but it is one which is fundamental to the discussions which follow, as will become clear.

2.4 WHAT ARE WE TRYING TO PROTECT?

In order to make security issues into a tractable field for consideration, we need to simplify the general discussion into one where we identify a manageable number of cardinal points that we wish to ensure are protected. There are many possible ways of defining these, but one possible set particularly relevant to electronic business is given in Table 2.3.

Confidentiality and privacy: It is generally considered a basic human right to be able to carry out legal activities without unwarranted observation by private individuals or commercial or official organisations. People ought to be able to purchase what they want, without being criticised or mocked, without being pestered to buy other products on the basis of what they are currently doing (unless they knowingly surrender the right), or without criminals being alerted to their possession of valuable property. Since that is what is believed, then any sensible vendor will try to avoid mishaps of this type. The need for confidentiality extends even stronger into areas where direct material damage can occur if security were to be compromised: credit card details, confidential contract bargaining, financial transfers, health records are areas that immediately come to mind. Businesses possess commercial data which would be damaging to their success if it fell into the hands of competitors.

Table 2.3 The cardinal points of security

Confidentiality and privacy	only the legitimate parties can observe what is be passed or stored
Integrity	only legitimate processes are enacted and data is not corrupted, or, if corrupted, the consequences are contained
Availability	systems are proof against attack intended to destroy their ability to provide service
Authority and accountability	only the correct people can use a resource, action a transaction and/or be reliably traced as such

Integrity: An attacker or an incompetent person or process can do damage to data without being able to read it. There is no system which can guarantee that a message can always be delivered, but there are ways to make sure that messages have not been lost or corrupted.

Availability: can be related to integrity, although it is generally concerned with the specific issue of denial of service. Without sensible security it is easily possible for an eCommerce activity to be rendered completely inactive, either by incompetence or malicious attack.

Authority and accountability: Processes should only be run by people entrusted with their proper operation. Only the authorised data-entry personnel should be allowed to change records. Incompetent operation can lead to loss or danger – perhaps failing to specify the arrangements for safely using a product. Malicious use can lead to deliberate destruction or corruption of data files and Web sites.

2.5 SPECIFIC SECURITY DIFFICULTIES OF COMPUTER SYSTEMS

Electronic business owes its origins to the development of affordable computers; to this also it owes its security problems. Computers accidentally infringe Bauer's law: '*Superficial complications can provide a false sense of security*'. Computers are inherently very bad from a security perspective. They are extremely flexible, and flexibility makes policing very difficult. They are complex and it is not easy to analyse the way that misuse of the system can be used to defeat standard controls. Computers now communicate over wide areas, in an environment that is open to anyone, using protocols which, by-and-large assume good faith on behalf of interacting systems. Assumptions of 'good-faith' are deadly to good security, and we probably need to add that, since most computer staff do think and act in good faith, they themselves represent one of the biggest weaknesses.

To aid legitimate use, computers have been designed deliberately to remove the complexities of their operation from their users and from application programmers who do not need to look at some of the more basic system components. It is therefore sensible in most circumstances to model a computer system as shown in Figure 2.1.

It is more often done to draw the model as a set of horizontal layers, with the user accessing only the top, 'application' layer, but here we find it more vivid to look at it as seeking access to rooms within a building. Once the user has gained access to the first 'room' of the system, all their other requirements are conducted, in isolation from them, in the deeper recesses. Negotiations between 'rooms' are provided by appropriate interfaces, which hide the detailed workings of the more basic software.

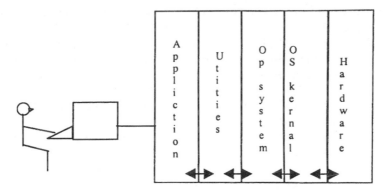

Figure 2.1   Computer architecture for 'good guys'

A similar approach applies to applications development, whereby the designer of applications does not need to know the details of the inner rooms and it can also make applications portable across a range of hardware and software platforms.

Unfortunately, as well as being a convenient metaphor for explaining 'good-faith' system design, it also provides a too-comfortable picture, when we try to draw an analogy with physical security, which progressively restricts access to critical resources, with checks and authorities on each door as one moves inwards. However, it is not a true picture. The reality is closer to that of Figure 2.2.

We first have to realise that the user does not access the system via some abstract 'application port', the connection is real, physical and very much

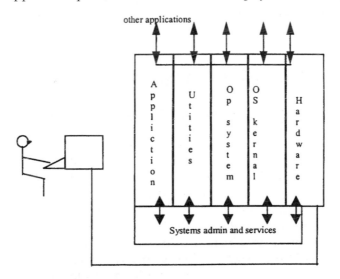

Figure 2.2   Computer architecture for 'bad guys'

into the heart of the system. How does the hardware 'recognise' the user, for example, even before we start discussing passwords and encryption? After all, as we can see, there is a tunnel that burrows right through our conveniently compartmentalised model: all systems have administrative and service functions than cannot really be placed within any one 'room'. What is to stop the user pretending to the hardware to be a system console? What if the user can get access to the root privileges of the admin system, even if only once?

Note also that applications themselves may not operate in isolation. Either by talking more-or-less directly to each other, or via any of the rooms (layers) in the model, they can create effects that were not intended by the individual application writers. We shall see examples of this later.

The big problem with computer systems is that the legitimate design model may be compartmentalised but the illegal attacker can attack the system at any point.

2.6 ATTACKING THE UNPROTECTED SYSTEM

Hopefully we have set off sufficient alarms to say that security should be looked at top down. To understand in more detail how technical attacks can be mounted, we now need to look at some specific technologies. In the material that follows, we describe a number of ways of attacking essentially unprotected systems, in order to give some idea of the complexity of the issues and some insight into the devious nature of attackers. It is intended to be instructive and fun, but it is by no means comprehensive or completely up-to-date. Readers should be aware that, although for most of the examples given, a fix can be found, there are many other ways to mount attacks.

2.7 EAVESDROPPING ATTACKS

When the potential reward is great enough, then attackers can be very sophisticated. Following on the lessons learned from military intelligence gathering, they can use eavesdropping equipment to gain access to the *plaintext* signals, thus circumventing the decryption problem entirely. For instance, every amateur spy 'knows' that the signals from computer screens can easily be picked up and decoded by a man in a van parked outside the building. In reality it is considerably less easy than that, but it is a fact that the military do project their secret computing data by using specially designed terminals which radiate a considerably reduced amount of electronic signal., thus suggesting that the threat is real. However, if you feel your data is sufficiently valuable to warrant this degree of security, then you should adopt a fairly parsimonious approach to the deployment of such screened terminals. They are expensive, perhaps 3–10 times as much as standard terminals. Also, bearing in mind that, by having your critical data handled on these machines, you are hoisting a flag pin-pointing precisely where your secrets reside, then you should treat them with care otherwise you will develop an attitude of false security that is worse than none at all. It is important to keep them in a secured area, making sure they are serviced in a competent manner and by trusted technicians. The risk is that someone accidentally or deliberately will damage the electromagnetic screening thus allowing electromagnetic radiation to escape. A light-hearted (true) story: a major diamond trading company conducted its highly secret price-setting plan in a 'screened' room, thus avoiding the possibility of being radio bugged by its competitors. During one of the meetings, one executive had to leave because he received a pager alert from his boss. A badly constructed or maintained enclosure is not a good idea!

Incidentally, it is not just video display screens that are at risk; it has been reported [80] that a number of smart cards that contain encryption software have some serious weaknesses. In many modes of use, smart cards that are inserted into terminals or readers, draw electric power from them during the transaction. It appears that the amount of power consumed changes slightly depending on the specific operations carried out by the on-card encryption process. In particular, it is claimed that some information regarding the private key settings used to encrypt and decrypt data can be identified by this means.

2.8 TAPPING THE INTERNET

It is quite difficult for an unauthorised third-party to mount a mass 'tapping' attack on all the users of traditional telephone networks. The lines that connect all the customers come under the control of the telecoms authority and they are also physically difficult to get at. This is not true of the Internet. Because of its federated nature – a deliberate design feature to provide diversity against point failure – the data packets can pass through a large number of domains that are not protected and are concentrated at the routers that connect sub-networks together. Moreover, these data packets carry with them their source and destination addresses (Figure 2.3).

That is how they are routed, and a router can be seen almost as an Internet tap in itself. Nothing is easier than to read and copy the traffic to a malicious attacker, using so-called *sniffer software*. All the data can be easily read, modified, or used to mount more complex attacks. The origin, the data itself or the application type can all be changed. Thus, unlike telephone taps, the process need not be one-way, receive-only: spurious

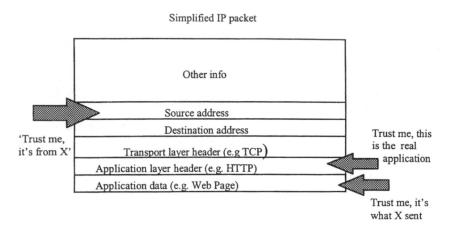

Figure 2.3 Simplified structure of an Internet data packet

messages, purporting to come from other sources than the real one can be injected into the network, in the process known as *spoofing*.

2.9 SOURCE ROUTING ATTACK

One example of generating spurious messages is the so-called *source routing attack*. On the Internet, the TCP/IP protocols are themselves very trusting. They are more concerned with ensuring the reliability of the connection than with its authenticity. This has opened up the possibility for a particular form of *source routing attack*, which allows a third party to pretend to be someone else, even without being able to observe the traffic across the network. Consider what happens when a TCP/IP session is initiated:

- A wishes to open a session with B. It sends a message 'from A to B: sync, ISN(A)' This message is addressed to B and tells it that A has chosen 32-bit number ISN(A), as the 'initial sequence number' for the session.
- In return, B sends 'from B to A: sync, ISN(B), acknowledge ISN(A)' which acknowledges A's initialising number and issues its own, ISN(B).
- A will then send back 'from A to B, acknowledge ISN(B)'.

This now means that A and B share a pair of sequencing numbers they can use to identify the connection between them. If these numbers are generated by a random process, then it would be difficult for a third party to guess what they were. But, often they are not. In general, they are generated by incrementing a simple counter, sometimes by a fixed

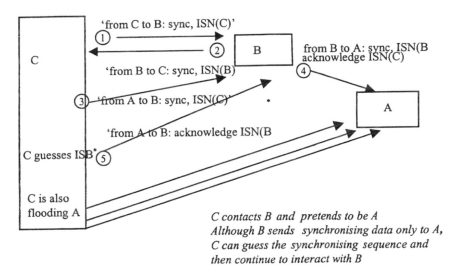

Figure 2.4 shows the attack described.

Figure 2.4 Source-routing attack

amount after a fixed period. It is therefore not always too hard to guess correctly at the value. What can happen next is as shown in Figure 2.4.

- C (the bad guy) first opens a connection with B, sending 'from C to B: sync, ISN(C)'.
- B replies 'from B to C: sync, ISN(B)'.
- C now *pretends to be A*, sending 'from A to B: sync, ISN(C)'.

That is, it uses A's IP address and also uses the same sync value as it did before. There is no reason why the latter could not happen in legitimate communication. The process continues as follows:

- B's reply (which C may not necessarily see as it is routed to A) is 'from B to A: sync, ISN(B)', acknowledge ISN(C).

Although C does not necessarily see this, (since the route from B to A may not pass through C's router) it can guess at the value of ISN(B'). Suppose it guesses correctly:

- C sends 'from A to B: acknowledge ISN(B)'.

This means that B now thinks it is in communication with A, rather than with C. If B is happy to accept the IP address as a way of authenticating the message, then it has just given C access to all of A's privileges.

From C's point of view a problem occurs if A is actively on-line and detects the messages from B. Then it will automatically generate messages that will tear down the connection. However, there are a number of strategies that C can use to get round this, including barraging A with traffic

so that it ignores messages from B. Note that C only requires to imperso-
nate A for a short period, just long enough to get into the system to plant
some bug that allows it easy access in future.

2.10 PACKET DE-FRAGMENTATION ATTACK

There are other attacks which make use of the security laxity of Internet
protocols. Some are effective even when even when security measures
have been put in place. One example of such an attack is the so-called
de-fragmentation attack. It is a way whereby attacks can be mounted on
Web servers and other servers behind them, right into the heart of supposedly secure, core corporate IT resources. It relies on being able to deceive
the receiver into believing one kind of application (usually an HTTP
request) is being requested, rather than, in reality, another, more dangerous one. Internal corporate networks are usually protected from the
outside, by means of *firewalls* (whose operation we explain later in this
chapter). However, as recently as four years ago, there were no widely
used countermeasures that could defeat a de-fragmentation attack against
a firewall. It is probably true that a significant number of installations are
still vulnerable. This is how it works:

We mentioned that the IP protocol has to cope with different networks
which set different values to the maximum size of packets of data. When a
long packet of data arrives at a router to a network that can only handle
smaller packets, the router breaks the packets up into *fragments*, and gives
them temporary labels that allow them to be identified at the ultimate
destination and re-assembled correctly (Figure 2.5).

Now, as shown in Figure 2.5, fragments are given a number, which
allows them to be re-assembled in the correct order, and the first fragment

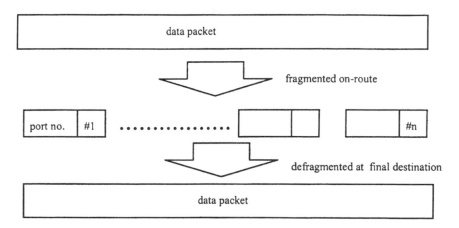

Figure 2.5 Packet fragmentation

also contains a *port identifier* which allows the destination computer (in this case, the server under attack) to know which TCP *port*is being accessed. Conventionally, certain port numbers are reserved for particular types of access. For example, port 80 is used for HTTP access (i.e. for accessing a Web server) and port 23 is used for telnet services (which allow access to, and movement of, files between the server and other computers. In general, it is pretty safe to let a third party through the firewall for an HTTP request, which will pull down a Web page, for instance. It might not be wise to let outsiders get access to files. So, the firewall is programmed to let through only HTTP packets and not telnet ones.

That is what is supposed to happen and many servers and firewalls trust the rest of the world to play by these rules.

But, suppose the attacker creates the situation shown in Figure 2.6.

The attacker has generated pseudo fragments, the first of which is labelled as being port 80, HTTP, a permitted request. The firewall there-fore passes the fragment, and all succeeding fragments that it knows to be part of that packet. Defragmentation only occurs when the fragments arrive at the destination server. (This is done for reasons of performance.) Here they are assembled according to their sequence label. The server places the first fragment (with the HTTP address in it), at the first place in the packet, the fragment with label 2, is placed after it and so on. But the attacker has cheated: there is a second 'fragment 1'. This over-writes the previous entry, and, notice that it is a telnet address! The attacker is in to the server's file structure.

Now that this method of attack has been recognised, some vendors of firewalls are producing new systems (and patches for old systems) which de-fragment the packets and examine them for danger, before passing them on.

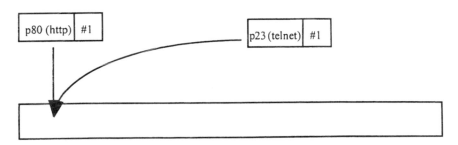

Figure 2.6   De-fragmentation attack

2.11   ATTACKING WEB USERS THROUGH COVERT CHANNELS

The unprotected nature of the Internet allows many other forms of attack to be mounted. Contributing to this vulnerability is also the increasingly active part played by the client terminal. As described on page 94, in the original mode of operation of the Web, the terminal's browser acted essentially as a simple down-loading mechanism for static Web pages whose only non-text information was that used by the browser to format the pages. This has given way to a situation where executable code, in the form of Java applets or ActiveX controls, is run on the terminal and can carry out a number of activities which can be potentially dangerous.

To give one example: Dean et al. at the University of Princeton have given details of a number of security holes in JAVA [81]. They distinguish between two and three party attacks, where, in the former, the Web server has to be involved, but the latter can be carried out without collusion from the server.

Three party attacks rely on a security weakness in the methods intended to restrict JAVA applets to communicating only with the server from which they were downloaded into the user's browser. In one variation of the attack, it is possible for the applet, once running in the browser, to request the server to send an eMail message to any computer elsewhere on the Internet. How does this malicious applet get into the server in the first place? Dean et al suggest that A could include it in an apparently innocuous utility that B found attractive and included in their Web page. C views the Web page and in so doing, loads down and enables the applet, which now communicates with A via the *simple mail transmission protocol SMTP* (e-mail) through B's server, unknown to B.

Another variation of the process can use the *Domain Name Service (DNS)* as a two-way covert communication channel: the applet sends a call to a fictitious name, which lies within the attacker's domain. The attacker's system recognises this dummy name. It is possible to use this method repeatedly to open up a two-way communication channel between the user's machine and the attacker's. Methods such as these have been used to extract credit card details from a client PC.

2.12   WEB SERVER ATTACKS

Many other attacks can be mounted on Web servers, with a variety of effects: at the simplest level, one can mount a *denial of service attack* simply by bombarding the server with requests. If this possibility has not been taken into account by the designers, it will probably lead to overflow of input buffers and can lead to random data being written into areas containing executable code. This will almost inevitably lead to a system failure. It is not always easy to detect at the server that a denial of service attack is in progress. As explained on page 105, the performance of a system, whether under attack or legitimate heavy demand, may require remote monitoring.

All but the simplest of Web server designs include the ability to invoke other application programs (for example, a data search and retrieve operation) in response to requests from user-clients by running executable code, based on parameters sent by the user's browser, using, for instance, the *CGI* or *Active Server Pages* facilities described in Part 1, *Retail Server*.

Too often for comfort, security investigations reveal that these processes are insecurely designed and installed and can be corrupted in order to run processes that the designers never intended to, when called remotely via a legitimate Web access HTTP command line. The extra instructions in the HTTP are passed on from the Web server to activate the various application services that make up the service and to supply them with data (Figure 2.7).

The application programs are set-up to receive data passed to them by the Web server, in the form of data strings included in the URL. For example, *www.myshop.com/cgi-bin/search.cgi?subject = cardigan*, connects to the *myshop.com* Web server, which in turn activates an application program called *search* and supplies it with the data string *cardigan*. Presumably, the program will search a database and return information on cardigans. That is the intention. But suppose a malicious user, observing that a correct URL has the form *www.myshop.com/cgi-bin/ search.cgi?subject = cardigan* sends the Web server something different, *www.myshop.com/cgi-bin/searchTEST.cgi?subject = cardigan*. What will happen? Perhaps nothing; perhaps, a reasonable guess will come off and *searchTEST.cgi* will run an early, faulty or dangerous process. What actually happens will depend entirely on the integrity of the design and

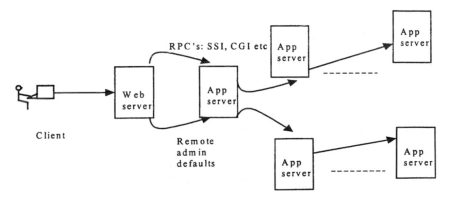

Figure 2.7   Propagation of effects of CGI, etc.

its resilience against attack. If it is vulnerable, it may be possible for the hacker to stall the server or piggyback on this application into other applications.

2.13 OUT-OF-THE-BOX ATTACKS

One does not even have to inadvertently install vulnerable software oneself, into the server. Sometimes vendors have left the doors open themselves. There have been reports that one popular UNIX server has nearly 30 possible illicit entry opportunities unless they are individually disabled. Another very popular operating system is reported to have more than 45 default passwords. If not removed, these can each allow unauthorised access.

One example cited by the US NIPC advisory body in 1999, (code named 99-027), is a remote database attack on a well-known Internet information server. Provided the default option of leaving the so-called *RDS sample pages* has not been overwritten, it allows Web browser to issue and receive SQL queries to a database and to alter entries. It is even alleged that there exists an attacking tool, which can automatically detect this vulnerability.

Another example, from another vendor, is even simpler. Provided a rather large number of / / / symbols are typed after the server URL, access to the root directory is obtained. The figure of 211 slashes is quoted, but varies with version of server.

Most of these *out-of-the-box* loop holes have been left in order to make it easier for systems administrators to set-up and configure new systems. It is asserted by vendors and, so far, not attested otherwise by the courts, that it is the responsibility of purchasers to take the necessary precautions to annul these vulnerabilities. Given competent – or perhaps better to say, 'perfect', systems administration, then they do not represent security problems, but who's perfect?

2.14 DATABASE ATTACKS

The potential sharing of critical data between different parts of the enterprise, typically across an enterprise extranet is seen as one of the major advances in business behaviour. (See Part 2, *Managing eBusiness Knowledge*.) Clearly this has advantages, but there are risks, too. Apart from variants of the attacks mentioned above, which allow attackers to access or sabotage files to which they have no legitimate rights, there are intelligence-gathering operations which can be mounted 'legitimately' by anyone having even restricted access to the database. One such activity is *inference from statistical queries*.

Table 2.4 A sales database

Company	Location	Type	Order value ($million)
BBLU	Washington	H/W	0.8
DEK	Frisco	H/W	1.6
HOGG	Seattle	GAMES	0.7
FRUIT	Frisco	H/W	1.0
KID	Washington	ROUTERS	2.5
MOON	Frisco	S/W	1.5
TINY	Seattle	S/W	3.0

Consider Table 2.4.

This represents the (confidential) value of contracts won by your company to supply a number of different types of companies, in different parts of the US. To protect the detailed knowledge contained in the table, but to allow (reasonably) trusted others to access some of the data, you restrict them to be able to carry out only a set prescribed set of statistical queries.

They can, for example, submit an SQL query which returns the average order value for the database.

They can also ask for the total number of companies in any location for which you have product type data.

What you do not want them to ask is any question relating to how much money you have taken from any individual company.

Suppose you allow them to ask for the average value of orders supplied to software companies in Seattle? (= $3 million).

Now, suppose you also allow them to ask for the total number of companies in Seattle that supply software? (= 1).

Oops! Although we did not allow direct access to queries of the type, 'how much money have we taken from Tiny', we see that a pair of queries taken together, has successfully found exactly that figure.

In general, it turns out that it is virtually impossible to stop a set of statistically averaged queries resulting in a specific answer. One part defence is to keep an audit trail of the queries and calculate backwards, to assess what users have been up to. But there is no easy total defence prior to the event.

2.15 TECHNOLOGIES TO SUPPORT SECURITY POLICIES

In order to understand security issues in eBusiness, we introduced the standard, three level model of policy, process and technology. We felt it best to begin with some examples of how this could be compromised in situations possessing little or no security measures. In order to see how such measures could improve things, we now need to look at basic technologies and then see how they can be built up into standard processes for wide area operation.

2.16 FIREWALLS

In planning a defence against an attack, the best thing to do is to rely upon the tried and tested security principle of cutting down the number of access points and then patrolling these points with rigour. This is the principle of the 'firewall'. Every entry to an internal corporate network from the public domain (in practice, usually an Internet connection) is made via a firewall, which inspects the traffic in order to determine whether or not it should be allowed to enter, and on what conditions. There are essentially two different types of firewall: those that act as *packet filters* and those that are *proxy servers* or *bastion hosts* as they are also called.They can be used separately or together. They can check-out data at the basic packet level or operate as *application filters*, refusing to pass data intended for specific applications or services.

2.17 PACKET FILTERS

We can have as many and as complex internal networks as we like, but they all connect to the outside world through one pair of filters, one for checking outgoing packets and one for checking the packets coming in. The filters are conventional routers enhanced to include the ability to inspect the addresses of every individual packet. The systems administrator (who must be someone who can be trusted – an issue in itself) sets up a table for each filter to check against, to see whether it is OK to forward the packet. If forbidden addresses turn up, they are rejected and the administration alerted. By means of an *service gateway*, filtering can also be carried out, not just at the packet level, but also on the specific contents and purposes of the data passing through (Figure 2.8).

For instance, it is possible to set up a free transit for eMail between the corporate network and the external world, but block off every other service. One of the most common selective filtrations is to let through packets that contain an HTTP request (i.e. for the retrieval of a Web page) but not FTP packets (which request access to files). However, as we saw earlier, simply introducing a packet filter is not necessarily sufficient in order to ensure security. In any case, there are some operational difficulties with packet filter fire-walls: firstly, since one is simply using a

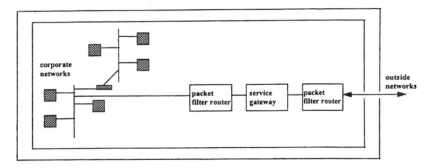

Figure 2.8   Firewall design

router which is set-up to reject packets that do not conform to its acceptable list, there is often no way of collecting logs on what has arrived at the
router. This means that it can be difficult to realise that one has been under
attack; equally, it is not easy to detect the rejection of people you would
rather allow through. It is also difficult to set up privileges for special
users, to allow their eMails to come through, or for field-force technicians
with securely identified terminals to enter the file system remotely.

2.18   SECURITY VIA A PROXY SERVER ('BASTION HOST')

Instead of allowing external, potentially malicious, client machines to talk
directly with servers and clients inside a corporate network, it is sometimes preferable to direct them to another *proxy server* or *bastion host*. The

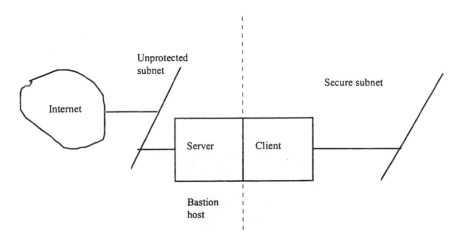

Figure 2.9   Operation of a bastion host

bastion host (Figure 2.9), is simply a standard computer, not necessarily particularly powerful (for example, a 486-DX66 with as little as 16 Mbytes of memory), that acts as a barrier between the outside world and the secure internal network. It has two network cards and two IP addresses. The server receives requests from the outside world but has then to request the client to vet the passing on of the request to the secure application. No IP packets pass through directly. Properly configured, this offers a high degree of security. It also can be configured to store a record of the messages passing through, thus establishing a full audit trail for on-line or off-line security monitoring. Bastions are particularly good for *application gateway* implementation, since it is possible to run complex checks on the contents of the messages that pass through them, for example, to avoid eMail viruses. One approach, *SOCKS*, developed by the Internet Engineering Task Force (IETF), provides for a standard way of permitting safe applications to operate through the firewall. It requires each such application to undergo a process of *socksification*. Originally this meant that the application had to be re-compiled and linked to a set SOCKS library functions. This was complex and time-consuming. More recently, dynamic link libraries that operate at run-time have removed this problem, at least for Windows applications. There is still debate, however, as to whether SOCKS is necessary and/or desirable in the longer term.

2.19 PROTECTING VIRTUAL PRIVATE NETWORKS USING FIREWALL TUNNELS

In the case of business-to-business operations the situation may some-times be different: the two parties may trust each other and want to do business over the Internet, yet preserve a secure channel between them. In this case, the firewalls can be configured to provide a virtual private network using a technique known as *tunnelling*. The tunnel provides a secure channel between the internally private networks of each company (Figure 2.10).

This makes use of one of a limited number of secure tunnelling proto-cols, such as the IPSec protocol, which allow for the IP packet to be encapsulated in a secure *wrapper*. This will be discussed later, once we

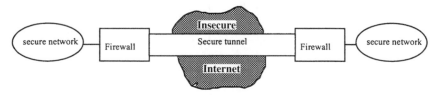

Figure 2.10 Secure tunnelling

have considered some encryption techniques which are used in its implementation.

2.20   BASIC PRINCIPLES OF CRYPTOGRAPHY FOR e-BUSINESS

Protecting access and exit from private computer systems into the wide area is only one aspect of system security. Data must also be rendered as resilient as possible against corruption. The usual way to do this is create a security process which relies on encryption algorithms. We make the point again, because it cannot be made too often, that we must make a distinction between implementations and algorithms. An encryption *algorithm* is a mathematical expression that describes how, for example, a piece of data is to be encrypted (and decrypted, if the method is different). The *implementation* is concerned with how it is to carried out in reality. This could be by software, hardware or manually and would include consideration of the network or other system used in order to transport secret keys and how agreements were arranged between users. The point to note is that security depends on both algorithm and implementation.

In Figure 2.11 we outline the basic components of an encryption process and its underlying algorithms, that are common to most applications, whether they are for keeping secrets, transferring money, proving you are who you say you are, etc.

The system is driven by an algorithm, which in some way protects the transaction by applying some privileged knowledge, held by one or both

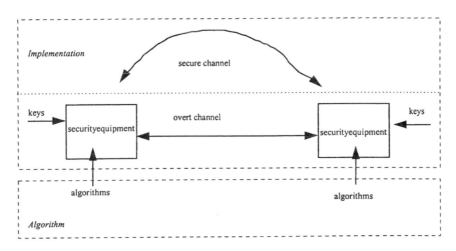

Figure 2.11.   Security implementation and algorithm

of the parties to the transaction. This knowledge could be a secret password, or it could simply be that the two parties know each other by sight and can therefore conduct their business over a video link.

There are always some hardware and/or software items of security equipment, which hold and operate the algorithms that provide the security – the video cameras, monitors and transmission links, in the example quoted. More often, there will be cryptographic machines which code up messages to prevent overhearing or manipulation. In this case, they are conventionally referred to as *cryptos*.

The final component of the system is the *secure channel*. This comprises some set of processes and technologies that allow the privileged information to be fully and reliably exchanged between the two parties. An easy example of this would be the exchange of keys used to set-up the enciphering and deciphering algorithms for secure message transmission. It could even be an exchange of photographs of legitimate participants in our conferencing example. Incidentally, it is sometimes claimed that security systems, such as public key encryption, make it no longer necessary to provide a secure channel. We shall discuss whether this is exactly correct, later.

2.21 'PERFECT' CIPHERS

The mathematical theory of encryption has advanced dramatically since the topic was first analysed rigorously during the last war, in particular by Claude Shannon at Bell Laboratories. When we encounter this research for the first time, we may find it surprising that it was not centred on making ciphers that are stronger and stronger. In fact, the major advance due to Shannon, was to prove that, although there did exist a 'perfect' cipher (in the sense that it could not be broken, provided it was supported by a strong implementation), in most cases, this implementation would be unworkable.

This perfect cipher is known by a variety of names, but, most commonly, as a *one-time pad*. According to occasional reports in public, it does find at least one practical application – as a means of communication between spies and their controllers. Here, conceptually, is how it is used:

The letters of the message (the *plaintext*) are first turned to numbers: A = 1, B = 2 and so on.

So, S E C R E T

becomes 19 5 3 18 5 20

The enciphering agent has a set of pages containing lists of numbers, within the range one-26, that have been generated at random: 21, 3, 13, 19,

5, 7, etc. These are added in turn to the numbers corresponding to the plaintext. If the sum exceeds 26, then 26 is subtracted. This gives the *ciphertext*:

> 19 5 3 18 5 20

> plus 21 3 13 19 5 7 gives

> ciphertext = 14 8 16 11 10 27

If the numbers have been truly generated at random, then there is no way that an intercepter can work back to the plaintext. However, if the legitimate receiver has a copy of the random numbers and knows where they start, it is simple to subtract them from the ciphertext (and then add 26 if the result is negative) to get back to the plaintext:

> ciphertext 14 8 16 11 10 27

> minus 21 3 13 19 5 7 gives

> plaintext 19 5 3 18 5 20

> that is S E C R E T

The 'key' for the method consists of the specific numbers on the pad; the 'crypto' is simply the operation of adding the key to the message.

Although it is the most secure system possible, the one-time pad has a serious problem: the size of the pad has to be the same size as the total number of characters in all of the messages that you want to send. It is not a good idea to re-use a set of numbers: an attacker can combine intelligent guesswork with some fairly simple statistics to successfully break the cipher. This may not be a problem where the messages we want to send are short and infrequent. The controller and the spy can, presumably, arrange to meet frequently enough to hand over new one-time pads. However, if we want to provide a convenient mechanism for inter-bank payment, involving millions of transactions per week, we can hardly imagine sending couriers around the world with reams of one-time pads. Or consider the case of trying to encrypt a TV programme so that only paid-up viewers can see it: we would require one-time pads providing several megabits per second to each TV subscriber.

2.22   NOT PERFECT, BUT PRACTICAL

Clearly, we need a more practical solution. This is achieved by sacrificing perfect security, in favour of a much shorter key and a more complex crypto. There are a large number of ways of achieving this. Some are

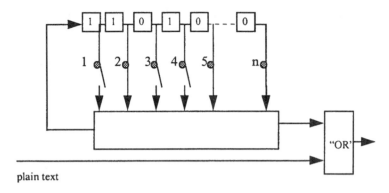

Figure 2.12 Keystream generation using linear feedback shift register

based on the principle of generating numbers that appear random, which are then used essentially in the same manner as the keys of the one-time pad. Others shuffle *(permute)* the data in the message in an order based on an apparently random key. A very simple example of the former, (which is cryptographically very poor), is the *linear feedback shift register* (Figure 2.12).

A shift register allows bit-patterns to move sequentially from left to right. In the device shown, some of these bits are also tapped off and passed by the closed switches (2, 5 and *n*, in Figure 2.12) into the 'OR' function device. The OR function performs a series of 'half-adding' operations (0 + 0 = 1, 0 + 1 = 1, 1 + 1 = 0) on the data, such that an even number of 1s will give answer 0 and an odd number will give 1. The answer is then 'OR'd' with the first bit of the plaintext, thus encrypting it. The answer is also 'fedback' into the left hand cell of the shift register after all the other bits in it have moved one place to the right. This is an example of a *stream cipher*: a stream of pseudo-random bits is generated by the crypto and added bit by bit to successive bits of the plain text.

It is possible to show that legitimate receivers of the data can decrypt it with a similar device, provided they know which switches are closed and the start sequence (110100, in this case). The 'crypto' consists of the shift register and the function generator (an OR function, in this case). It has to be assumed that the attacker can have at least some knowledge of the number of stages in the shift register and the nature of the function generator. The only security that can be assumed is that contained in the key. There are two components to the key, in this device: the start sequence, and the switch settings. These must both be exchanged between sender and receiver, using a secure channel (for example, a trusted human messenger), and kept secret until a new set can be used.

Practical systems use functions other than a single shift register with feedback and a simple OR function. While the single shift register provides some security when the attacker has no knowledge of the contents of the message, it is extremely weak when some of the text is known or can be guessed. (For example, many letters begin, 'Dear Sir', contain the date, or end equally conventionally.) In this case, it can be shown that the system can always be broken, given less than 10N successive bits of plaintext, where N is the number of stages in the shift register.

There are two important points to be drawn from this: firstly, crypto systems must consider practicalities, as well as theory, and in particular the possibility of the attacker starting with more than zero knowledge. Secondly, a critical design factor is the ratio of attacker effort to that required by the legitimate users. What we must aim to achieve is presenting a degree of difficulty to the attacker that increases faster than any increase in the complexity of the crypto. This is not true in the case of the shift register, where the complexity of the system and its susceptibility to attack both go up 'linearly', that is, with N. We would like a system whose complexity could be increased by N, with a resulting increase in difficulty of attack going up by (at least) N^M, where M is some large number.

Some practical alternatives offering better security include cascades of shift registers, whose modes of operations interact. For instance, we can use the output of one register to control the clock controlling the following one; that is, make it move on more than one shift at a time.

2.23 BLOCK CIPHERS

An alternative to stream ciphers are *block ciphers* which operate on a block of data, which is a fixed number of bits long to produce an equivalent, encrypted output. Rather than generate a separate stream of pseudorandom bits and combine them with the plaintext, block ciphers often carry out 'scrambling' operations on the plaintext data alone, reordering the bits in the block and combining them with other bits, for example using an OR function. Alternatively, we can simply take a large block of data and scramble the order of the bits, according to some *permutation* formula, known to sender and receiver, but, again, kept secret from others.

Several examples are based on *Feistel Ciphers* which repeatedly pass the text through the same scrambling function ('F', Figure 2.13).

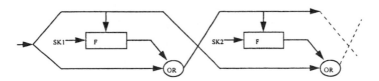

Figure 2.13 Two stages of a Feistel cipher

A block of plain-text data is split in half. One half of the data is operated on by the function, F, which carries out a re-ordering (shuffling) of the bits in the block, according to a 'subkey', (SK1). The output of the re-ordering is then 'OR'ed with the untreated half. The halves are then swapped over and the same re-ordering function now applied to the other half. Note that the re-ordering function is still the same, but its effect on the data may be different because it is controlled by a new subkey, (SK2). The process can be repeated over and over. Feistel ciphers are convenient to decrypt, because the process of encryption is simply reversed, applying the subkeys in reverse order. The re-ordering functions are chosen such that their detailed operation on the data is difficult to decipher without knowing the sub-keys. Thus the re-ordering functions can be made public.

2.24   BLOCK AND STREAM OPTIONS FOR CIPHER SYSTEMS

The Feistel cipher description given above implies that that the data needs to exist in blocks which are of a certain size, commensurate with the size of the key. This is not actually a requirement; it is possible to use the Feistel principle to generate encryption on each individual bit of input data. In the example of a linear shift register encryption scheme discussed earlier, Figure 2.12 shows the shift register operating in *feedback*mode where the register generates a *stream* of pseudorandom data which is OR'd with the cleartext data. We can do exactly the same thing with Feistel ciphers, priming the re-ordering box with an initial sequence, allowing it to scramble this with the key and thus producing a set of output bits some of which can be OR'ed with the data to be encrypted. The output bits from the re-ordering box can then be fed back into the box, to generate a new sequence.

There are also a number of ways in which the basic *block encryption* of a Feistel cipher can be implemented. For instance, it is possible to OR the previous output of encrypted text with the current block. This has the benefit of making it more difficult for an attacker to take advantage of repeated blocks of identical data. However, like all variations, it also has disadvantages. We have to remember that attackers can, in theory, inject their own messages in order to initiate a known plaintext attack, or can introduce errors that may propagate across several blocks of data, making the system unusable and the legitimate parties liable to want to revert to an insecure alternative.

2.25   THE DATA ENCRYPTION STANDARD (DES)

One published and extensively used example of a Feistal encryption system, is the *Data Encryption Standard (DES)*. This uses a key typically at least 56 bits long. This provides nearly 10^{17} possibilities that an attacker would have to try in a brute force attack. A well-programmed, general purpose computer, in today's technology would take a hundred years or so, to test every possibility. Doubling the key length to 128 bits increases the encryption time by the same order, but makes the brute force attack 10^{20} times longer to do. This would require an impressively large amount of processing power if it were to attacked using a single conventional computer, but, beware, there are other ways to go about it.

One proposal envisages a very large population of patriotic home-workers and a level of not very expensive, but widespread technology. The emergence of satellite broadcasting in China has suggested a particular technique and the name the *Chinese lottery* (Figure 2.14).

The attacker has intercepted the link and transmits the signal on a satellite broadcasting channel into many million homes, each of which is equipped with a TV set and a satellite decoder, in the form of a set-top box. The latter is, essentially, a modified computer, and could simply be further modified to allow the crypto algorithm to be included. Keys selected at random by each household are tested out to see if a meaningful message appears on the TV screen. One of the many million attempts will soon be successful.

It has turned out that one of the earliest successful attacks on an encrypted message, using this technique, was achieved not through any great act of patriotic solidarity, rather by a volunteer force of 14,000 Internet users, one of whom managed to find the 56 bit key. A previous success

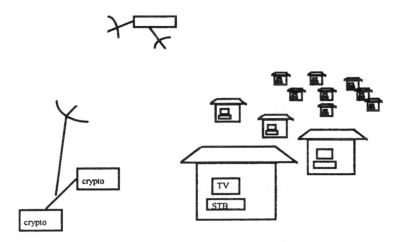

Figure 2.14   The Chinese lottery

using 3,500 personal computers broke a 48-bit key in 13 days. The serious implications are that attacks based on heavily parallel computing, will always pose a threat to any encryption techniques. It has been estimated that special purpose machines, costing less than $1 million, that carry out a large number of parallel attacks on 56 bit DES, could achieve a success within several hours. This is well within the means of government security departments of many nations (and probably of several global criminal syndicates). There is also a continuing conspiracy theory that DES contains a weakness that was intentionally introduced by the National Security Agency, but despite there being many attempts by leading experts opposed to government 'interference', no successes have been published, and the story must be treated with some scepticism.

Other attacks have been proposed, that use *plaintext attacks*, where the attacker manages either to get the sender to encode selected messages chosen by the attacker, or simply using messages whose plaintext is known. One example of the latter successfully recovered the key in 50 days using 12 workstations, but this required 2^{43} known plaintexts, and thus is probably not practical.

2.26 TRIPLE DES

Whether or not DES does have weaknesses, it is still better to rely on it than on many other systems that have not been extensively attacked. In any case, the balance between attacker and defender does not stand still. One of the problems of coping with advances in attack technology, is in still preserving backward compatibility with legacy systems. Rather than throw away DES, introduce a new system and re-key all the data, there has been a tendency to extend the working life of DES by introducing what is effectively a longer key. In fact, three 56 bit keys rather than one have been proposed, under the name of *triple DES*. A particularly favoured option is to use only two distinct keys, applying one of them twice. It happens that this allows the decryption to be essentially a reversal of the encryption.

Whatever the real or imagined weaknesses of DES, in a world where nothing is certain, it is probably true that triple DES is more secure than is the practical management of its keys. We now look at how some of the key management problems can be alleviated.

2.27 PUBLIC KEY SYSTEMS

The systems we have looked at above, have all been concerned to achieve a high degree of security with a manageable arrangement for key hand-

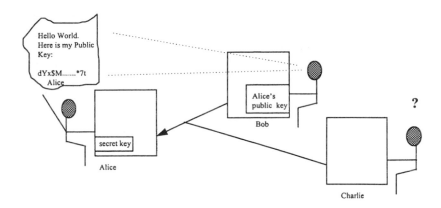

Figure 2.15 Using public key encryption

ling. But they all share the common problem that there still has to be a secure channel for delivering the secret keys. This is an example of a relatively secure algorithmically secure system being potentially compromised by a cumbersome security process. At first sight, this seems an inevitable requirement; if you find out the key that I use to encrypt a message, then surely it is easy for you to decrypt it? It is true, if the key for decrypting a message is either identical to, or easily derivable from the encryption key and the decrypt method is a simple reversal of the encryption. But suppose this is not so? In this case, it turns out that there are a range of techniques which open up a surprising possibility.

The ability to publish openly an encryption key for anyone to use, that does not make it easy for anyone, other than the issuer of the key, to decrypt any messages encoded with this key.

To understand what this means, consider the example shown in Figure 2.15.

Bob wants to send a secret message to Alice. He notices that Alice has told the world her *public key*. He uses this to encrypt his message, according to one of the algorithms we shall discuss later. He then sends his message to her, over an open channel, knowing that, even if Charlie can intercept it, he will not be able to understand it. However, Alice, and only Alice, who possesses a *private key*, which operates as the inverse to the public one, and which she carefully guards to herself, can be used to decode the message. Because of a special relationship between the private and public keys, it is extremely difficult for Charlie to decode the message and work out the private key,. What is perhaps more surprising is that no-one, not even Bob, can work back from knowledge of the message and the public key, to the private one.

Public key encryption systems rely on the fact that some mathematical operations are easy to perform in one direction, but difficult in another. One example of this is forming the square of a number – an easy operation of multiplying the number by itself – compared with being given the result and asked to find the square root. The number of steps required to do so, is much greater than for the squaring operation.

Another, more relevant operation, is that of 'factorising products of prime numbers'. We remind ourselves that prime numbers are whole numbers that cannot be 'factorised' into the product of a number of smaller whole numbers. 1, 2, 3, 5, 7 are primes, but not 4 ($= 2 \times 2$), 6 ($= 2 \times 3$) or 9 ($= 3 \times 3$). It is known there are an infinite number of primes, about 10^{150} of them with 512 bits or less, but there is no known formula for proving, in general, that a number is or is not a prime, that is significantly quicker than trial or error. The time taken to do this increases very rapidly with the size of the number.

A similar problem is that of 'factorising the product of two prime numbers'. (i.e. finding p and q, given only p time q.) It is easy to multiply two prime numbers together, but difficult to reverse the process.

2.28   THE RSA PUBLIC-KEY CRYPTOSYSTEM

The differing difficulties between the multiplying of prime numbers and the reverse process, is the basis for one of the most successful public-key systems, *RSA*, named after its inventors, Ron Rivest, Adi Shamir and Leonard Adleman. A full proof of the technique requires some complex mathematics, but the principle is simple. (Details are given in Table 2.5.) Two large prime numbers of approximately equal size are selected and multiplied together, to produce a number several hundred bits long. This is called the 'modulus'. Two further numbers are calculated from the two primes. These are called the *public* and the *private keys*.

The owner must keep the choice of the two primes and the private key secret from anyone else, *but can make the modulus and the public key openly available*. Thus there is no need for the complexity of setting up a secure channel for passing out secret keys.

Anyone wishing to send a message uses the algorithm described in Table 2.5 to encrypt the message, operating on it using the modulus and the public key. Once the message has been encrypted, the task of decrypting it is equivalent to trying to factorise n, and this is known to be 'difficult', unless one has access to the private key. Knowledge of this private key allows the legitimate recipient to decrypt the message using a similar algorithm to the encryption, except that the private key is employed instead of the public key.

Table 2.5 The RSA encryption algorithm

The algorithm is based on selecting two large prime numbers, p and q, whose identity must be kept secret. Let $n = p \times q$ and chose a number e, which is less than n. e need not be a prime number, but it must not have any factors in common with either $(p - 1)$ or $(q - 1)$

We now need to find another number, d, such that $(e \times d - 1)$ is divisible by $(p - 1)(q - 1)$ without a remainder

Thus we have three numbers: d, e, n. We make e and n public. (They are the 'public keys'). We keep d (the 'private key') secret

Suppose Alice wants to send us a message, m. She looks up our value of e in a public list. She then raises m to the power e. (That is, she multiplies m by itself, e times. If m was '345' and e was '7', she would calculate $345 \times 345 \times 345 \times 345 \times 345 \times 345 \times 345$)

She then finds n from the public list and repeatedly divides it into the previous calculation until only a remainder R is left. (e.g. suppose m to the power $e = 10n + R$. In mathematical notation, this is know as finding $R = m^e$ mod n). R is the cyphertext that Alice sends to us over an open channel. It can be shown that any technique that allows m to be found from R and knowledge of n and e, is computationally difficult, equivalent to trying to factorise the product of two primes

We take the cyphertext R and raise it to the power d (the secret key). We then find the remainder after repeated division by n (i.e. we have calculated R^d mod n). Surprisingly, this turns out to be m, the plaintext message

A typical public key value, coded as an alphanumeric string, would look something like this

IQEVAwUBNOW/58UCGwxmWcHhAQG + Mgf/RRBnjaC5ir0QKuD8 + tgNhhvdoi2ajZRkKgaBzsd0PDRy/3W2UJTuIR/Hq4k7EoEE + Z5KYl + bcXLJuIPHdBKpNbDZ9IPrs5z2S0gyc1IFPZkj + EjO8oNXXA9H9nqzjh-F5aHvLv5XZrxvb08iLAMl6iepBQz7qdh617CPm7tu7A8VD9G + 46kGXN4SnjOjRUXgKnoQGMTs1TBdMWuse6qjipSMLxXVQXCvAOoD-8VScg3lI3EQB7mjX + P2nVchVYVUBfpxgYJsvtw5zqBBd3f1tpeLCeI29Sg7/Fzwi

---

2.29 COMBINING PRIVATE AND PUBLIC KEY SYSTEMS

It turns out that, although public key algorithms have this great advantage of much simpler key management, in practice they have not completely replaced private key ones. Private key systems have other advantages, notably speed of computation. Consequently, one possibility is to use a public key system as the secure channel for passing private key information (Figure 2.16).

As shown, A can use B's RSA public key to encrypt a DES private key that will be used for a secure session between them and B. B can do the same. The DES keys are used to prime DES cryptos for use throughout the session.

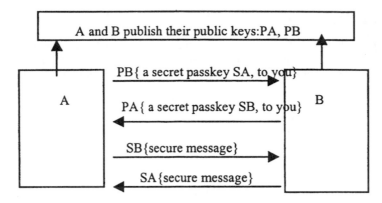

Figure 2.16 Combining public and private keys

2.30 OTHER PUBLIC KEY SCHEMES

Clearly the security afforded by a public key scheme is dependent on the 'reversibility' of the process from public key to private key. We saw that RSA relies on the difficulty in factorising large numbers into their prime factors. One other commonly used scheme, published by *ElGamal* [82] makes use of the *discrete logarithm problem*, the difficulty of finding x, given the final remainder after repeatedly dividing a^x, where a is a known number, by prime number p. The ElGamal algorithm has been used for digital signatures. It has a slight disadvantage that it generates a ciphertext that is twice as long as the original message.

2.31 ISSUES AND STATUS OF PUBLIC KEY CRYPTOGRAPHY

Optimists and commercial opportunists will tend to imply that the encryption problem has been more or less 'solved' through the development of public key cryptography over the last quarter century. Of course, near-perfect security is never achieved, except in the most trivial of cases, but there are some specific issues that still remain and which mean that we must not place our entire trust in public key systems.

Nevertheless, process is being achieved in developing a consensus and set of standards for public key useage. (A comprehensive overview is given in [83].)

One issue is the question of 'strong primes', which is not just relevant to cryptographic technology, but also acts as a reminder as to human propensity for criminal behaviour.

Earlier we explained how the RSA algorithm was based on the difficulty of factorising large integers which are the product of two prime numbers. This difficulty is not uniform: some products are easier to factorise than others. For example, let p and q be the two prime numbers and suppose $p - 1$ can be factorised into prime factors, of which r is the largest. Then, there is a method (*Pollard's P − 1 Method*) for factorising pq, which will take around r steps. Obviously, the bigger r, the longer it will take. P is said to be *strong*, if r is large.

It would seem therefore that strong primes are a good idea; but subsequent discussion and the development of alternative factorisation methods has cast doubt on the advantages to be gained. However, in recent years, strong primes have regained some popularity, for an interesting reason: it is speculated that unscrupulous users could *deliberately* use weak primes, in order to give themselves a loophole, in the form of the opportunity to repudiate a digital signature. After all, the user could claim, 'I did not mean to chose a weak prime; it was just my bad luck (or, really, yours, chum) that I chose one that someone else could crack and then use in my place'.

The point to note is the lack of a common, long-term viewpoint with regard to some issues of public key cryptography. Remember that provability is critically important to the acceptability of any security algorithm.

2.32 THE KEY-VALIDITY QUESTION

Another aspect regarding public keys may be worth briefly touching on: how do we know that a key is valid? A major claim for public over private key systems is that key management becomes simple: you do not need to set up a secure or covert channel for passing key information; you simply publish it or provide it on request. True. But what happens if I am given a public key that I believe belongs to someone, but in fact, belongs to someone else? (Figure 2.17).

Figure 2.17 The need for public key validation

I may be sending large amounts of electronic payment across the network, secure in the knowledge that no-one can intercept it, or channel it off to a third-party account, but how do I know that I am making a payment to the correct person and not to an impostor? I may have even been trapped into using the criminal's spoof key to transmit the key details for a private key system, DES, for example, that opens up my whole bank account or my company computer system.

This highlights again the difference between the security algorithm, which may be secure, even provably unconditionally secure, and the implementation and the application, which may lead to serious breaches.

Public key systems may have removed the need for a secure channel for transporting keys; that is, they have removed the need for a *confidential* channel. But they have replaced it with the need for a channel which can *authenticate* a set of keys. Various approaches have been adopted, but the basic principle is the same. In every case we need to have some person or organisation competent and willing to guarantee that the keys used to sign any document are valid, in the sense of belonging to the person who claims them, and that all reasonable care has been taken to ensure this guarantee is of value. Moreover, this person or organisation has to be trusted by all parties in the transaction. It is in how this trust is developed that differences arise.

Treating specifically the issue of who guarantees the authenticity of a public key, we now look at two alternative approaches, which are based as much on different cultural attitudes, as on technical approach.

2.33   CERTIFICATION AUTHORITIES

As we have repeatedly said, the issue resolves around trust, and around policies that make that trust justifiable. One useful definition that addresses this is *certification*: 'assessing whether a product is suitable for a given application' [84]. A *certificate* in our case is some reliable evidence that an identified person has the right to a set of privileges. The certificate is provided by a *certification authority*. Let us take a simple example of how this can enable a transaction. Suppose someone, purporting to be 'Alice' publishes a public key. Bob wishes to conduct business with the real Alice. How does he know that she and 'Alice' are indeed the same person? What Bob requires is an unforgeable document, from a publicly known body, that binds the public key of 'Alice' to the real Alice.

As shown in Figure 2.18, Alice approaches a *certification authority* and asks it to validate her public key. After due diligence, perhaps even a live meeting between Alice and a member of the authority, the latter agrees that Alice is who she claims to be. Now Alice has to take something on trust: she has to assume that the certification authority is widely known and monitored, to the degree that, when 'its' public key is published, then

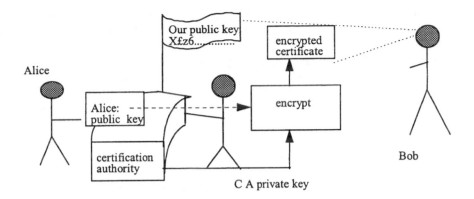

Figure 2.18 Using a certification authority

she can be sure that 'it' really is it. When this is done, the authority then creates a message encrypted with its private key, which includes Alice's public key. Alice sends this message, the *certificate*, to Bob, before they enter into a business negotiation. All that Bob has to do, is to use the certification's public key to decrypt the certificate and thereby establish 'Alice's' identity. Bob knows that the authority's key is genuine and he knows the authority know 'Alice's' key to really be from Alice.

We mentioned *trust*. The fact that the first and second parties to the transaction are prepared to trust a third (the certification authority) is critical to the process. The trust must operate in at two ways: the parties must believe that the authority is who it says it is, and must also trust that its processes are fair and competent. To ensure both of these conditions is difficult for a single authority. It is very unlikely that one organisation could carry out a reliable audit of all the parties that wanted to enter into transactions great and small, across the whole of a continent.

One way round this scalability problem is to create a multi-tiered hierarchical structure of certification authorities [85]. Higher, more general authorities would authenticate the public keys of lower, more specialised authorities. It is even possible to envisage a network, rather than a tree, of trust hierarchies, with several independent authorities validating a certificate.

The certificate needs to validate the public key of the party it has audited, but it can, even should, hold other information as well. It needs to specify the encryption algorithm to which the authenticated party's public key refers, and it should also include an expiry date for the certificate. The latter is a very important item of information and it has to result from a value-judgement on behalf of the certification authority as to how long the authenticated party can maintain the integrity of the private key. (Notice the usual trade-off between security and convenience: the more often the key is re-certified, the more hassle, but more likely that it is valid.)

The certificate can also contain other information relevant to trading relationships in general; for instance, it could hold information on licences to deal with certain information and carry out regulated processes (doctors, hazardous waste disposal and so on). It could also hold contractual terms – insurance limits, authorities within an organisation (e.g. 'finance officer') and so on. Some of these are particularly applicable in business-to-business transactions, and begin to extend the role of the certification body beyond that of simple financial certification. The parallels and integration possibilities of this with PICS rating authorities and P3P (Part 3, *Trust*) should be clear.

2.34 CERTIFICATION AUTHORITIES, TRUSTED AND DISTRUSTED THIRD PARTIES, PGP

In principle, anyone can be a certification authority, but in practice, a very large degree of credibility (trust, again!) is required. Large, well-known and well-founded companies with reputation for security (for example, banks and other financial institutions) can take on the role, as can post offices and telecommunications companies. Some large companies have initially set up certification authorities in-house, but a tendency has been reported for these to 'spin-off' joint venture certification authorities which trade under the trustworthy aura of the parents [86]. Government or quasi-governmental bodies (the assessment is subjective) can also be involved in certification.

Sometimes the term *trusted third party* is used synonymously with certification authority. This is probably not a good idea, as it can conflate the idea of authenticating someone's good faith and identity, with a the concept of *key access* or *key escrow* which is tied in with the need to fight crime and to support state security. Governments, notably the US and, to a lesser extent, the UK, do try to restrict the use of strong encryption to activities that they define as legitimate. However, they realise that this may not completely feasible. Instead, they have been trying to persuade their citizens that anyone using strong encryption methods must lodge the decryption key with some *trusted third party* who, under appropriate warrant, will produce it to the authorities.

Governments and other official bodies tend to favour the trusted third party and or certification authority approach. This worries some people, who see this as a potential avenue for exploitation by a not-necessarily benign state. Maher [87] believes that such 'licensing approaches which overtly constrain the freedom of open markets' will be destructive. Lomas

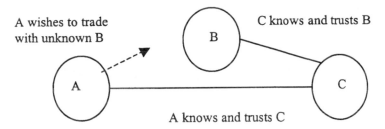

Figure 2.19 Setting up a web of trust

and Crispo [88] caution that: '… it is important to note that 'trusted' and 'trust-worthy' are not the same'.

But if we are to remove the officially sanctioned third party, how are we to be able to generate the necessary level of trust between parties to a transaction?

One option is to use a *web of trust*, a decentralised network, relying on trading parties themselves to authenticate others (Figure 2.19).

Suppose A wishes to trade with B but knows nothing about B. In particular, A does not know whether a digitally certificated message from B can be trusted to be non-repudiable or even actually from B. Fortunately, A does know C, and C, through personal dealings with B, can vouch for the validity of B's certificates. (Perhaps C has previously made physical visits to B's site, and has got hold of B's public key information in a guaranteed manner.) It is then possible for the claimed details of B to be encapsulated in a wrapper from C, using standard encryption hashing, for example, so that A knows they are getting valid certification details that they can use safely in trading with B (Figure 2.20).

The web of trust thus always relies on a trusted intermediary peer, rather than a 'higher authority'. This removes the bureaucratic complications of the hierarchical tree of certification authorities as well as putting to rest any worries regarding their probity in use or misuse of your key (a major civil rights issue, in some countries.) On the other hand, there are issues of scalability (or, rather, lack of it) and opportunities for bad practice on behalf of the trusted peer.

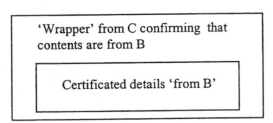

Figure 2.20 Validating the unknown party

The most widely deployed scheme based on webs of trust is *Pretty Good Privacy (PGP)* [89].

2.35 TECHNOLOGIES FOR KEY MANAGEMENT

In constructing a corporate network, it will almost be inevitable that some thought will have to be given to the various security keys are managed. In doing so, it is necessary to consider whether this is going to be a completely closed network, in which case, an in-house solution can be adopted. However, with the growth in business to business operation, it may be perhaps be wiser to consider choosing a widely deployed system. (There is no one 'standard' – see [42].) One good general principle to take into account is the possibility of integrating the key management task into the rest of one's resource planning and process control software, in particular, using a *corporate directory* of employees, as the hub for the control of access rights and authorities. On page 138 we describe how the *LDAP directory protocol* can be used for this purpose. One common practice is to combine LDAP with the *X.509 standard* for digital certificates. Since X.509 is primarily concerned with binding a person's name to a digital certificate, we can see that it makes sense to combine it with the natural source of corporate names – the directory service.

2.36 APPLICATIONS OF ENCRYPTION TECHNIQUES

So far, we have talked fairly generally about encryption techniques without any real attempt to categorise their applications. We have looked at individual aspects: cryptography as a way of processing messages in order to make them resistant to misuse by an unauthorised person, and the creation of a digital certificate by a certification authority. There are, however, other cases: perhaps we want to sign the message to prove that it did indeed come from us; maybe we want to make sure that it cannot be deliberately corrupted by the person who received it and later produced to justify a fraudulent claim. In the following sections we will look at a number of the most common applications of cryptography that would be found in eBusiness.

First we must remind ourselves that the model for a military security system is not an adequate way to describe the commercial situation (Figure 2.21).

We do not have two unconditionally 'good guys', fully trained and fighting fit, one at each end of the transaction, under attack from a 'bad guy'. There are many other alternatives.

Figure 2.21 The military security model

2.37 NON-REPUDIATION

One extreme opposite to the military scenario idealised above, is the situation which occurs when the partners to the transaction cannot be assumed to be acting in good faith. A contracts with B to supply goods under certain conditions – or at least that is what B thinks. They have traded across continents without meeting physically and can only exchange agreements electronically. B is disappointed because it feels that A has not delivered to the terms of the contract they hold, but confusion arises because A appears to have a differently worded one. How can this be avoided?

Long before electronic contracts were exchanged, physical contracts on paper were deliberately torn in two and each party held one of the parts. The tear was jagged and difficult to forge. Putting the two parts together provided proof that this was the original of the agreed contract.

A somewhat different approach is used to seal sensitive documents against unauthorised opening – envelopes containing instructions 'to be opened in emergency only', are sometimes signed on the flaps and the flaps stuck down with adhesive tape, which defaces the signature if attempts are made to remove it.

These are part proofs only. We can do better electronically. What would be even more convincing would be some way of protecting every single bit in the message from tampering. Suppose someone changes $100 to $1000, we want to be able to identify that a change has been made and, preferably, in what form. We want to be able to demand that the sender attach some form of *Digital Signature* that authenticates the data (and the sender).

2.38 CRYPTOGRAPHIC HASH FUNCTIONS AND DIGITAL SIGNATURES

The basic principle is shown in Figure 2.22.

A part or even the complete text of the document is passed into a

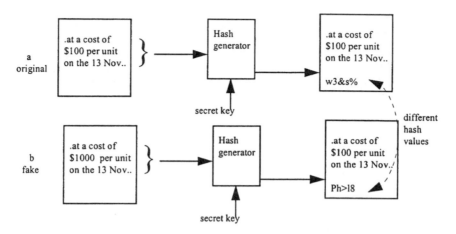

Figure 2.22 Using a hash function to provide message integrity

cryptographic device, a *hash generator*, which is primed with a secret key. This device then takes the given text and converts it to a *hash value* which is an encrypted summary of the document. The algorithm used to hash the document is designed so that, in most cases, a change to the original will produce a change to the hash function. There are a large number of ways we could do this. One simple approach might be to count the total number of characters in the document, encrypt this value in some way, and use this encrypted value as a hash function. Most changes to the document would result in a difference between the original hash function and the hash function calculated for the new document. Of course this is a dreadfully insecure example; if someone found out that you were using this hashing algorithm, they could easily generate document changes that would fool it. Practical hash algorithms have to satisfy a number of more stringent conditions.

Let the text we wish to hash be x. Let hashing operation be h and the result of hashing x be $h(x)$. Then, a good hash function, h, is one which has the following properties:

1 *Computability*: it should be feasible to compute $h(x)$.
2 *One-way resistance*: given hash value y, it is, in general, infeasible to find value x such that $h(x) = y$.
3 *Weak collision resistance*: given x and $h(x)$, it is infeasible to find x^*, where x^* not equal to x, such that $h(x^*) = h(x)$.
4 *Strong collision resistance*: it is infeasible to find different inputs x and x^*, where $h(x^*) = h(x)$.
5 *Compression*: irrespective of the length of x, h(x) should have a fixed length, n.

Conditions 1–4 imply that the hash function is *one-way*, in that it is easier to create one than to work back to the original text. (Note the similarity to public key systems.) The conditions also imply that the hash function tends to maintain a 'good separation' between different texts: although it, in general, maps them into a reduced number of bits than in their original form, it tries to find as different a solution as possible for each. Suppose we think of each bit in the original message as being written into a different 'dimension' in a message space. Then a message M bits long can be considered to be a point in M dimensional space. (For example, the three bit message '1,−1,1' has coordinates {1,−1,1} in three-space.) The hash function operates on the message to produce another one in a space of reduced dimensions (of fixed dimensionality n). A well-designed hashing function will tend to space out the answers for the set of all messages, to be as far apart as possible, in the reduced dimensional space. It also has the property that it takes two messages that are very similar and creates hash values that are, in general, not particularly close together, thus making it difficult to work backwards, and also avoiding the possibility of accidentally generating feasible, similar messages, when a spurious decode has been attempted.

Hash functions can be generated in a manner similar to encryption methods. One method, *discrete exponentiation* is the inverse of the discrete logarithm problem described earlier. However, it is rather computationally intensive and other, speedier methods are sometimes used instead. Examples include the *secure hash algorithm* (SHA-1) which is used as part of the US Digital Signature Standard and the MD5 standard, which is very common in the case of Internet applications [90].

To use a hash function to protect the integrity of a set of data, the sender has to share the secure private key with the receiver. This can easily be done if the receiver has first published a public key, which can then be used by the sender to encrypt the authentication private key.

Obviously, as well as confirming that no malpractice has occurred on behalf of the two parties to the contract, the hash function can be used to protect the integrity of the document 'in transit', e.g. to protect against a third party modifying an agreement, perhaps to 'cream-off' surplus value from a financial transaction.

2.38   SECURITY PROCESSES IN NETWORKS AND COMPUTERS

Now that we have discussed the principles of digital encryption, we can look at one or two standard processes that employ it as a way to provide security services in computer networks.

Figure 2.22 made the point that security measures must take into account the fact that attacks can be mounted at different levels within a networked computer system. The problem with providing security at the application is that it makes it difficult to ensure that no trap doors to lower levels have been left open and it also places responsibility for security with each application developer, which can be inefficient probably inse- cure. We begin at the packet level.

3.39 IPSEC

We have seen that the basic packet transmission protocol of the Internet, *IP*, is a hostage to fortune for correct delivery of the packets in the correct order and without error, and we discussed how this can be vulnerable to attack. For applications requiring security, there is a protocol *IPSec* [42], which lies within the IP layer and tries to ensure that the packets are authentic, uncorrupted and confidential (Figure 2.23).

IPSec is described by the Internet Engineering Taskforce as a 'frame- work', which may evolve with time in order to provide enduring network security [91]. One of its advantages is that it is being refined in concor- dance with the IETF's developments for later versions of the IP standard, notably IP version 6. The details are complex and will not be covered here, but the basic principles are simple: within the standard IP packet, IPSEC defines a couple of headers; one of these defines the authentication details and the other one specifies the encryption used. Public keys with digital certification (to prove that they genuinely belong to the communicating parties) are first used to encrypt and transmit secret keys. Some of these keys are subsequently used to encrypt the packet traffic, to stop it from being read without authority, using, for example, DES. Other keys are used for hashing algorithms which append hash signatures to the data packets to authenticate them against corruption. IPSec has been designed to allow the incorporation of new encryption algorithms as and when these are developed and proven.

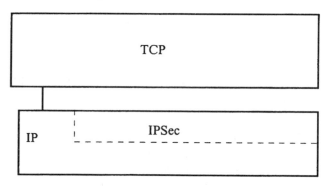

Figure 2.23

In addition to providing protection against overhearing and corruption of data, IPsec can also protect against someone mounting a *man-in-the middle attack* (intercepting legitimate packets and replaying them at a later time). It does this by adding an encrypted *sequence number* to the packet, and increments it every time the same packet is retransmitted. This can also defeat attacks such as the de-fragmentation attack mentioned earlier.

2.40 SECURE TUNNELLING

In Part 2, *E-business Systems Architecture*, we discuss the key role of extra-nets and intranets in the construction of business-to-business and intra-business logistics of the electronic enterprise. Vital to these services is the ability to provide communication that is confidential and guaranteed between sites connected over the public Internet. So far, our discussion of IPSec has been concerned with the first part. But there is little point in constructing a network that we feel might be attacked, and which can carry secrets, if these secrets are not delivered to the right place because someone has streamed them off elsewhere. Nor is there much point in the network if someone can flood it with traffic which appears to have originated at legitimate points, but is actually injected by a third party, to confuse or disrupt.

One of additional requirements, therefore, for IPSec and other security protocols is in the guaranteeing of point of origin and point of delivery of IP packets. IPSec handles this problem in one of two ways, depending on whether the transport of data is from the original host (*transport mode)* or one of the ends of the secure path is only a gateway (for example, a fire-wall, Figure 2.24).

Again, the details are complex, but the main difference is simply in how visible is the end-to-end routing of the packet to a third party observing the link. The security of the link is defined unidirectionally, that is, there are separate security arrangements for traffic passing in opposite directions. Tunnelling is the fundamental security feature of Internet *Virtual*

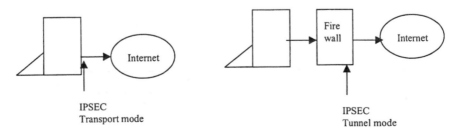

Figure 2.24 The two modes of IPSec

Private Network (VPN) technology, as described in Part 2, *E-business Systems Architecture.*

2.41 SECURE SOCKETS LAYER – ENSURING CLIENT-SERVER HANDSHAKING

IPSec and the like protect the end-to-end transmission of data across the Internet. In addition to this protection we also require methods for ensuring that applications can be conducted in a similarly secure and validated manner. One of the simplest, but most commonly used, applications requiring security, is a simple transaction between a client PC and Web service, for example when passing over details to authorise a payment. What is required is a security process that sits between the application layer and the generic TCP/IP connection that goes on below it. One very common method is the *Secure Sockets Layer (SSL)* protocol, developed originally by Netscape (Figure 2.25).

The details of the transaction are quite complex, but the mechanism is relatively simple: data between client and server are encrypted, as part of an *SSL record layer.* It is necessary for the client to lay down the rules for this encryption and the server to agree this and act on it. This is done during an *SSL handshake* (Figure 2.26).

- The client begins the transaction by sending a random number to the server.

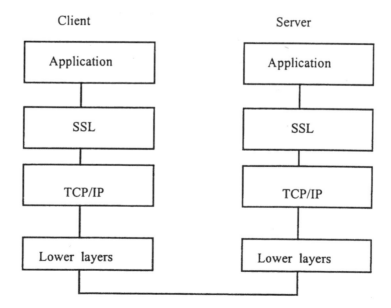

Figure 2.25 The secure sockets layer lies between applications and TCP/IP

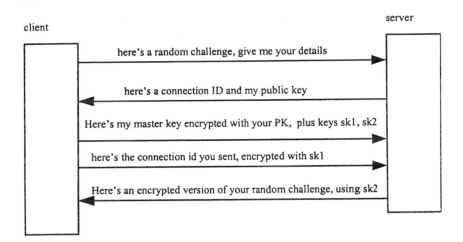

Figure 2.26 SSL secure handshake

- The server responds by issuing a connection number and its public key.
- The client responds with a message, encrypted using the server's public key, which contains key information (that will be used for encrypting transaction details, e.g. using DES), plus a couple of keys sk1 and sk2 which will be used to validate the messages between client and server.
- Using its private key, the server decodes this information, and waits.
- The client uses sk1 to encrypt the connection id originally issued by the server and sends it to server.
- The server has sk1, so can decode the message to find a connection id that is indeed valid between it and the client.
- The server now encrypts the random 'challenge' sent by the client at the beginning of the handshake, using sk2, and sends it to the client.
- The client can decrypt this and verify that it is indeed the correct server it is talking to.

Both parties now share secure keys and can continue the transaction.

SSL conforms very much to the good-security principle that little is required of the application designer: all that needs to be done is to replace a call to a TCP utility with one to an SSL one. This has been one reason why SSL has become the most widely used browser security protocol.

We should mention one further refinement that turns the security technology of SSL into a secure process: the authentication of the server's public key. Only the clever (or slightly sharp?) amongst us will have thought about the possibility that the Web server we are dealing with is not actually who it says it is. Yes it declares a public key, but not necessarily one that belongs to the company it purports to be. The people who designed SSL have thought of this: it is not actually a public key alone that can be offered by the server; it can be a digital certificate, issued by a certification authority, providing proof of the server's identity.

2.42 PROTECTING COPYRIGHT

There are a number of issues of a security-related nature that we have not yet discussed. Of these, one of emerging importance to electronic retailing is copyright. In particular, how can we protect against unauthorised use of valuable content – music, video, whatever, particularly that which has been loaded down over a network? First we have to recognise that there is unlikely to be a complete solution to the problem. Just as for the protection of secrets and for authentication of participants, there are theoretical and, perhaps more to the point, practical difficulties that stand in the way. All we can hope to do, is to make the theft too expensive to all but the most determined and resourceful attacker.

2.43 RE-PUBLISHING OF COPYRIGHT MATERIAL

Not only is digital electronic encoding of copyright material vulnerable to local copying, it also provides a very easy means of making it available on a world-wide basis. Currently lawsuits are in place to try to prevent this happening – and may be successful – but the technology is so suited to this kind of thing, that we are not likely to have heard the last of it. The particular example currently at law relates to the provision of services for acquiring copyright music that has been encoded in *MP3* format (the audio portion of the multimedia standard, *MPEG3*) and stored on the PCs of individual music fans. We look at two examples of this, *Napster* and *gnutella*.

Gnutella operates on the principle of providing application software that allows these people's machines to behave either as client or server (or both). The hybrid machines are known as *servents*. To use the software, all you require is the IP address (and port, but this is detail), of any other servent. Your machine announces your presence to this servent and it in turn passes this on to all the other servents it is connected to. Each of these servents then responds to you, giving information on whatever part of their file structure they want to make public. In announcing yourself you can issue a search request, for example '.mp3', which is obviously requesting information on MP3 files. The response from any servent is rather arbitrary, depending on how it has been programmed to interpret '.mp3' and how it has been set up to respond, but, in principle, you could expect to get sent back to you, a number of names of files matching the search string. If you ask your machine to retrieve one of these files, then, your machine simply issues a standard HTTP request to call the file down, as the distant servent has been set up to respond appropriately [92].

Napster, which is currently fighting and losing battles (but not quite yet the war) against music companies, works on a similar principle, but requires a centralised server. This is why it is easier for lawyers to complain that it is complicit in copyright theft.

2.44 PRINCIPLES

One of the principal differences between electronic media and the more traditional methods such as paper and photographic film, is that the former are considerably cheaper and easier to copy and to distribute. That is the origin of both their success and weakness. Moreover, the digital nature of much electronic media means that copying can be *lossless* – no degradation in quality need be experienced either in copies taken from a master version, nor need the master itself suffer any degradation in the process of being copied an infinite amount of times.

Digitisation does provide one level of protection to the legitimate owner – it is amenable to highly secure encryption, using the techniques we discussed earlier. Therefore, it can often be secured in transport, and, if it is required to be viewed promptly, (for example, in the case of a sporting event), it is often possible to charge for the timely delivery of the decryption key. This is the case with broadcast TV, where subscription and pay-per-view channels are regularly encrypted and keys are purchased for use on the set-top box decoder. It is worth remembering the special aspects of key management in the case of TV set-top boxes. Each box owned by a customer has to be different, in terms of some key settings inside. There also has to be a way of distributing additional keys which will only operate with that box and which can be disabled remotely or by time expiration, in the event of termination of the contract to view. These are not trivial process complications and must be considered as potential weaknesses against attack.

Nothing can stop the recording of the finally viewed material. Large amounts of money have been wasted on schemes that claim to make it 'impossible' to copy a piece of recorded music or a video film. It stands to reason that if you can display it, you can copy it. Of course, what is displayed may not be quite as good as what is stored on the original source and there may be an intermediate *lossy* process in the way, too – copies of analogue tape copies, even if themselves digital, cannot restore the quality of a digital master. There are also *spoiler* systems available that interfere with information that is not subjectively observable, but necessary for the reproduction equipment. For instance, video signals for TV sets and video cassette machines contain synchronisation that is required to scan the picture across the TV set at the correct rate. If, as is done with some scrambler systems, this synchronisation is dithered in time away from its normal position, then it is impossible to view it on a normal screen, without using a descrambler to return the synchronisation to its correct value. This is effective against casual attempts to copy and replay the content, but it is relatively simple and cheap to build 'pirate' descramblers to defeat the effect.

2.45 WATER-MARKING

If we cannot make it impossible for people to make copies of valuable content, then we can at least find ways of discouraging them. The law is quite clear on the matter – it is illegal to make unauthorised copies of copyright material (and other material, such as that supplied on condition it is not copied, state secrets, etc.). If we can show who carried out the copying (or who lent them the copy) then examples can be made, that might discourage others.

One effective method would be to mark each copy with the name of the person to whom it was issued, with that person under a duty of care to make sure that it was not copied. Provided we could see this information carried over onto the illegal copy, then we could know who to blame. But, of course, we can hardly expect any copyist with much common sense to allow these details to remain on the copy, if they know they are there and know how to remove them. If we are to make this impossible (or at least very difficult) we have to do two things:

- Hide the information in the material.
- Or, at least do it in a way that it cannot be easily cut out from the material itself.

Such a scheme is known as a *watermark* in analogy with the way that banknotes and other physical materials, principally, paper ones, are imprinted across their surface with an inconspicuous imprint. Where the imprint is invisible and not intended to be easily found, the technique is known as *steganography*. The author developed a very simple scheme several years ago for doing such a thing to documents produced on a word-processor, to provide an audit trail in the case of unauthorised 'leaking': each time a copy of the document was printed off, the proportional spacing of the words was modified in such a manner that a hidden 'unique copy number' was effectively put into the text itself. Obviously this copy number is not lost after photocopying, photographic copying, etc.

In general, although there are some counter-examples where the fragility of a watermark can be valuable, most watermarks are intended to have precisely the opposite properties of cryptographic hash functions.

That is, they must be resilient against a number of transformations of the original material, for example, the compression algorithms used for reducing the number of bits required to code an image, cropping of the image into smaller parts, and so on.

In distinguishing between copyright watermarks and other watermarks which provide authenticity by providing a signature which cannot be forged, we have to decide what it is we want to encode in them. For copyright watermarks we shall probably want to have a number of variants that can be recognised despite degradation (as in the case of our document copy number) and we shall consequently need to decide how we recognise this variant. One way is to compare the recovered watermark with an original; for example we might 'hold the two up to the light' and measure the differences between them. This method implies holding an original for each set of variants, and having a process that allows us to compare original with variant, despite subsequent distortion and degradation. Furthermore, the original thereby essentially becomes a key in our cryptographic sense as used earlier, and we have the beginnings of a key management problem. What is preferable is a method of embedding at least some of the encryption process within the watermark itself, so that authenticity and other parameters can be recovered from the copy itself, without reference to the original.

In order to get a clearer idea of what is meant, first let us look at a very simple example of how this could be done for a speech or music signal (Figure 2.27).

The coder has the choice of adding a short or long delayed version of the audio signal to itself (an 'echo') depending on whether it wants to code a binary one or a nought. Provided the delay times are chosen carefully, there is little perceived degradation of the signal and by switching between them, a rate of a few bits per second can be achieved. The decoder works by 'correlation': the signal is split into two channels and each one multiplied by a version of itself, delayed by one of the delays. The bigger signal is obtained from the correct multiplication, thus the secret bit stream containing the watermark can be recovered.

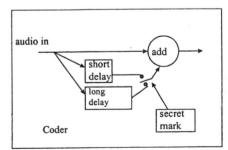
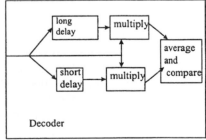

Figure 2.27 Adding a secret mark to audio signals

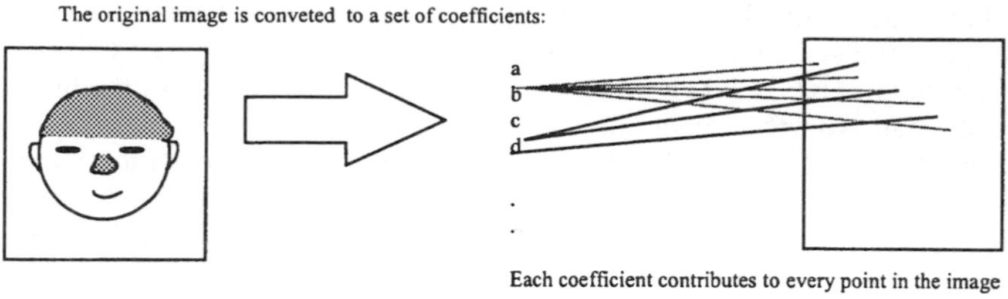

The original image is conveted to a set of coefficients:

Each coefficient contributes to every point in the image

Figure 2.28   Transform coding of images

For various reasons this is not a very secure way of coding (and it does require sophisticated coding to get round some of the special properties of speech and music), but it demonstrates a number of the basic principles. First of all, it inevitably damages the original signal. The echo that is introduced was not there in the first place. Any watermark that is spread through a signal must damage it to an extent; the skill is in choosing a method that is resilient without being obtrusive. Secondly, the water-marking is achieved by adding something to the signal that can be extracted from it, without needing a separate recording of the original.

We next look at a much more realistic example involving the water-marking of video images. The raw signal from a TV camera is digitised and then compressed; this compression is required in order to reduce the data rate to an affordable level for transmission. Most coding/compres-sion techniques in common use (JPEG, MPEG, H261, etc.) make use of *transform coding,* which represents a digital picture by a set of numbers (*transform coefficients*). These have the peculiar property that each number describes (in a limited way) every point on the image, rather than just a single point on it. Individual points on the reconstituted image are there-fore formed by the combined effects of all the transform coefficients (Figure 2.28).

So, if we arrange to modify even a single coefficient, the effect will be spread across the whole image and will not be immediately obvious. In practice, digital coders split the picture up into smaller blocks and calcu-late sets of transform coefficients for each block. This allows us to make fairly large changes to a selected subset of the coefficients, which makes the watermarking resilient to further codings and distortion to the image. It is possible to produce watermarking schemes which do not require an original set of images to be used for recovering the watermark.

Of course, there has also been activity in the attacking of watermarks, too. Some of them have been reported as highly successful [93]. For instance, one can simply average across a large number of copies in order to confuse the unique per-copy watermark. Introducing slight distortions into the pirated copy of the image can make it difficult for the owner to reverse the watermark and claim their rights. Digital watermarking is, as with other technical measures of copyright protection, only partially successful. There are major problems, perhaps insoluble ones, in this area.

A good survey of the subject of watermarking is contained in [94].

2.46   PORTABLE SECURITY – THE SMART CARD

We saw earlier in this chapter that providing a secure communication service relied on holding at least some information secret. Trying to keep a long, complex private key in our heads is not feasible. It is possible to store this information in a PC, but there is always the risk that someone will use the machine illicitly when you are not there. There is also the inconvenience of not being able to work on a different location. Ideally, it would be better to be able to carry the secret information around with us, in a secure and compact form. One way to do this is to use a *smart card*. As shown in Figure 2.29, smart cards are computers encapsulated within credit-card sized plastic bodies. The processors are designed to run rather slowly but thereby consuming little electrical power even when running at full speed, and the entire card consumes pico amps (for memory retention) when not in use. The card has fixed, read-only memory, which contains programming instructions. For security applications these could comprise the currently favoured encryption algorithm or algorithms. The read-only memory may also hold data including the user's private key. An alternative approach is to implement the algorithm on special purpose hardware, such as an encryption chip. A number of

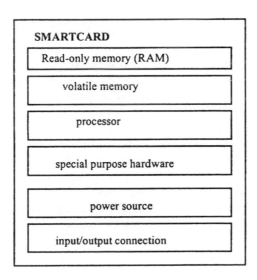

Figure 2.29   Smart card technology

input/output options are possible: current cards usually make contact with other devices by means of electrical connectors on the edge or surface of the card, but magnetic, capacitive or radio connection can also be used, as can optical methods.

A study carried out in 1995 identified nearly 200 organisations actively involved in developing equipment for smart cards or their integration. There is vigorous activity in the area of international standards for card architectures and basic parameters such as working voltage and specification of the electrical connection, as it is seen that a major cost-reduction opportunity exists if the number of variants can be reduced. Smart cards can store generally useful information; for instance, they can hold a list of telephone numbers that we can invoke by a short code dialling code. (1 = 'mum', 2 = 'work', etc.). GSM mobile telephones contain a smart chip, the *Subscriber Identity Module, (SIM),* that holds the user's own identity code. Without one, no GSM phone will operate. With it the mobile can send out user identity to the mobile network, giving access rights and allowing charges to be attributed to the user's bill. In Part 3, *Electronic money,* we discuss how this can act as the basis for a source of electronic money. Cards can also be programmed to hold biometric information, in encrypted form, so that any instance of using a biometric for access can be checked securely.

One drawback of smart card characteristics has been the lack of good standardisation, at the physical and application level. Progress has occurred in recent years [95], but is still not complete. One major development has been the emergence of a layered standard from the *Personal Computer/Smart Card (PC/SC)* group [96].

2.47 PROOF OF IDENTITY - BIOMETRICS

In the previous section we mentioned biometric information. One vitally important aspect of security in a networked environment, is proof of identity. This was not a problem when people dealt face-to-face, but how do we identify someone at a distance? We have seen with digital certificates that a public/private keys can be validated by a third party, but they can thereafter be stolen. There is still no way that these can be uniquely associated with an individual, without something extra, some unforgeable characteristic, measurably unique to the person seeking access, at the time that access is requested.

What we need is something that identifies a person through a unique biological characteristic. These 'biometric' measurements rely on the identification of some pattern of habitual behaviour or physical characteristic that is more specific in some way to an individual than to people in general. Not only must the measurement be able to identify individuals, it must also be possible to prove that this is so. This tends to favour physiological parameters, such as fingerprints or iris images, which are precise and relatively unchanging throughout time or situation, rather than 'habitual' measures, like voice or handwriting, which are learned habits not genetically 'burned in'. Nevertheless in practice, the measurement must be cost-effective and its administration acceptable to the user and this sometimes means that habitual features have an important place.

Identification by biometric means is a particular example of the general class of *pattern-matching problems*. Given an example from an unknown subject, there are number of techniques whereby this can be compared with existing patterns from a number of subjects, in order to find out which subjects' patterns it is most like, and hence provide a best guess at the identity of the individual. Different techniques have been applied to a variety of biometric types, using a range of measurement technologies. In all of the techniques and biometrics involved, however, there are a set of common features that must be satisfied in order to produce a system which is satisfactory in practice.

2.48   REJECTION AND ACCEPTANCE

There is always a trade-off between the performance of a system in incorrectly rejecting someone who should be allowed access privileges and in its susceptibility to impostors. Each scheme has a performance curve whose general form is that of Figure 2.30.

Most schemes allow us to set recognition thresholds that determine where on a graph of false recognition/false rejection we wish to operate. Usually if the access is to critical procedures, we will choose to operate with a low false recognition rate and thus sometimes will expect to reject, wrongly, access attempts by legitimate parties. They may be asked to try again. However, most of the common biometric schemes in current use tend to be both rather poor at rejecting impostors and at accepting legit-

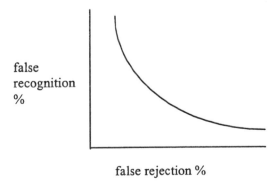

Figure 2.30   False recognition and false rejection errors

imate users. Consequently, they tend to be set in favour of acceptance, rather than rejection and used in conjunction with passwords and PINs as secondary security measures.

Another problem with many biometric schemes is in provability. Remember that we said that one of the important criteria for the acceptability of a security scheme was that it was 'demonstrably secure.' Later we look at one measure that can be used to select schemes with high 'provability', but first we consider some basic classification theory.

2.49 BASIC PRINCIPLES – LINEAR CLASSIFICATION

Mathematical methods of pattern matching can be very complex and difficult to understand. Perhaps the most approachable method is 'linear classification'. Since most of its principles can be applied to other, more complex techniques, we shall look at it in a little detail.

Suppose we decide to use a simple handwriting recognition scheme. Consider the problem of recognising the authorship of the hand-written letter 'W', in Figure 2.31.

For the purposes of example, consider that we have decided to measure two features of the letter:

*   The average slope of the down stroke AB.
*   The speed of the upstroke CD.

Suppose we have several examples of W's written by each one of N people. They will not all have the identical value, neither across all writers nor even for each individual. Suppose we plot the values on a graph as shown in Figure 2.32.

Each of the five ellipses represents a ring drawn around most of the values for any one writer. We see that, taken across all writers, the value of slope ranges from about 45–70°, and speed between about 1 and 1.5 cm/s. We also see there is a mystery writer, 'W'. To make Figure 2.32 easier to see, we have deliberately chosen a position for W that is far from the

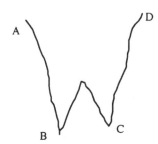

Figure 2.31   A letter 'W'

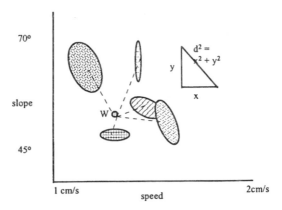

Figure 2.32 Distribution of individual variations

centre of the values for any one writer, but we are going to insist that W must be one of them. But which one?

Our first thought is simply to measure the lengths of the lines between W and the centres of each writer's spread of values and chose the known writer who is nearest. On a two-dimensional graph,

Distance = square root of (x distance squared + y distance squared)

according to Pythagoras' theorem. Note that if we had three different measured parameters, rather than two, the formula can be easily extended by adding a term in z-squared. In fact, it is possible to extend it to cover as many 'dimensions' as we like, adding in a term for each.

But it is not quite as simple as this. Returning to our two-dimensional example, suppose we simply changed the horizontal scale from cm/s to m/s, or the vertical from degrees to radians? The relative distances would change and we might find that we now identified W with someone else.

Rather than use arbitrary scales, we have to select a scale that take into account the relationship between the spread of values for *each feature across all writers*, compared with the average spread for *each feature within any one writer*. We define a ratio, conventionally called the *F ratio*:

$$F = \frac{\text{range of values of the measured feature, across all writers}}{\text{average variations of individual writers}}$$

A little thought will explain that this calculation, applied to each dimension ('feature'), will give an estimate as to the discriminating power of the dimension, a big F ratio implying that the specific feature is highly effective in discriminating between writers.

In practice, the F ratio is used as a weighting factor in the calculation of the distance between unknown example and the various possible writers. It can be shown that this removes any influence of specific units (such as centimetres versus metres). There is another, very important aspect of the F ratio: it is usually easier to establish confidence in a component of a system if its F ratio is high, as we shall see, in the following examples.

2.50 HANDWRITING RECOGNITION

Authorisation by means of a signature is almost as old as writing itself. We have previously briefly mentioned measurements of direction and speed and there are a large number of patents that employ variations of these. Some only operate on the static signature (i.e. direction), whilst the newer approaches increasingly emphasise the dynamic approach. Handwriting possesses some of the problems associated with other 'habitual' measures. The F ratio for any particular measurement dimension always turns out to be rather low, principally because there is quite a lot of variability within each individual's way of writing. There is also a slight technology problem: to capture the signature, we either need a relatively expensive tablet to write on, or a generally cheaper, wired pen which however, is more liable to be broken (or stolen!). One reason why the tablet is expensive, is that it may need to be sensitive to 'air strokes', the movement of the pen when it is not in contact, as this has turned out to be an important distinguishing feature of signatures.

2.51 FINGERPRINTS

This is a biometric measurement that antedates electronic technology. There is little doubt that fingerprints offer an extremely high effective way of identifying individuals uniquely, provided the print can be captured reasonably completely. The theoretical F ratio for fingerprints is almost infinite because individual changes are approximately zero, unless fingers are deliberately or accidentally profoundly damaged. It is not necessary to have this property of individual invariance, but it does make it easier to justify the security of the biometric.

In principle, an image of a fingerprint could be captured by a TV camera, transmitted over a communications link and checked against a database. In practice, however, there have been a number of problems with automated, remote access systems. Because of the high density of lines within a print, a large amount of data is required and there is a non-insignificant amount of processing power required to align the data if the print taken is only partly captured or poorly outlined. Fingers must be pressed against glass screens in order to position them correctly for scanning. The screens can easily become dirty. Recently, commercial products claiming to have got round some of these problems have begun to come onto the market.

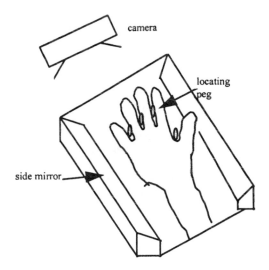

Figure 2.33   Recognition of hands

2.52   RECOGNITION OF HANDS

A very simple, but quite reliable, biometric is the 3 dimensional shape of the hand. This can be measured quickly and accurately, using basic technology (Figure 2.33).

   A system like this is already proving its worth with the US Immigration Department, for immigration control to the US. People already having legitimate residential rights in the US can acquire smart cards that hold an encrypted version of their hand measurements. If they leave the country and subsequently return, they can be automatically monitored against their card to see if they are who they claim to be. One of the major advantages of this technique is that only a small amount of data is required to specify the biometric – one system quotes 9 bytes.

2.53   SPEAKER VERIFICATION – 'VOICEPRINTS'

The name *voiceprints* is sometimes given to the technique of *speaker verification*, because it has been claimed that each individual's voice has unique characteristics that are analogous to the uniqueness of fingerprints. This claim is highly misleading and has never been convincingly demonstrated. There are wide differences between the two, as outlined in Table 2.6.

   The real problem that the advocates of speaker verification have to deal with is the fact that the F ratio for speech does not appear to be anything like as great as that for fingerprints. At the very least, it is extremely

Table 2.6 Comparison between fingerprints and voiceprints

Fingerprints	Voiceprints
Are genetically determined	Genetic determination has not been proven. There is a degree of correlation between voice and physiology (e.g. length of vocal tract) but the former is not rigidly determined by the latter. Voice may be learned; certainly, accent is acquired, not inbred
Cannot be altered by will; does not change with age	Voice changes with situation; voice can change dramatically with age, state of health, situation
Individual's prints are identifiable, discrete lines that can be seen to retain the identical pattern under all circumstances	Individual's voiceprints comprise complex 'features' that are flexible, vary from utterance to utterance, and have values that are sometimes shared with others

difficult to measure, for any of the features that go to make up the various systems on offer. Thus, it will be difficult to estimate the reliability of speaker verification schemes, and so it is unlikely that they will be trusted on their own, for critical applications. Nevertheless, in communications, speech is one of the easiest data sets to collect and there are indications that speaker verification does give a modest degree of protection. We could say, for example, that from a population of a 1,000 voices, we could reasonably claim that we could narrow the possibility down to 50 or even less, with a 'good chance' of being correct. If, for example, the authentication procedure comprised asking someone to speak their personal identity number (PIN), voice recognition would thereby add significantly to the security of the process over that of simply keying in the PIN.

Regarding the details of voice recognition, there are a large number of parameters that have been tried: voice pitch, frequency spectrum, vocal tract resonances (*formants*) and derivations thereof. Some are more resistant than others to the noise and distortion problems of communication networks, or require less computational activity.

2.54 KEYBOARD IDIOSYNCRASIES

Another example of a statistical parameter that is a 'natural' candidate for use in an IT-related environment, is *keyboard statistics*. Increasingly, at least for the next few years, keyboards will be the predominant way of personally passing information across the network. It would seem logical to investigate whether we can be distinguished, one from another, by the rhythm with which we use the keyboard.

There is a respectable history behind this: during the 2nd World War,

British warships would set off from port after landing the Morse code operators they had used when they were in harbour. These operators would continue to transmit messages from shore, knowing that the German interception units would recognise their 'fist' and therefore believe the ship was still at anchor.

Application of this to automatic recognition of keyboard style has been achieved in a number of ways: one method measures the press and release times of the keys when users typed names. From this was calculated the time between successive presses and also the time the keys were held down and these were used as classification parameters. The system was trained using a neural network although in principle other classification strategies can be used.

2.55   IRIS AND RETINA SCANNING

Recently, significant success has been reported in practical measurement of the pattern of veins on the surface of the iris or retina and the use of these as a unique 'eye print' for individuals. Indeed, eye scanning is one of the front-runners in the quest for a reliable, non-invasive biometric The user simply looks at a TV camera and an image is captured and processed on a medium-range computer workstation costing a few thousand pounds. The technique has the obvious advantages of fingerprints – a very high F-factor – as well as being non-contact and providing a better quality image. One system quotes a stored pattern of 256 bytes and a 'cross-over error rate' (that is, when the false acceptances equal the false rejections) of better than 1 in 130,000.

2.56   SECURITY – FINAL WORDS

In our tour of security problems and solutions we have gone through the technologies and operational detail of only a small subsection of what is possible. None coming to the subject for the first time should consider themselves now skilled simply by having read this chapter. In many ways, security is the most difficult side to eBusiness design, although it is still often neglected or forgotten. It is not perhaps something that can just be taught; it has to be learned, perhaps the hard way and a large number of organisations are beginning to do just that. There is, in fact, a race on between organisations that are converting their critical operations to eBusiness mode and the criminals and others that are attacking them. Today's computer security situation is poor for the simple reason that networked computers are inherently complex and there is little sign that this will undergo much change in the foreseeable future. Therefore, process control must carry the burden of good security and this requires thoughtful and mature design. It will pay well because the failure costs will be high. It is also intellectually challenging, for the good guys as well as the bad. Let us hope that the industry begins soon to realise this.

3

Electronic Money

In other chapters we discussed the technologies for carrying out secure transactions at a distance. In the area of electronic trading, undoubtedly the highest profile application is in the transporting of credit and money, in electronic form. Prudent financial dealings the world over are conducted on the basis of minimising risk, in an environment where everything has to be done to avoid fraud and greed. Moreover, it is a multi-participant activity, with each player trying to minimise its own loss, without putting too much in the way of making a sale. It is difficult for most technologists who, naturally, are interested in promoting their favourite solutions, to keep this in mind. Banking law and practice is however, keenly aware of these issues and financial institutions have a justified reputation for caution and their rules often limit the opportunistic adoption of technology. In this chapter we look at some of the processes that apply secure techniques to minimise risk in on-line financial transactions. This is an extremely volatile area and the intention will be to describe enduring principles rather than all of the current instantiations, some of which may turn out to be only of historical interest.

3.1 PARTICIPANTS, LIABILITIES AND PROCESSES

In a purchasing transaction there are a minimum of three participants: a buyer, a seller and a bank or other financial body. Very often there are four: the buyer and the seller may use different financial institutions. The bank which issues the credit card is known as the *issuing bank* and the bank which receives payment on behalf of the seller is the *acquiring bank*. A variety of intermediaries to the transaction may also be involved. For on-line transactions there are at least a network provider and a service provider to be included, and they may or may not be sue-able if a transaction goes wrong. If things do turn out badly, the participants (or their lawyers) will apportion blame according to well-established rules which were written before eCommerce began. The new situation will be mapped onto the old order until it too has acquired a body of legislation and case-law. One such example is 'cardholder not present', which occurs, for example, when a merchant accepts credit card payment over the telephone or the Internet. In this case, the risk lies with the merchant, not the bank, which would generally be otherwise if the card was signed in the presence of the merchant.

Whilst having to take regard to the risks involved, many and varied are the ways of parting a customer from their money, and this variety has developed in order to make it easy for all levels of price and location. Cash and credit are different ways of paying, with the credit agency creaming off some fee as a reward for accepting risk. Authentication processes such as signatures and other methods, whose complexity varies with the amount of money, are involved and, finally, inter-bank processes have to be carried out to complete, or 'clear', the transaction. The mechanisms can be quite complex and vary from country to country and are frequently discussed in texts covering the commercial aspects of eCommerce, including [97-100].

3.2 DIFFERENT LEVELS OF TRANSACTIONS

There are a number of different approaches to electronic money and its transfer. The World Wide Web consortium [101] lists at least 28! The differences are partly because of commercial competition but the fundamental differences are mainly due to the relative size of the transaction and its cost. Clearly, where very large sums of money are involved, it is not unreasonable to spend quite a lot in protecting the transfer. If, however, the sum is very small, then the corresponding transaction charge must be even smaller. It is for this reason that a significant number of transactions which to the shopper appear like real-time charging, are actually batch processes. Consider, for example, the case of paying for something in a store, using a credit card. Although the owner of the store will swipe the card through a reader and it appears that this transaction is being cleared over a telephone link (or it may be over an X25 packet network connection), all that is actually being dealt with is the card verification. The actual payment for the goods is held against the card reference in the store machine, usually until the store is closed, and then all transactions for the day are downloaded to the store's bank. This is done because of the cost of carrying out the individual transactions one at a time would be too expensive. The consequences are at least twofold: firstly, the store does not get instant clearance of the money due. Secondly, a cardholder who carries out a large number of small transactions may appear to the system to have exceeded his or her credit limit. This occurs because, often, the card company's security systems will count each valindation transaction as being set to a nominal maximum figure allowed to the card and the retailer (e.g. £100). In this scenario, eleven validations against a card rated at £1,000 credit limit would be rejected as 'over the limit', even though the total cost for the day was considerably less than £1,000.

It is generally believed that replacing the telephone connection with an Internet one would significantly reduce the transaction cost and might make it possible to clear every individual transaction in real time.

3.3   ELECTRONIC CASH

Most financial transactions comprise amounts less than a few pounds; many are even less than a pound. Any such transaction that involves handling by a bank is going to incur an unacceptable surcharge. Just in the same way that you would not write out a cheque for a pound and would use cash instead, remote payments of this type require *electronic cash* rather than credit transfer. With 'real' money, you carry about coins and notes that you have previously withdrawn in a single transaction from your bank, or have been given by a third party. Similarly, you load electronic cash into an *electronic wallet*, at some stage prior to paying some of it out when you buy something.

One of the simplest forms of eCash is the *pre-payment card*, for example, a telephone card. The early versions of this were purely mechanical: the card had a metallised strip that was progressively eroded by the mechanism in the payphone, thus lowering its value each time it was used. Today cards can be more sophisticated, but the principle is the same: customers make *pre-payments* to the card issuer or draw down funds from their accounts into the cards. Once in a card, the money behaves like cash – it can be lost and the loss falls to the cardholder. It is essentially untraceable and can be spent without any third party, including one's bank knowing anything about it. Actually, the anonymity part is controversial; some authorities see this as way of making it easy to carry out crime, payment for drugs or other criminal activity. There have been some calls for the cash to be less than completely anonymous and some implementations do indeed label the cash tokens with a unique reference. Political and social arguments regarding the desirability or otherwise of this anonymity are beyond the scope of this book, but, clearly, are of some significance to its acceptability [102].

One difference from cash is the degree of security attached. eCash is usually held in an encrypted form in the electronic wallet and it requires the owner to enter a secret password in order to open the wallet and effect a withdrawal. Microsoft offer an electronic wallet of this type on their newer browsers. The Mondex service [103] offers a small, portable wallet.

One interesting development is the realisation that the cash itself does not need to reside in a user's wallet, if you only want to deal on-line, only the unlocking mechanism does. You can simply purchase a supply of pre-payment 'money' in the form of an account with an eCash vendor, and make payments from this cash by releasing it from any terminal, using a password which you key in as an encrypted message.

3.4   WEB MICROPAYMENT STANDARDISATION

As reported by *WWW.org*, there is no fully standardised approach to micropayment for goods or services provided on the Web. They have recently issued a public draft [104] which tries to lay down guidelines for such an approach. The intention is to provide a description of a *Per Fee Link Handling* process which will make it easy for developers to mark-up Web pages in a standard manner and provide users of these pages with a simple 'click to pay' mechanism. The principle is outlined in Figure 3.1.

The Per Fee Link Handler resides on the user's client machine. It is intended to communicate with the merchant's server, using HTTP proto-col. (Strictly, the secure variant of SSL/HTTPS is recommended, we discuss this further in the next section.) The messages which pass between the client and the server must be compatible with existing browsers. In these messages, a number of mandatory, recommended and optional fields are specified. These include:

- The on-line location of the merchant's site (mandatory).
- The merchant's name (optional).

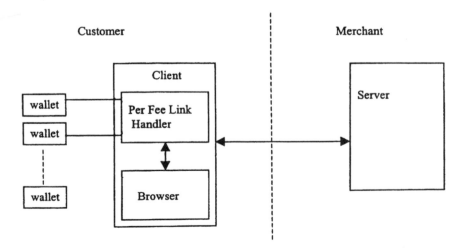

Figure 3.1   Principles of W3C micropayment mechanism

- An on-line reference to what is being bought (mandatory).
- (short)textual description of what is being bought (mandatory).
- (long) description (recommended).
- Image/graphical description (optional).
- Price (mandatory).
- Persistence in time of the selling site (recommended).

(There are other, less obvious terms. Refer to the site for more details.)

The W3C activities recommend ways of embedding this information into existing HTML browsers, using a simple 'plug-in' mechanism, or as a JAVA Applet. (Note that the latter course is not recommended for HTML version 4.0 and newer, where the OBJECT element is the preferred solution. Again, see the full text for details.) The emerging use of XML (see page 179) is also covered. The group have also recommended designs that make it clear to users that they are following a *fee link*, for example by displaying a link title of the form *'I wish to purchase it'*.

3.5   ELECTRONIC TRANSFER OF CREDIT/DEBIT CARD DETAILS

Although we are now going to discuss new methods for taking higher value payments, we have to recognise that, today, none of them have really taken off, compared with the simple mechanism of typing in one's credit card details. Many people do appear to believe that the convenience of doing so outweighs the level of risk. Of course, surveys also indicate that there are a large number who do not shop on-line precisely because they do not trust the system.

One way of lessening the risk of someone stealing your card details and reusing them later for fraudulent purchase, is to encrypt the card transaction using, for example the *secure sockets layer (SSL)* method, whose technical details are described on page 266. As shown in Table 3.1, the operational principles of SSL for financial transactions are quite straightforward: the eCommerce server issues the customer's client machine with a public key, whose ownership is guaranteed by a trusted third-party, and which the client uses to encode the card details.

This at least protects the link between the customer and the vendor; what happens thereafter may be another thing. The customer's card details can be decrypted by the merchant and thereafter used as the merchant sees fit. A number of cases have occurred of merchants storing the card details in an unencrypted form in a database which has subsequently been hacked into.

Notice also that the onus is on the customer to decide either to trust that the merchant's key is genuine, or to invoke a procedure which checks out the validity of the digital certificate. There *should* be no problem with this

Table 3.1 Using SSL to protect card details

Participant	Stages in transaction			
	Prior to transaction	Intent to purchase	Provide secure method	Send encrypted card details
Trusted certificate authority	Issues a certificate and public and private key to merchant			
Customer		Notifies merchant of intent to purchase		
Merchant			Sends customer (certified) encryption method (public key)	
Customer				Uses public key to encrypt card details and contacts merchant

process, but it is open to possible misuse. Nevertheless, SSL is quite widely used as a means of protecting transactions as it is built into Web browsers as a standard and easy to use feature.

One of the problems with all the proposed alternatives, is the need for some form of validation token (for example, a smart card) that can be connected to the network. This would usually require the purchase of additional equipment, if used for home shopping, and there would need to be a pre-established standard for its connection.

Purchasing transactions obviously require to be protected against basic fraudulent attack, for example, theft of PIN and its reuse. But there are other things that need to be protected too: how you spend your own money ought to be something that you can do without everyone else knowing. From a list of your transactions, a competitor might be able to guess at your product strategy; knowledge of your share dealings might give someone some insider dealing; the state of your health might be known by an employer from the treatment you purchase; information on purchasing details may enable a more elaborate fraud to be perpetrated, and so on. So, only those parts of the transaction that are required in order to effect it, should be revealed to the person with whom you transact. A credit agency that handles the credit, should not necessarily know what you have bought, but simply enough information to know that a payment transaction is valid.

3.6 SET – SECURE ELECTRONIC TRANSACTIONS

A mechanism has been defined by card providers VISA and MasterCard, under the name *Secure Electronic Transactions (SET)* [105], which addresses these more complex aspects of transactions. It is intended to become an open standard, although progress in this direction has been, so far, slow. It is based on a chain of digital certificates that are provided to cardholders, merchants and acquirers by their sponsoring organisations. The issuing of the digital certificates that are required by the process is highly hierarchical, requiring a top-level *SET Root Certification Authority.* The mechanism is a conventional version of the security processes we have described earlier, with message data being encrypted using a randomly generated key that is further encrypted using the recipient's public key. This is referred to as the 'digital envelope' of the message and is sent to the recipient with the encrypted message. The recipient decrypts the digital envelope using a private key and then uses the symmetric key to unlock the original message. Users of SET effectively encrypt their credit card details in a manner than cannot be unlocked by the merchant, only by the card-processing centre.

3.7 DETAILS OF A SET PAYMENT

Prior to a SET payment operation, a number of steps first have to be taken: the client machine (belonging to the purchaser) must be equipped with a SET wallet, which contains a certified *card-holder certificate* with an issuing organisation. Similarly, the vendor (merchant) has a digital certificate and account with an acquiring bank, who will settle up later with the purchaser's bank.

1 When the time to pay is reached, the merchant server sends a message to the purchasing client, containing details of goods and a request to the client's SET wallet. This message has been digitally signed, using the merchant's private key.
2 The client can check that it has not been tampered with, by decoding it with the merchant's public key. Included in the message are certificates for the merchant and for the payment gateway operated by the

acquiring bank. These can also be checked for authenticity by decoding with public keys.

3   Now the purchasing client creates an encoded message describing the order details and payment instructions. These two parts are encoded separately, making it possible for the purchaser to hide payment details from the merchant. The client signs the message and sends it to the merchant.

4   The merchant checks the authenticity of the message, decodes the order details and sends the (still encrypted) payment instruction to the payment gateway.

5   The gateway validates this payment request, including authorisation from the customer's issuing bank, and then sends a token to the merchant, which can be redeemed for payment from the issuing bank.

6   The merchant creates a digital receipt and sends it to the purchaser.

3.8   ELECTRONIC TOKEN HOLDERS

Earlier we mentioned *electronic wallets* as a mechanism for holding electronic cash. Although we referred to the wallet residing on a PC, the concept can be extended much further. The wallet can be a smart card, a personal organiser or even a mobile telephone, and device can be used for credit as well as cash transfer, and also the purchase and holding of non-monetary tokens such as tickets and other permits. The advantage of using these devices is obvious – portability.

The simplest form of wallet is probably the smart card, which we already discussed in Part 3, *Security*. Cards can hold personal keys and other permanent data and can be loaded with temporary tokens, e.g. electronic cash. As well as the card, three other prerequisites are probably required: a terminal for loading the tokens, a terminal for extracting them and probably some way of reading the contents to see whether they need replenishing.

One early example of this is the *Mondex stored-value card*. This comprised a credit-card sized smart card which held the tokens, a small reader which clipped over the card and allowed the contents to be read (with password protection), a larger, pocket-diary-sized wallet, which allowed for more complex interrogation and programming, and a variety of loading and unloading terminals. The last included special purpose countertop units for use by shopkeepers, and which could be connected by dial-up telephony or via X25 packet services, for periodic download of the cash to the merchant's bank. For Mondex trials, telephones using *CLASS* signalling (see page 323) were also modified to provide a card-slot, into which the Mondex card was plugged, and similar modifications were provided on a number of payphones. The Mondex card is not just a memory device with encryption; it actually includes a single chip micro-processor and an operating system, *MULTOS*, which allows it to be loaded with a number of applications, perhaps from different vendors or service providers. MONDEX is also unusual in that the tokens are genuinely cash, in the sense that they are deducted directly from the owner's bank account as a cash withdrawal. Thereafter they are anon-ymous 'coins' which can move from buyer to seller without audit trail and, if the card is lost, are lost with it.

3.9 MOBILE TELEPHONES AS TOKEN HOLDERS

It is easy to see that the functional components of an eCommerce financial transaction can be realised in the form of a modern (e.g. GSM) mobile telephone and its associated network infrastructure: the channel is reason-ably secure (more on this later), there is a keypad and a display, proces-sing power and memory. Less obvious is the personalisation provided by a GSM Subscriber Identity Module (SIM) and the relative ease with which attachments such as smart card interfaces can be provided. Each GSM telephone contains an SIM, itself a smart card which contains access rights and other details for the user of the mobile. Note that it is associated with the user, not with the telephone. Users can remove a SIM and insert it into another telephone if they wish.

So far, experiments with using mobile telephones for banking services have moved cautiously. Most consist simply of providing balance-check-ing, the ordering of statements and viewing recent transactions, as well as payment of predetermined accounts (e.g. to utilities), essentially the same services permitted to customers who use wire-line tele-banking. However, some trials have been carried out whereby electronic cash can be loaded down into a banker's cash card connected to the mobile tele-phone. A call is made to the issuing bank and the customer uses the telephone keypad to input the required amount and a PIN code. This information is encoded by the phone, using a private key and sent over the network, to the bank. The bank decrypts and verifies the request and then sends the cash tokens, which it deducts from the customer's account.

A major issue here is the use of two smart chips, and the distribution of the application between them. In at least one of trials carried out, the mobile telephone's SIM had to be modified to be able to handle the process for a specific bank; to enable the same service for another bank would have required re-modification.

There is quite a serious issue here which is in some sense a re-run of the issue regarding client-centric and server-centric applications. Does the majority of the application (banking, obtaining tickets, etc.) reside on the smart card of the application provider (bank, ticket agency), in the mobile telephone, or on a server?

The mobile's SIM can provide transmission level security and also application security, plus a degree of authentication. The external application card can contain user rights to the application's features and authentication information. Thus cellular operator's SIM and the bank smart card offer duplicated functionality. That is, they (or rather, their issuing companies) compete.

Alternatively, would it be better to maintain rights and privileges on an application server, merely using the customer-end smart chip(s) for security? But whose application server? That of the cellular operator or the bank? These are issues which can only be resolved once practicalities and business politics have been dealt with first. It will depend, amongst other things as to whether eCash is to become important and the relative importance of interaction with banks in pure account transfer terms against the purchasing of eCash or non-monetary tokens. Networking issues will play a part: suppose one purchases in advance a right to a place a car in a car-park. On arrival, you must present this right to an unattended barrier. If the barrier is connected to a wide-area network, e.g. GSM, then there are a variety of, thin/thick, client/server strategies possible. Alternatively, if the terminal is not connected, then you must pass to it a token that you hold locally and which it can 'understand' and remove from your terminal. Thus some of the application will be in the client.

3.10 PAYMENT SERVERS

Currently, most payment schemes are based on taking credit card details, although most of the principles involved can be carried through to other schemes. A very simple approach is for the merchant to simply collect the payment details in its own database (hopefully, in encrypted form) and carry out negotiation with the payment agency, off-line, either re-keying information or by electronic data transfer. A more advanced approach is to transfer the payment details automatically to the authorising agency almost immediately. This involves providing an *on-line processing gateway* or *payment gateway*, which transfers details of the credit card and the payment details and sends them, via a secure link, to the bank. Notice that this is to the merchant's ('acquiring') bank, which has arranged terms and conditions with the merchant, principally regarding the size of the transaction that can be accepted without checking. Systems such as I/ Net's Merchant/400 [106] enable a response from the card processing centre 'usually within one minute'. The acquiring bank can also use such gateways to upload into the merchant's system, a list of 'hot' cards, or other security details, that can be used by the merchant's processes to screen transactions before accepting the order. All this data traffic must pass in a secure way, either over secure lines or, increasingly, over the Internet, using security protocols such as SET.

Whatever method is chosen, the most fundamental architectural decision to be made is how directly involved will the vendors be in designing and implementing the service. One option is to buy it in from a specialist supplier. Sometimes this may be integrated into a shopping cart design, which is then installed on the vendor's server. Alternatively, vendors can place a Web link on their site that links to a payment server run by the payment-service supplier. Actioning this link delivers a form to the customer for completion. This is then sent in secure form to the payment service provider. For larger vendors or those requiring to carry out additional control on the payment, it is also possible to purchase software which runs on the merchant's server and operates as a secure client to the payment service provider's site. See for example the services offered by [75].

With the continuing uncertainty over the ultimate 'winners' in the electronic payment race, it would not be unnatural to expect that the vendors of payment servers are hedging their bets. Payment servers and gateways are being designed to handle a variety of protocol options.

3.11   PUBLIC KEY INFRASTRUCTURE

It can be seen that the majority of payment mechanisms described above are based on a *public key infrastructure (PKI)* being in place. This is slowly becoming a reality, with some emphasis on 'slowly'. There have been some interoperability problems with legacy systems and there is a real issue with how to deal with ensuring digital certificates can be revoked promptly but without incurring too much bureaucracy with legitimate users. There is also some concern about the shear difficulty of the technology. However, progress on all fronts is being made and, by and large, the industry is beginning to look optimistically at the wide-scale adoption of PKI within a very few years. [107].

3.12   THE GENERAL OUTLOOK FOR PAYMENT MECHANISMS

Although we have discussed a number of payment mechanisms and associated technologies, it should be clear that there is no obvious winner. Perhaps we should not expect there to be any one single solution. Perhaps, the process of parting with our money is so sensitive that we will take time before we unconditionally settle on a preferred way. Maybe there exists already a satisfactory solution – the credit card – and all we need is to be assured that on-line companies can be trusted to use its details in a sensible and secure manner. This, more than anything, may be the deciding factor in the ultimate choice: the performance of a technical solution will be judged on the operational surroundings rather than on its own merits. Several times in the recent past there have been security breaches involving credit details, which have compromised because of frailties and failures of process rather than technology. On far too many occasions have credit card details and accounts been susceptible to hacking simply because of bad system design or operation, rather than inherent weaknesses in encryption. We make no apology for yet again warning about this. If the level of professionalism in this area does not improve, then eBusiness will receive a significant setback which will have a major impact on revenues and profits. There is a real possibility of some companies going bust.

Part 4: Service, Supply and Marketing

Chapter 1: Service and Support

We explore how comprehensive customer service can be provided by the integration of voice, Web, eMail and paper sources. In particular, call centre architectures are analysed and their probable evolution discussed. Automated alternatives to human agents are considered. Electronic support for help-desks, maintenance and repair crews is detailed, including the potential of telemetry as a way of removing the need for site visits.

Chapter 2: Supply Chain Management

An integrated supply chain, 'from melon to customer' is widely seen as a key differentiator for successful enterprises. Traditional Enterprise Resource Planning (ERP) is discussed, particularly in relation to an on-line environment. The demanding nature of logistics is discussed with respect to exception handling and the robustness required from processes and hardware. Warehouse radio LANS, bar coding and radio picking systems, fleet management and scheduling schemes are explained. The potential for automated purchasing, using intelligent agent software is surveyed, including the need for standard data descriptions and security issues.

Chapter 3: Electronic Marketing

Although electronic commerce opens up a new channel to customers, it can be fitted into the traditional models of marketing. However, there are unfamiliar aspects that need to be critically appraised. On-line methods of branding, attracting and retaining customers and market research are considered. Issues of advertising, getting listed on search engines, and other ways of gaining footfall in cyberspace are examined. Creating a corporate portal and using its customer-facing nature to drive organisational thinking is explained. A variety of push-services for proactive marketing are considered.

1

Service and Support

1.1 PROVIDING SERVICE

We have said repeatedly that, in many ways, the simplest part of eBusiness is the creation of the catalogue and the taking of customer orders via an on-line form. Even fulfilment is not always a problem, if the goods are tightly specified and well understood by the customer. But most business is not like that: customers often have to make further enquiries about product specification, perhaps talking to an expert, and there are often after-sales queries to be dealt with. Some products go faulty after installation or have to be maintained on site. Even at the very beginning of the sales cycle, it is often not sufficient to wait until people come to your real or virtual store of their own accord. 'Financial services, ' (it is said), 'are sold, not bought'. Proactive, outbound calling is often required. Often customers prefer to deal with a human being rather than a machine. With frictionless global trading becoming available on-line, products themselves become differentiated only on price and availability, but one of the things that is most difficult to commoditise in this way, is service. Provision of all the supplementary activities described above becomes not just an ancillary to product selling, but a critical differentiator. Drawing the attention of the customer to new products, providing advice at point of purchase, and after-sales support, seven days a week, 24 hours/day, will become increasingly important. Doing so will cost considerable amounts of money and distributed automation will be key to keeping down those costs.

1.2 CALL CENTRES

One of the biggest growth areas in new types of employment has been the *call-centre* whereby customers can telephone in to a set of lines dedicated to handling their requests and proactive selling can be achieved by out-bound calling to customers. To call this 'new', is perhaps misleading: for many decades, such services have been provided by a number of companies and, of course, the entire customer-handling services of telecoms companies themselves have been almost entirely based on a mixture of telephony and post. However, in recent years there has been a dramatic shift in the scale of these services and in their efficiency. This is overwhelmingly due to automation in public telecommunication networks and in private exchange equipment, and in the introduction and integration of computer systems into the query-handling environment. This is a rapidly changing scene, where the rules have not yet been firmly defined. A traditional approach has been through voice telephony, but the impact of the Internet is now beginning to be felt at all technology levels.

1.3 CALL ROUTING AND THE INTELLIGENT NETWORK

Starting with the telephony background, we note that telecommunications networks essentially perform two functions: *transmission*: the transporting of information across distances, and *switching*: the routing of the information, so that it does in fact go to the right place. In our studies of retail and inter-business communications, we have, so far, emphasised the former, in particular the speed of transmission of bits of information. However, at least as important for today's call-centre operations, is the capability of the network to provide flexible options for *call routing*, to meet customer patterns and organisational capabilities. One method for providing this flexibility is the so-called *intelligent network*.

Probably the most intelligent networks ever developed for telecommunications were deployed over 100 years ago, in the form of the telephone operator! Up to today and even into the foreseeable future no machine has ever been envisaged that can really replace the human operator's ability to respond to implicit, ambiguous and down-right incompetent enquiries, and this is a significant lesson for all who would automate the call centre. But human operators are very expensive – and were so even a hundred years ago. This was one of the reasons that automated dialling was introduced. In its original form, customers used a rotating dial which operated as a switch, interrupting the flow of electric current on the line. Each time the current was interrupted, the interruptions triggered a stepping switch in the telephone exchange, which moved round one place for each interruption. Dial '5': five interruptions, five movements of the switch, connecting you to a different outlet connection.

This is a fully *hardwired* relationship between the number dialled and the line to which you were connected. With the growth in the size of the telephone network and the increasing desire to make national, then trans-national and intercontinental calls, solutions were required to make this less unwieldy. At the very least, it was inconvenient to have to dial a different number, even though you wanted to call the same person, just because you were making the call from different locations.

The answer was *number translation*: the numbers dialled no longer acted directly on the switches, five interruptions invariably leading to five movements of the switch. Instead, the five interruptions were fed into a *register-translator* which had been previously set up with information which converted the '5' to the number appropriate to making the connection from the caller's exchange to the exchange of the called customer. Routing now used *stored programme control*: the telephone exchanges became like simple (but ultra-reliable and real-time, multitasking) computers.

From the beginning of the 1970s there has been an ever increasing interest in the introduction of computer processing into switching systems, in order to provide a cost effective and flexible intelligent network, because it was soon realised that simple number translation for trunk call dialling was only one of the benefits that could be expected. Once the network becomes intelligent and built to respond to standardised commands, it becomes possible to give a very wide, potentially global, set of services offering maximum flexibility to customers.

For instance, suppose a business needs to handle a large number of in-coming calls from customers. It may want to re-route these calls to a number of enquiry points right across the world, to give 24 h service and to achieve economies of scale by selectively locating specialist problem solvers. There may be a requirement for customised recorded announcements, in a range of languages selected on the basis of the call-er's location. The service may even benefit from the integration of advanced, perhaps experimental, services such as speech recognition.

What many businesses do not want is an intensive capital investment and installation programme, particularly if the virtual business is to be flexible, ever-changing and, perhaps, rapidly dissolving. They will prefer to buy in a service package. This is one major market for the Intelligent Network, which can provide all of these services. The technology to deli-ver these services is outlined in Figure 1.1.

In this design, telephones are still connected to local exchanges, some of which will still be 'unintelligent'. However, at some point in the path, the call will reach a *service switching point* which is responsible for determin-ing the route to be followed by the voice call (shown as a continuous line),

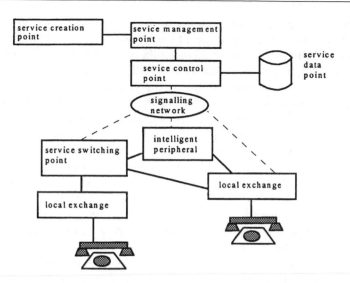

Figure 1.1 The intelligent network

under the command of signalling instructions (the broken lines). The precise nature of these instructions (which may set up different paths at different times of day, for example) is dictated by software in the *service control point* and its associated database. A simple example of its operation is the providing of number translation services: for instance, a business may supply a single number for customer inquiries, but want to handle them at different locations depending on the time of day:

- The *service switching point* will have been instructed to signal to the *service control point* when it receives any such call.
- Then the *service control point* checks the dialled number, realises that number translation is required, retrieves the translation details from the database and returns them to the *service switching point*.
- This, in turn, connects the call to the appropriate destination exchange.

The service control point is a high performance, reliable, multi-tasking computer that can handle real-time interruptions from the exchanges requesting services. Apart from providing high-speed decision-making, it also performs another important function: it acts as a buffer between the management and service creation elements and the detailed complexities of switching the calls and of network monitoring and alarms. This means that the 'application programs' that set up the various services to be performed, can be written on other computers used for service management and service creation, in high level scripts that do not require detailed understanding of the switching, etc. For example, we could imagine a programme called 'time of day routing', which simply requested the business customer to complete a table of destination address against time of day.

Also shown in the diagram, is the 'intelligent peripheral'. Callers and called parties can be connected to such a device, which could, for example, provide a network answering service. A service could be created that, after normal business hours, switched all calls to the recording machine. In the morning, the business could call into the answering machine and retrieve the calls.

The intelligent network is a major departure from traditional telephony in a number of ways:

- It replaces calls to fixed numbers by calls to flexible 'names'.
- It treats calls in two parts, in particular, a caller can be connected to an intermediate platform such as an intelligent peripheral.
- It allows flexible configuration of the network, to deliver a range of customised services.

The Intelligent Network therefore provides a capability to businesses to develop flexible working services and interfaces for their customers, on a platform that is provided by the telecoms operator and without incurring significant capital expense or uncertainty in the amount of equipment to purchase. It is however, limited in its flexibility, is very voice-oriented and with little Internet capability; it may also be more expensive in the long-run. Some organisations may therefore prefer to build their own in-house capability; most organisations will want to do at least some of it for themselves, perhaps leaving to the telecoms provider the basic time-of-day traffic management. This has led to the phenomenal growth in the *call-centre private network*.

1.4 CALL CENTRE REQUIREMENTS

Call centres are not simply places where people answer the telephone in response to customer enquiries. Their prime distinguishing requirement is that they do so in as cost-effect a manner as possible and their approach in doing so is, at every point possible, to speed up the enquiry handling and to minimise the number of human agents and their skill level. This implies extensive automation, integrated, as far as is possible, into a continuous work-flow process. What then are the various elements in the process? A simple model is given in Figure 1.2.

This shows a basic in-bound call-centre, set up to receive calls from customers, although essentially the same system could be used to conduct outbound sales. The heart of the operation is the *automatic call distribution (ACD)* system which is similar in many respects to a conventional PABX and acts as an interface between the multi-line inputs from the public

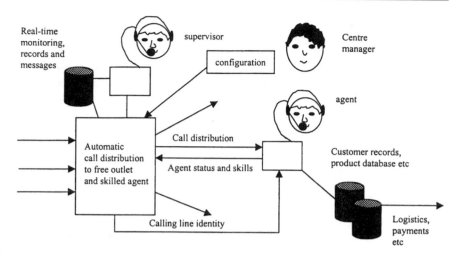

Figure 1.2 Typical call-centre functionality

telephone network and the call-centre agents. Its function is to monitor the agent positions and direct the incoming call to a free position, where there is an agent competent to handle the query. If there is no suitable agent free, it can also put the customer 'on-hold' to a recorded announcement. Some systems allow agents to log-on from any desk position, using a personal identity code, thus informing the ACD of where they are. The more sophisticated installations, with networks of ACDS installed on different sites, can also transfer the call from a fully occupied site to an alternative, via the installation's private network. They can even be transferred to other parts of the organisation, to the shipping department for example, to allow the customer to deal direct with the person in charge of handling the order.

Agents also have access to computerised resources such as customer and product database information. On systems where the information system and the ACD are well integrated, these databases can be automatically updated with the date and time of the call and caller's name and availability parameters can be passed to the logistics and supply systems. Alternatively, supply and delivery information can be retrieved from these databases and appointments booked between customer and the supplier of the product or service.

Payment details can also be taken. Currently, people appear to be willing to give credit card, etc. details over the phone to a live operator, without inquiring into what happens with this information thereafter. This does, however, pose a potential security problem, which astute companies ought to consider: what happens to these details once the operator receives them? Are they entered 'in-clear' into the database at the agent's terminal (risky), or are secure transmission protocols used to pass this information to a payment centre? If the latter, how are agents authorised (and de-authorised) to make these transactions?

Security and trustworthiness of employees is only one aspect of call-centre performance monitoring. Since, as we said, today's call-centres are designed to provide service at minimum cost, we have to introduce checks and balances into the system to make sure that the level of service is reaching acceptable targets. Again, we would like to do this in a low-cost automated manner, if possible, and some aspects of the process can be carried out in this way, by *call-centre management systems* that monitor and record system parameters such as percentage busy time, agent time-to-respond, agent call holding time, etc. Once this information has been collected, it can be analysed to provide performance against target and to plan workforce routes and demand. It can also be made available to management and to the individual agents themselves, in order to allow comparison with group performance.

Perhaps fortunately, though, it is still not possible for all aspects of agent performance to be measured soullessly by the machine: customer satisfaction and agent integrity must still be assessed by a human. Most call-centres are equipped with *silent listening* facilities, whereby supervisory staff can listen in to conversations between customer and agent without themselves being overheard. If this is deemed by some to be an invasion of privacy, then it at least has a long pedigree, for this is how manual telephone exchanges have always operated! In any case, it is difficult to see how otherwise one can avoid situations similar to that once experienced by a major telecommunications company which exhorted its directory enquiry operators to minimise the duration of the call – customers did get numbers, all right, but since it is quicker to make them up than look them up....

In some call-centres where word-of-mouth transactions need to be backed up by more permanent records, (financial services are a case in point), tape-recordings of calls are also made. The most convenient method for this is still in the form of compact cassettes. Tapes can run out and recorders can go wrong, more frequently than is convenient. There is only one highly reliable way to avoid this: fit a playback head on the downstream side of the record head and arrange to monitor some form of comparison between the input signal and the playback signal and for an alarm to ring if they vary beyond limits. A twin track recorder, with a tone recorded on the non-voice track is one easy way to do it. Measuring the signal from a down-stream playback head allows the system to detect any recording failure. Tapes of real transactions are also very convenient for training purposes and it is perfectly legal to make these, although it is generally considered courteous to inform customers that this might happen. Again, it would be wise to consider what security precautions are taken to protect against misuse of these cassette tapes which can be considered attractive items to pilfer, by some members of staff.

Call-centres do not necessarily only handle calls originated by custo-mers; many centres also handle outbound marketing. The principles are very similar to the inbound case: agents are provided with scripts and order-taking screens and so on. One difference is the ability to generate outgoing calls automatically, from information supplied by the marketing department. An agent is provided with a list of calls, which is passed out to them on the basis of the length of their current work queue. As soon as a call is terminated, another one is dialled. It is even possible for some systems to be setting up calls before any agent is free, on the expectation that one agent will be available by the time the call is answered. In this case there is the possibility of a delayed response to a customer, and traffic levels and response times must be closely monitored in order to avoid this happening.

1.5 COMPUTER-SUPPORTED TELEPHONY INTEGRATION (CTI)

Call centres have been around for a long time, in the form of internal telephone networks, controlled by a human operator or by a 'private automatic branch exchange' (PABX). Telephone extension numbers were listed in a phone directory and the PABX was configured by the company's telecom engineer to route calls to designated extensions. Until twenty years ago, this was mainly done by hardwired connections; since then, software programming of the PABX, including the provision of ACD functions, has become much more common, but ACD/PABXs even today tend to have specialised, limited functionality, more appropriate to exclusively conversational services. When a customer does get through to an enquiry agent, the latter is often supported only by paper files of instructions, or, more recently, by having access to a computer terminal on a local area network which is totally independent of the telephone and ACD.

However, in recent years there has been the recognition that the two networks – telephony and computer LAN – can be integrated to provide useful synergy. To give two very simple examples: it creates a much better impression if the enquiry agent has instant access to the customer's files and knows what purchases the customer has made, the current balance in the account and so on. The need to ask the customer for a name and then key it into a terminal can be done away with, if the caller is equipped with calling line identity and the ACD can extract this information, passing it to the system computer even before the call is answered. The agent who answers the call, can therefore be already primed with all the details necessary. The time of the call can also be automatically entered into the computer, for example to maintain fault records or for scheduling a visit.

Another application is *out-bound calling*, in a *telesales* operation. Computer analysis of sales and marketing data can lead to the creation of lists of prospects' telephone numbers, a short-list of products that each might want to buy, and a third list of sales agents, skilled in selling these kinds of products, their rotas and their extension numbers or log-on passwords. Co-operation between the ACD and the computer can lead to automatic call generation, distributed to the correct agent extensions together with a suitable *script* on the screen that they can use when the prospect answers the telephone call.

These are examples of *computer/telephony integration*, commonly known by its initials, *CTI*. Instead of building ACDs with inbuilt, specialised intelligence, the ACD is designed to consist of a simple, but reliable, switch for setting up and closing down calls, which is connected to a general purpose computer which carries out all of the intelligent call-control as well as its normal computing functions. Because the computer is a standard product, its hardware and software are considerably cheaper to develop than that of a specialised ACD and much easier to integrate with other application packages.

CTI can be achieved in a number of ways. At the bottom end, a small system can be constructed using a PC and a single telephone (Figure 1.3).

Two configurations are shown: in one, the telephone is used to interface to the network; in the other, the computer is the dominant unit, with telephony relegated to handset functions. The choice of configuration is partly dependent on the desire to minimise the disruption to any existing equipment, but it is also 'cultural', depending on the attitudes to telephony and computing within the organisation.

On a larger scale, CTI involves combining traditional ACD functions and hardware with traditional computer networks (Figure 1.4).

In Figure 1.4, there is no direct physical connection between the telephones and the workstations. Instead, functional integration is achieved by connecting the corporate LAN to the ACD via a CTI server, which can communicate switching requests from the computers to the ACD and call

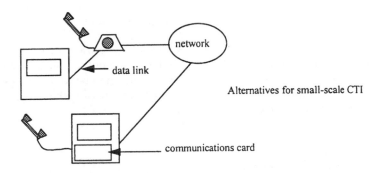

Figure 1.3 CTI-a small installation

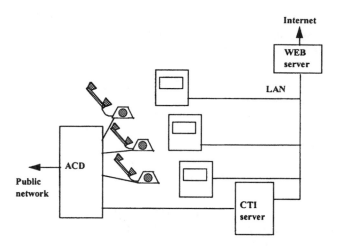

Figure 1.4 CTI-large scale installation

data from the ACD to the computers. It uses a set of communications protocols which can be proprietary but are increasingly becoming standardised. To give some idea of scale, Table 1.1 gives a specification for a typical large installation.

The intelligence embedded in CTI private networks such as these is considerable, and allows independent organisations to construct a number of features which, only a few years ago, would have been considered the domain of the 'Intelligent Networks' offered by public telephone companies. Moreover, shown in Figure 1.4, *but not always shown when the diagram is drawn by telecom engineers*, is the gateway to the alternative to the telephone network – the Internet. We need to look at how Internet (and

Table 1.1 Functions and sizing of typical large ACD

Function	Size
Public network connection capacity	$2 \times T1/E1$ (i.e. 60 simultaneous voice call connections from local exchange 30 high speed dial-up modems from public network
Private network capacity, e.g. to connect to the company's regional office, in order to pass on calls or data to functional units	$8 \times T1/E1$ (i.e. 240 simultaneous calls can be passed out to regional office (or the same number of 64 kbit/s data circuits are available)
Number of agent terminals	256
Levels of priority	64
Number of agent teams	64
Number of agent identities	2000
Agent skill levels (for prioritising who is connected)	250

corporate intranet) services are likely to modify current CTI practice, which has, so far, been driven by a predominately telecoms view.

1.6 CALL CENTRE EVOLUTION

The architecture described in the previous section is the most common today and reflects the historical need for handling customers whose normal means of access is by the voice telephony network. However, the growth in Internet traffic and on-line shopping introduces the need to expect traffic to arrive via TCP/IP networks as well as traditional telephony ones. From customers, we expect that the Internet activity will mainly be as a result of looking at Web pages or in the form of eMails. But there is also another source and destination of TCP/IP traffic that the call centre has to deal with: that which occurs on the company's internal network. Some of this is data. Increasingly, we shall also see a significant amount of Internet (or, at least, IP-based) voice traffic.

All this introduces two elements of complexity. At the network level there is a need to achieve integration between telephony-based equipment and TCP/IP; at the application level, consideration needs to be given as to how telephone calls, independent of whether made over the traditional network or by an Internet telephone, can be related to the Web application the customer is interacting with.

('Look, here's my bill. Can't you *see* how much you've charged me under this heading?')

Looking first at the issue of network level integration, we show in Figure 1.5 how this is commonly achieved.

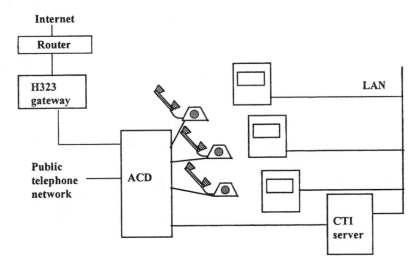

Figure 1.5 Call-centre integration of IP and PSTN telephony

Here the traditional access from the public switched telephone network ('PSTN') terminates on the ACD as normal, but the IP telephony comes across the Internet to a gateway. This gateway runs the *H.323 protocol*, which manages interworking between IP and traditional telephony [14]. Note that the IP telephony may come from individuals using modem to Internet connections (probably via an Internet service provider) but may more likely be from the internal company network or from corporate customers with direct LAN connection to the Internet. Most likely, the initial users of IP will be other departments *within* the company, who are connected to customers via the call centre.

This approach requires minimal modification to the existing call-centre design, whilst allowing the service to migrate swiftly to the handling of Internet telephony.

However, this solution does not really integrate the customer-agent interaction with the customer's use of an eCommerce application. Taking our example, above, of a customer querying a bill, it would clearly be desirable for the customer and the agent to be able to view it page by page, in synchronism. What might also be desirable would be for the agent to 'push' pages onto the customer's terminal, perhaps to lead them through a complex catalogue. In any case, the balance between voice and data is bound to shift in favour of the former, as eBusiness data becomes more readily available at every point in the supply chain. For all these reasons, the call-centre architecture moves from one which is telephony-centric to one which is Web-centred.

Figure 1.6 gives a very simplified illustration of this. In principle, the PSTN telephony is converted to IP traffic, the ACD hardware disappears and the call distribution and system management is all done via the call-centre server(s) which also handle Web applications. The agents interface with the system entirely via their workstations, which also have voice

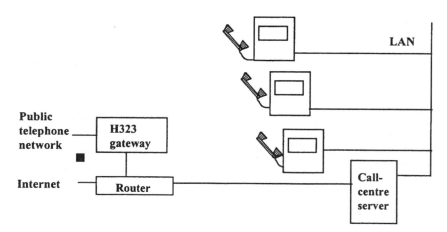

Figure 1.6 Simplified Web-centred call-centre

(and perhaps) video capability. Some have suggested that the above architecture puts into question the whole concept of a call *centre* [108]. Converting to a TCP/IP approach means that, in principle, customer handling can be delivered to anywhere in the organisation that is connected to the corporate intranet.

1.7 QUALITY OF SERVICE

In what sense is this a naïve view, or, at least, one which is futuristic rather than immediate? The principal issue, as ever with Internet technology, is quality of service. As it stands, Figure 1.6 is not likely to be able to guarantee good quality voice performance and low waiting times. What is lacking is some mechanism for prioritising traffic which is a) delay sensitive (such as voice) b) customer-generated. Current practice employs the telecommunications public and private networks which provide the organisation with guaranteed, low delay, almost always available voice networks, already installed within premises and running on already installed and proven public or private circuits between them. The new architecture as shown above introduces the uncertainty of unspecified performance across LANS on corporate sites and intranet virtual private networks, at traffic volumes often far in advance of the purely delay-insensitive data traffic that was carried before. To implement a large call-handling service, particularly one which is distributed rather than 'centred', may often require a serious overhaul of corporate networks. In an organisation which is prepared to set this up and introduce priority routing on its corporate intranet for delay sensitive traffic, then the truly distributed and integrated call-centre may be possible. Today, however, it cannot rely just on TCP/IP principles. Figure 1.7 gives one possible implementation.

This fairly ambitious layout is worth a little study. It represents a local site, perhaps one large headquarters building, which is connected to the rest of the company's premises by a high-speed Wide Area Network, making it one part of a distributed call-handling service that runs across a number of sites. Initial enquiries, received at one agent station can be fed with customer details pulled down from anywhere on the network and customers can be transferred from this agent to any other agent anywhere on the Wide Area Network. At the top, the voice calls from customers from the public network come in as part of the digital stream delivered by the corporate intranet, as does other traffic from within the corporation. On the right, the public Internet traffic and some direct telephony voice/modem traffic also come into the call-centre network. This will comprise mainly traffic from individual customers and consolidated traffic from any Internet Service Provider with whom the company does not have a quality of service agreement. All of the traffic moving around in the

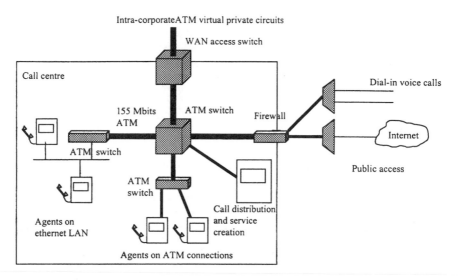

Figure 1.7 An advanced distributed call-centre architecture

diagram is, or has been converted to IP format. However, it is being connected through the network using *Asynchronous Transfer Mode (ATM)* switches rather than routers. ATM is switching technique, developed specifically to minimise end-to-end delay, whilst being able to carry relatively economically, a mixture of traditional voice/video telephony and packet (usually IP) data [3]. This allows quality of service traffic grading, so that voice calls from and to customers, for example, get priority and do not experience poor quality audio performance or network congestion.

It may not be strictly necessary to use an ATM approach, at least in a few years time. There is a great deal of activity underway to produce router protocol standards that can provide a reasonable level of quality of service and some proprietary solutions already exist. However, these are still some unresolved significant issues and given these fairly formidable requirements being demanded of IP-based platforms, we have to ask whether organisations should still continue to put their faith in traditional telecoms PBX and ACD hardware, rather than instantly abandoning it in favour of the newer architecture? This is a difficult question to answer, but it is one which clearly frightens the telecoms equipment manufacturers. For instance, Mitel [109], whilst cautioning against a rush to IP, have recently announced an 'IP Investment Protection Plan', which publicly declares that 'up to 90 percent of their current Mitel infrastructure [on certain of their core products] will be supported on an IP-based voice solutions.' The correct answer may be to build any new systems with this level of IP-readiness, but not, at this exact moment employ any too-radical solution that might screw-up this critical area of customer handling.

1.8   CTI DEVELOPMENT AND IMPLEMENTATION TOOLS

One of the main developments in call-centre design, indeed, of any installation requiring that telephony and computing be integrated, has been a
move over to standardised interfaces [110]. Until a few years ago, PABXs
and ACDs, even if they did have computing interfaces, all used their own
proprietary specification. This has been gradually eroded by the work of
the computer manufacturers and, in particular, one of their standardisation bodies, European Computer Manufacturers Association (ECMA),
which realised that promoting an open standard would be beneficial to
their industry (and incidentally perhaps adversely affecting that of the
PBX manufacturers). They took on board the *Telephone Server Application
Interface (TSAPI)* originally developed by Novell and AT&T, and portable
across a number of operating systems. Of course in the computer world,
not everyone agrees to this single 'standard': Microsoft offer an alternative
in the form of *TAPI*, which is centred on a PC architecture rather than
TSAPI's server bias and integrates Microsoft's ActiveX components. The
original positioning of the two alternatives was clear: TAPI was intended
for the smaller end of the call-centre market, with TSAPI covering the
larger, but the differences have been reduced with time. There is now
also a Java-based JTAPI environment available. As the simple, basic
call-centre requirement extends to encompass more of back-end processing such as scheduled follow-ups to telephone calls in the form of fax,
eMail or even physical mail envelop-stuffing, the range of customised
design will increase and more flexible tools will be required. Both ActiveX
and JTAPI are object-based and a large number of CTI objects have been
specified in class libraries which make it easier for designers to put
together a bespoke application. Many of these have been further integrated into toolkits which allow non-experts to construct their own application using menus.

1.9   RELIABILITY

What must not be neglected in the considering the setting-up of a call-
centre is the issue of reliability. After all, one is creating the company's
most critical interface with the customer. In some cases, especially
perhaps with support/complaints desks, one may be dealing with customers who may not be in the best frame of mind to tolerate further failure.
   Telephone networks and products have a deserved reputation for reliability; computers and computer networks do not. Solutions hosted on
standard PCs are risky, unless duplicated and on hot-standby. The telephone network routinely operates with levels of reliability only found in
the most expensive *fault-tolerant computers* used for mission-critical tasks

such as air traffic control, achieving 99.99% availability and correct operation. Although PC reliability is frequently quoted at 99% or higher [17], translating this into downtime reveals a salutary figure of nearly 90 h of service failure per year.

Without going to the full extent of providing fully fault-tolerant machines, however, it is possible to achieve outage levels of below 10 h/year by employing machines with some simple ruggedised and duplicated components. These are usually rack-mounted, which means that firmer connections can be made on the back planes and, more mundanely, they are less likely to be left on the floor of the call-centre where they are vulnerable to accidental (and, sometimes, non accidental!) kicks and knocks. Expansion cards are also properly connected to the main processing boards via reliable, passive back-plane wiring rather than by being plugged directly into the motherboard, as happens with most basic PCs.

Providing reliable source of power to the system is an obvious need. It should be possible to plug-in a replacement power supply and remove a faulty one from a system without interrupting service. It is preferable to have system which automatically switches over to a new power supply when the current one fails. Do not do not forget to have an alarm go off to let you know this has happened, or you may inadvertently run the system on one supply in ignorance of the failure of the other. Check periodically to see that the alarm is working! *Uninterruptable power supplies (UPS)* should also be used, in case of failure of the public supply, but again, beware: the author has known these to fail themselves, causing loss of service.

One of the least reliable components in integrated electronic systems is the cooling fan. Systems that incorporate fans presumably require them for a purpose, which may not be clear for some time after the fan has failed. So, fans should also be fitted with alarms and checked and replaced promptly.

1.10 LAYOUT AND WORKING CONDITIONS

The offices where the agents are located also require good management. Data and mains wiring should be neatly conveyed to the PCs, so that cables do not get entangled or strained. Since over 80% of call-centre costs may be in salaries [111], the agents themselves should be treated with consideration: full-day use of computers can be mentally and physically stressful. Apart from the morality issues surrounding employment conditions, call centre operators should give thought to the economic future of their business: it is likely that human operators will process less and less routine traffic and will increasingly be called upon to handle the more complex tasks. That is, operators will require to be good and to employ a higher degree of tacit knowledge. Paradoxically, human performance may become the principal differentiation factor in the electronic support business of the future. It would clearly be sensible to treat them well, to avoid expensive re-training costs, in the event of them handing in notice. Furniture, air-conditioning and lighting should be appropriate to the task. Training and team building may also become increasingly important. One advantage of modern, programmable call-centre equipment is that it can be configured to run with dummy traffic or with selected, simpler query handling tasks, and temporary teams can easily be constructed and trained to work together.

1.11   INTEGRATING PAPER AND eMAIL INTO CUSTOMER SERVICE

There is little point in providing access for customers to your help-desk or order department if the responding agents have nothing to say, or if what they say is nonsense. Agents must not only have access to obvious items that are likely to be queried, such as bills and delayed orders, they will increasingly have to be fully integrated into the entire business process and their jobs supported by *workflow* systems which mirror the organisational working practices – and those of customers. You may have integrated your telephony and Web forms ordering processes, by using the call centre architecture described above, but what about the letter of complaint to the CEO, from the important customer? Paper-based systems are going to be around for a long time and, perhaps will still represent the most important communications between customer and company. (The on-line writ is probably a long way off!) They may become the exception-handling medium of choice for people with a complaint or urgent request that has not been satisfied electronically. People will therefore expect their written correspondence to receive at least the same level of priority as eCommunication. Document-scanning may even become more necessary than before, with the scanned document 'travelling' through the response process accompanied with electronic annotation until it is ultimately archived with any further correspondence from the customer and images of written responses by the company.

Another major paper-based activity is in *loyalty cards* and *product coupons*. Many companies make heavy use of these to encourage repeat purchases by customers. In many cases, customers have to fill in their personal details in their own handwriting and perhaps tick a box for a particular offer. This information is extremely valuable for loyalty or affinity marketing schemes, but once the coupons are collected, they must be entered into the system. In the past, this was done manually, but recently high performance scanning and character-recognition software has been developed that can process the information with low error rates. In one such system, it is reported that 15,000 such cards are processed in a single day [112]. This combined clever algorithms for character recognition with post-code data to produce error rates of the order of 4% or less.

What about eMails? Even if the latter can be captured automatically by the Web server and, as is the emerging common practice, answered by call-centre agents during lulls in voice traffic, are they the right people to do it? A good 'voice' manner does not guarantee the ability to write. Perhaps it is better to pass this information on to the line unit responsible, but it is necessary to tie them into a workflow task that makes sure the job is done within a specified time limit. Thus when the operational unit responds to the customer, this is flagged up to the call centre agent so that they can make a follow-up call, also having on-line access to the customer eMail and the response.

Although eMails do look like conversational activities that occur at only one point in time, rather like telephone calls, their electronic nature allows them to be treated in a rather more procedural and permanent manner. Several corporate knowledge management systems, such as are discussed in Part 3, *Managing eBusiness Knowledge*, specifically include eMails as just one more class of documents to be managed. In this way, they can be indexed just as any other document can be, and used, for example, for customer studies, product fault investigations an so on. The strategy for eMails, in general, should be to try to handle them with as little cost as possible, whilst treating them as requests for human response and squeezing as much useful information out of them as one can. After all, they represent primary contact with customers and exist in electronic, machine-processable format.

1.12 PRE-DELIVERY MAIL PROCESSING

One quite recent refinement to integrated mail processes is the *pre-delivery agent* [113]. This can be conveniently used to convert the rather person-oriented eMail into one which fits into a process. Specifically, mails addressed to a person, for example, can be converted into those that can be handled by a duty, comprising a number of people.

To understand how this is achieved, it is first necessary to understand a little about how eMail systems operate. Almost invariably, they consist of a mail router and a set of mail boxes. The mail router resides on a mail server and is principally concerned with proper input and output of mail from other servers. In the case of Internet eMailing, it uses the *Simple Mail Transfer Protocol (STMP)*. In order to facilitate the queuing of messages, the conventional solution is periodically to transfer incoming mail from the mail router into local mail boxes, using, for example, the *POP3* Protocol.

Traditional eMail management provided *post-delivery agents* which were associated with each eMail recipient's mailbox (Figure 1.8).

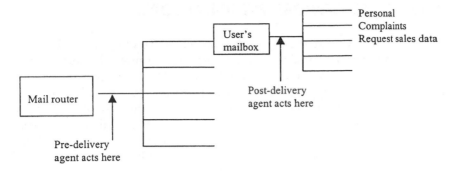

Figure 1.8   Pre- and post-delivery filtering of eMails

These agents could be configured to file automatically the incoming mails into folders that the user had set-up. The filling could be done, for example, by sender's address, subject line, and so on. The agents operated on the eMails after they had been delivered to the user's in-box. That is, storage space in the in-box was used up, even if the user wanted to refer the mail on to someone else. This can be a problem if large attachments are provided. It also meant that users were alerted every time a message came into the in-box. This can be a nuisance in a true process-based operation, where it is desirable to keep a user working at handling mail within the files at a pre-set order. ('Work on the urgent box until it is empty'.) Also, process-based activities are usually assigned on a role, rather than personal, basis. Consequently mail should be routed on role rather than individual, but this is difficult to achieve through a post-delivery agent. Carrying out this kind of activity is enhanced by moving some of the agent functions to *pre-delivery agents* that operate on the mail router, rather than on the individual's mailbox.

These agents can be programmed to directly transfer eMailed problem reports direct to the problem-reports files of a number of users within the organisation, without going via individual in-boxes. In this way it is possible to arrange to transfer mail to a 'duty' address, rather than that of an individual attendant. Mail gets handled as part of a collaborative process, rather than as a task which is wholly the responsibility of one person. Other tasks can include automated junking of *spam* messages, on the basis of sender address, the transferring of large attachments to central files and the sending of a notification to interested parties that the attachments had been so treated. With pre-delivery agents in place, it is also possible make better use of 'intelligent' software that can attempt to deduce whether an eMail is a complaint, a purchase request, etc. and then direct it to the most appropriate person or persons within the customer-handling operation.

1.13   CALL-CENTRES WITHOUT PEOPLE

Given that today call-centre growth is an established fact and call centres have been hailed as a major source of new jobs, it might be seen as irrational to claim that there may be trouble ahead. However, even employing all the automated support that we have described above, call-centres are still a very expensive part of a company's costs and one where they will be looking for continued operational improvements. In fact, there is a potential for cost *increase*: centres are not now just asked to process telephone enquiries; they also receive faxes and, increasingly, eMails. Traditional call-centre equipment can be up-graded to log these new enquires and, in the case of eMail, distribute these electronically to answering agents, but all at increased cost and complexity. Elsewhere [3], the author has pointed out that the wages paid to basic call-centre staff (in some cases not much more than £5/hour) tend to indicate that the skill levels are unlikely to be high and it is fairly confidently predicted that these jobs will, in the not-too-distant future, be replaced by machines. One answer appears to be in moving towards fully automated handling of simple queries.

Figure 1.9 explains the principle: the cost to handle queries entirely manually varies with complexity along the line A–B. Today, automated support given by ACDs, etc. has meant that people can handle the calls at a generally faster rate than before, reducing costs to line C–D. However, the cost of queries is still related to someone's time to answer them. Even relatively simple queries are labour intensive and complex ones more so. Suppose, instead, it were possible to leave the customer to interact with a

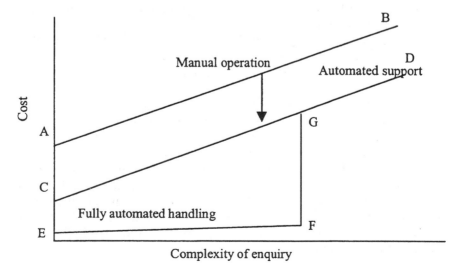

Figure 1.9   Relative cost of manual versus automatic processing of enquiries

computer, at least in the case of simpler queries. In general, the basic cost is lower and the marginal increase with complexity not so severe (line E–F). There may be a need to provide slightly more resources to meet the longer customer sessions involved and connection charges may be an added cost but basic processing costs are essentially independent of complexity. Of course, there does come a level of complexity which is too great for fully automatic handling and, at this point, we need to re-introduce a live agent, (F–G–D) before the customer loses patience.

If we can indeed replace most of our agents with automated response, then we can radically reduce our call-centre costs. Just how practical is it? Let us now look at how this might be done.

1.14 AUTOMATED ENQUIRY-HANDLING SYSTEMS

Simple alternatives are already in-use for handling in-bound telephone enquiries. The simplest way is to segregate calls on the basis of the number dialled: we have seen that the Intelligent Network architecture has freed the digits of the telephone number from any rigid destination. Instead, we can use the number to access a specific service – a recorded announcement, for example. This functionality can be provided either via the Intelligent Network facilities of the public telephone network or on a modern, *direct dial-in (DDI)* PABX which can be configured to provide a different announcement on each number. A variation of this principle is used for information retrieval via fax machines.

These basic services can easily be extended where *calling line identity (CLI)* is available. A computer-controlled response system which processes the CLI can customise a reply according to the CLI customer's profile.

A further extension, employing telephones with multi-tone dialling is also in common use: voice announcements inform the caller as to which numbers to press in order to be connected to the appropriate service. It is even possible, with appropriate recorded announcements, to get custo-mers to enter product codes and quantities, as well as credit-card details. Good dialogue design is required, as is the provision of the ability to delete and re-key erroneously sent or received data. As mentioned on page 46, multi-tone dialling is not available in all countries. Moreover, it is sensitive to telephone line quality and the tolerances permitted on tone levels and purity vary from country to country. (Perhaps predictably, US systems are keen to recognise tones even when there are none or different ones there; European networks are predisposed to treat legitimate tones with suspicion, if not delivered in a high quality manner!) Notwithstand-ing the basic, if not primitive, nature of these methods, it is possible to use them very effectively and at low cost as an integrated part of ones end-to-end eBusiness strategy.

1.15   AUTOMATED VOICE RESPONSE FOR ORDER-TAKING

Because of its virtual ubiquity, the telephone, increasingly the mobile telephone, offers the potential for other, more ambitious services. One of these is *automated voice recognition*. Claims have been made for many years regarding the ability of computers to recognise speech and there are a number of profitable products on the market that can do just that. Unfortunately, these claims can be rather misleading. There are usually severe constraints surrounding systems that claim 'near-perfect' results: they may operate only with one speaker, on whom they have been trained, their recognition vocabulary may be to within a very restricted vocabulary, words must be spoken carefully and in isolation, and so on. Performance of voice recognition systems is improving, but it would be extremely risky to base an order-taking system for use by the public in general, that was intended to handle unconstrained queries, in arbitrary vocabulary and accent, on a totally automated system.

A partial answer is to use a process that constrains the likely set of utterances that a customer is likely to make at any stage in the conversation. This is known as *dialogue construction* or *design*. With a skilfully designed dialogue, it is possible to use a relatively simple and error-prone speech recogniser but still achieve high overall accuracy. Consider the process of taking an order for some goods, in the case where we have a set of registered customers whose names and account codes are known. Figure 1.10 describes such a system.

The speech recogniser has been trained to recognise the numbers 0 to 9, 'yes' and 'no' and the spoken names of the customers who have an account. (The last can be done by arranging for them to speak their names over the line when they register for the service.)

After the user has responded to the machine's request for a name, the system then tries to identify the name by comparing the spoken name with the examples it has already stored. It forms some numerical measure of 'closeness' between the spoken word and these examples and retains the values for the top few. It then asks the next question, 'Account number?'. Again it calculates the best few guesses and their closeness to the spoken string of numbers. It can then look at the joint closeness of name and account number and provide a composite value that will allow it to make a good judgement.

Even then, there is the possibility that it is wrong, and it speaks its decision back to the customer. The customer can confirm ('Yes') or refute ('No') the decision and the process can be repeated, if necessary. In the event of complete failure – which will happen sometimes, with some speakers – the system can simply pass these occasional failures to one of a very small number of human 'exception handling' operators. Note

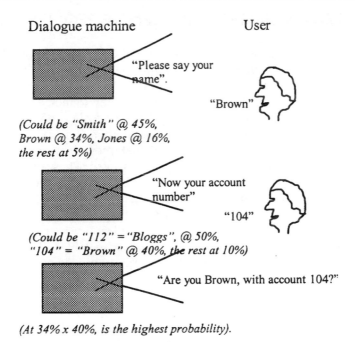

Dialogue machine   User

"Please say your
name".

"Brown"

(Could be "Smith" @ 45%,
Brown @ 34%, Jones @ 16%,
the rest at 5%)

"Now your account
number"

"104"

(Could be "112" = "Bloggs", @ 50%,
"104" = "Brown" @ 40%, the rest at 10%)

"Are you Brown, with account 104?"

(At 34% x 40%, is the highest probability).

Figure 1.10   Speech recognition for automated order taking

that the dialogue design can be constructed independently of the speech recognition algorithm used to measure 'closeness'. This means that it is possible to incorporate any advances in algorithms without needing to change the dialogue.

1.16   THE MARKET FOR VOICE RESPONSE

It has to be said that the take-off of automated voice response has been rather slow, if indeed 'take-off' is the right word. The problem has been that, given the limited performance (quite high error rate, limited vocabulary), there are often alternative solutions available. However, performance is improving, gradually rather than spectacularly, and it also might be reasonable to speculate that a new market for voice response might arise in the area of mobile terminals. The message capability that can be displayed on a typical mobile device screen is rather constrained because the screen is very small and sometimes only able to handle a few lines of text and primitive graphics; users often have to respond via primitive keyboard and control buttons. Perhaps these communication limitations might be ameliorated if the server could also understand spoken commands? There are problems with noisy background and with the limited voice encoding, but it may be possible to overcome them, in time.

1.17 REMOVING THE CALL-CENTRE REQUIREMENT

The automated response systems described above are ingenious ways of using machines to mimic human call-centre behaviour. They can require signal recognition technology that is quite complex and they are not always successful. Perhaps this is not the best way to tackle the problem of providing automated query handling. There may be a way which is at the same time simpler *and* more useful to customers. One possibility is to structure the customer interactions so that customers themselves discover the solutions to their queries and can, perhaps, even act as solution providers to others. Much of this can be achieved by a simple extension to the online shopping catalogue. Behind the pretty pictures of the products should lie a well-structured database and a help interface which puts a series of structured queries to it. Simple pull-down menus offering a range of prices can initiate queries on database tables. These can be combined with other parameters, starting place and destination, dates, etc. to provide a very well-focussed selection. This is already widely used in the cut-price air-ticketing market in the UK, to the extent that it is severely frightening the independent travel agent industry. Perhaps, just as business process re-engineering, EDI, etc. were as useful as formal methods for identifying how businesses operated as they were as implementations, the recasting of ones sales channel in terms of a Web site, might give organisations a great deal to think about!

There is no need to restrict interactions of this nature to selling. Customers could, for instance, be provided with access to on-line manuals for operation or repair of equipment. For many years, automated support systems have been available for fault diagnosis on electronic equipment. Some are designed using 'expert system shell' tools, whereby it is relatively easy to construct logic rules and humanly understandable messages of the 'IF mains switch is on AND red light is not lit THEN check fuse' variety. Some organisations provide these systems for their help-desk operators and it is only a matter of time before many more will be available on-line, without human operator mediation. In many ways, an on-line form with limited input fields, which forces the dialogue to follow a logical path, might be better than a live operator, who cannot block off unprogrammed statements from the customer.

Customer: 'There are sparks coming out the back and there is no picture'

Agent: (who has nothing about 'sparks' in their database but does have pre-determined set of scripts to talk through): 'Have you checked the fuse?'

Construction of these scripts does require some skill. They have to be driven from a phenomenological (how the product behaves) rather than from a functional (how it does it) approach. On-line simulation for faulting and servicing can be provided, with graphical models of the product.

Interestingly, these help facilities might be best done by the manufacturer of the product, rather than by the retailer, and might represent differentiated added value on the former's behalf. It might also be better for manufactures to handle maintenance, with a direct link from the retailer's Web site, for example. The consequences of ownership of the customer, and manufacturer versus retailer brand, are obvious.

Some companies, particularly those with reconfigurable products such as computers or custom cars, or those associated with sports, leisure and hobbies, might want to offer a Web site where the customers themselves can discuss issues and problems with the product. The simplest way to do this is via an eMail *user-group*, where members can mail a central point which distributes the mail onwards to all members. More advanced groups can be divided into subgroups and can archive past messages in organised files [59]. This sort of information can be invaluable to marketing departments: you have an ear on what your customers are saying about your products. It can also be scanned to identify recurring problems, leading to a posting on a *'frequently asked questions' (faq)* page. But it is probably not a good idea to be completely passive or only maintain a casual interest: eMail is an excellent way to propagate gossip about a company or its products and it is also an effective way to distribute viruses in attached documents. Professional management of brand image requires a similar level of professionalism and attention to such user-groups. Also, in creating an on-line help service, whether simply a set of static Web pages or a more elaborate search through a service centre database, it is important to consider data security. There is obviously merit in being frank and responsive to problems with your product, but remember that salespersons are frequently recruited from rival firms on the basis of the 'dirt' they can bring about their products. It could be all too easy for someone, whether an external customer or a disenchanted employee, with access to your fault reports, to mine the data for such damaging information.

1.18   PROVIDING PRODUCT MAINTENANCE AND REPAIR

Even the best products go wrong and this failure cost is often borne by the vendor. Minimisation of the cost is therefore important to the bottom line.

Although, for obvious reasons companies are reluctant to report bad news regarding their operational processes, there have been some very disappointing experiences with automated work management for repair teams. Many problems appear to be caused by naïve assumptions in oversimplifying the repair process. One has to remember that even the basic scheduling task of trying to optimise the route to be followed by a single person visiting a relatively small number of places, is mathematically a 'hard' problem, without bringing in issues of different skill levels, absenteeism, location of spares depots, etc. (Software for this is discussed on page 352.)

In scheduling the work to be carried out, an early approach was to get the technicians to download their entire next-day's work-plan into a terminal in their homes on the previous evening. This could mean equipping them with expensive PCs and modem connections, but some companies used interactive voice equipment which could be accessed using multi-frequency tone keying from a conventional telephone. In the light of experience with this approach, there has been a tendency to prefer the alternative of allocating a job only after the previous one has been completed. This allows a more dynamic approach which can cope with unforeseen events, such as one job taking longer than expected, or a customer reporting that a fault has been cleared and no longer requires attention. Obviously, in this case, the technician must have continual access to a communication channel. Freephone services can be used, if it is felt that customers will allow repair staff to use their telephones – this is probably more acceptable to business customers than domestic ones. An alternative is, of course, the mobile phone. There is clearly a market for WAP mobiles and other wireless terminals. Because of their ability to down-load and up-load data, there is also the possibility of using them not just for work scheduling but also for remote test and diagnosis. One could imagine test-gear being operated in client-server mode. Using a set of probes and, say, a voltage measuring device, a semi-skilled service operative could be guided through a test routine, with each new measurement being transmitted to the server which, in turn generates and sends a new test script to the hand-held terminal. One limitation in most customer terminals is the inability to make real connections between them and customer products. It is possible that this limitation may be removed in the future and the WAP telephone, for example, being provided with a connector to plug into a faulty piece of equipment. Alternatively, a Bluetooth radio connection (see page 39) could be used.

1.19 REMOTE FAULT DIAGNOSIS AND SERVICE METER READING

Even though today's customers regularly report faults remotely, usually by telephone, rather than calling at a store, almost invariably this is followed up by a visit from a service technician, much of whose time is taken up by on-site diagnosis, rather than repair. Clearly, this is expensive and, as we saw above, not always does the right person turn up, at the right place, with the right spare parts. A preferable option would be to carry out the fault diagnosis remotely, prior to a repair visit. For many years this has been the regular practice of telecoms companies, who have equipped their central exchanges with line testing capability that can be used by telephone operators and technicians. Many payphones are connected up to management networks to detect 'coin-drawer-full' or acts of vandalism. More recently, large computer installations are frequently tested from remote service-centres. However, it is only in the last few years that other organisations have begun to think about remote diagnosis. Vending machines, which are essentially similar in their requirements to payphones, can be connected up to their own management centres. Apart from fault reporting and vandal alarms, remote indication of out-of-stock conditions can significantly reduce service costs at the same time as improving service quality and availability. In the past, this was usually by private wire or dial-up modem calls to the centre over the PSTN, but increasingly the better option is to dial to a local Internet service provider point of presence and then operate over the Internet. Another approach is to use ISDN: the full ISDN service delivers (128 + 8) kbit/s of bidirectional data, which is certainly great enough to permit even the use of limited quality moving video, if one wanted to monitor users of a kiosk, for example. This comes at the cost of around that of two analogue telephone circuits, but has the advantage that the connection can be made in about half a second, if that is required. A more practical and lower cost service can be achieved by using only the 8 kbit/s data channel available from some ISDN providers. ISDN has at least one useful property that PSTN/Internet connections do not usually have: it is possible to address more than one device on the premises over a single telephone line without 'ringing the bell' on other devices connected to the line. This allows *silent monitoring* of equipment to be carried out without disturbing anyone in a building by generating false in-coming calls.

One application of this technique is in remote reading of service meters, for electricity, water and gas, but it can obviously extended to cover any remote monitoring or control operation. An alternative approach is to make use of *PSTN CLASS services* which provide a data service between a customer and the local exchange. The principle is shown in Figure 1.11.

CLASS signalling is provided as a service by the telecoms company. Its original purpose was mainly to provide a means for sending *calling line identity* data to a called customer's equipment. (Although this information is delivered to the called party's local exchange, there had not, prior to CLASS, been any widespread solution for carrying it the extra distance to the called customer.) CLASS is instantiated by means of a modem in the local exchange which sends a relatively slow data-rate signal (200 bits/s) using a conventional frequency-shift keyed modem. A reverse data channel, from customer to exchange, is also provided. This uses touch-tone, multi-frequency keying, for example, from the telephone keypad. Thus

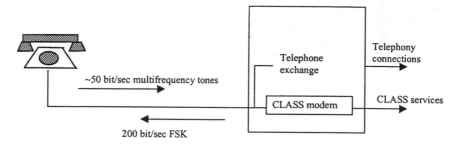

Figure 1.11   CLASS services

we have a (slow) two-way data channel between the customer and the exchange.

In the original Bell Laboratories specification for CLASS signalling, it was intended that the calling line identity information would be inserted as a data burst in the gap between the ringing current that is sent from the exchange to fire the 'bell' (nowadays almost invariably a tone generator) to indicate an incoming call. One of the reasons for putting the burst after the first ringing current is that the latter, being a fairly high voltage signal tends to reduce the high electrical resistance that sometimes develops across badly made cable joints, which would otherwise corrupt the modem signal. Similarly, the customer to exchange signalling by means of the multitone keying has been preceded by a direct current which results from lifting the handset and connecting to the exchange circuit.

An alternative strategy was pursued by companies such as British Telecom, (BT) whose version of CLASS puts the modem signal before the first burst of ringing current. In order to make sure that the dry joint problem does not occur, the BT solution precedes the data signal with a short burst of direct current (Figure 1.12).

This relatively minor variation opens up the possibility of providing silent monitoring, in a simple way: as shown in Figure 1.13, the telephone and the telemetry terminating unit are both paralleled across the incoming telephone line in the usual manner adopted by extension telephones. The telemetry terminating unit detects the data-burst and 'loops the line' (i.e. 'answers' the incoming traffic) before the ringing current is sent. A

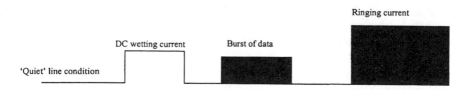

Figure 1.12

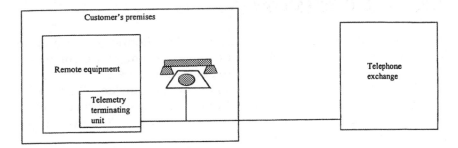

Figure 1.13

normal, but silent, incoming call is therefore set up between the local exchange and the telemetry unit, allowing any measurement and control data to pass back and forward, without disturbing the household.

The system was deployed some years back on trials for utility monitoring, but, it has to be said, that large-scale roll-out plans do not appear to be active.

Another option is to use the Internet. Without any modification to Internet standards or practices, we could envisage a PC being used as a house controller, constantly polling a range of devices within the house and reporting their conditions to the relevant companies, for example, product failure, utility power reading and control, even replenishment of intelligent larders and freezer cabinets (Figure 1.14).

The PC could do this by periodically dialling up a service provider or perhaps the market for wire-based telemetry systems will be re-invigorated by current moves to offer unmetered access to the Internet. One has to realise that the reliability of such services are dependent on the reliability of the PC and that the whole system is driven by client push rather than server pull. Thus is could not be used for critically important services. Another alternative is for each of the devices to have a simplified browser connection themselves, without the intervening PC, or, perhaps, the development of a network controller of very simplified but robust and reliable design.

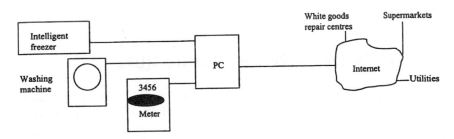

Figure 1.14   PC as telemetry interface

1.20   RADIO TELEMETRY SYSTEMS

Yet another alternative approach to remote monitoring is to use radio. Trials have been conducted for remote meter-reading that involve vans equipped with short range transmitters and receivers that drive along streets requesting readings from radio-equipped meters. Propagation difficulties are always a problem with radio systems, particularly when one unit is located in an out-of-the way place, perhaps below ground level and or screened by metal structures.

One possible solution is to use a local radio link, such as *Bluetooth* (see page 39) to connect devices to an Internet-connected PC or other home terminal and allow the latter to communicate with a company's service-centre server, perhaps by a standard wired Internet connection Alternatively, the Bluetooth connection could be to a WAP-enabled cellular terminal. A wide range of opportunities opens up because of the low cost of such a system. With Bluetooth transceiver components targeted at around £10, it becomes feasible to imagine locating them in many household goods. There they can be used as the diagnostic interface between the product and the remote service centre; they can be the channel for downloading software updates; one could even imagine a refrigerator or freezer box which notified a retailer to send replenishment stock.

The development of low-cost cellular data systems, using perhaps the WAP specification, is a possible solution where radio propagation is not a problem. One application that has been studied in detail [114] is a telemetry system for vending machines. It turns out that WAP is not at present ideal for this application, for a number of reasons. Firstly, there is still a need for dial-up telephony-type connection. This will no longer be necessary once GPRS or similar connectionless services becomes available. Also, WAP is currently only available built-in to mobile telephones. It does not yet appear possible to get hardware that provides interfaces between the radio-telephony and any external data-port. Again, this may be overcome in time. Finally, the early WAP implementations do not really possess a true 'push' architecture, as promised in the standard (page 390). This means that the client has to request information, rather than being directed from the server, a situation that is not ideal for a reliable, server-driven telemetry service.

1.21   TELEMETRY SERVICE ARCHITECTURE

Whether the connection between the device and the server is by wire or by radio, consideration has to be given as to how the service requests are to be handled. Take the example of remote meter reading: there are a number of parties to such an exercise, each with rights and limitations, as well differing views regarding identity. A utility, such as an electricity company will have an agreement with a named person or other legal entity. The communications company providing the reading service will know the utility customer by name, by telecommunications network account, by telephone number or IP address, and by physical details of the location of the meter to be read. The meter may be in rented premises and shared occupancy. The telemetry device must be accessed even if the network customer has diverted their telephone to another number.

What happens if the customer wants to change utilities, or has meters provided by several different ones? How does the telecoms company make sure that this information is kept separate and confidential? Does the telecoms company provide a logically direct channel to the utility's server, allowing it to read the meter whenever it likes, or does it interpose a server of its own?

What looks at first sight to be an extremely simple requirement turns out to be more complex. Nothing in it is insurmountable, but the effort required for the design and implementation is not trivial. Again, radio solutions appear to have the edge over those offered by the fixed network; unique line-to-meter mapping is not a relevant issue and the flexibility provided by the mobile network's separation between telephone number and user identity is also beneficial.

2

Supply-Chain Management

Around Christmas 1999, the Internet book-selling business was taught a salutary lesson that needs to be studied by all on-line retailers: out of sight of the smart corporate offices and the seduction of the shop-front lies the messy business of warehousing, packing and delivery. In the excitement of the new marketing opportunities of eCommerce, it is all too easy to forget that when you have taken the money, you have to deliver the goods. But this is not forgotten by the customer. Figure 2.1 shows the effect on Internet ordering of books, approaching the Christmas period of 1999.

Before the beginning of December, customers were ordering more

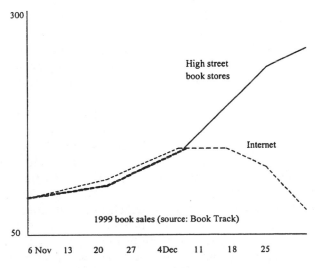

Figure 2.1 Orders for books, Christmas 1999

books over the Internet than they were buying from high street retailers. However, around 11 December they began to realise that orders to be fulfilled by post could not be guaranteed to arrive in time for Christmas Day, and a dramatic drop-off in sales was the result.

One element in successful fulfilment is *logistics*, a term that originally comes from the military world. In order to triumph in battle or in business, the important issue is not just to have the best army but also to bring it to battle in the best shape, at the right place, at the earliest possibility. In the battlefield of commerce, the hardest bits are those concerned with getting the existing processes to come in behind the new on-line developments, and in effectively handling and delivering to the customer the goods and services, particularly physical ones, that have been sold. It is on end-to-end supply chain management, that many organisations must concentrate, if they are to benefit from the heavy investment in communications and computing technology. According to consultants PricewaterhouseCooopers, 'Direct access to global markets will drive product suppliers to add value through customised products and services' [115]. Increasingly, this will be a key differentiator in the market place.

Successful running of an end-to-end supply chain, 'from melon to customer', is partly in the hardware for the movement of physical items, but, just as much, in high quality management of the information that is required to make them move effectively. Nearly always, this information must flow between a number of groups that are geographically remote and organisational distinct. This is true, whether we are talking about exotic, virtual organisations, or traditional ones. In Figure 2.2 we show a generally accepted view of how information flows within modern *head-office based replenishment retailing*.

In Figure 2.2, we see that there are a number of large flows of data within and outside of the central organisation. Interestingly, information flows between store and customer are rather modest, but there is significant, two-way flow between HQ and store. Undoubtedly, in the eCommerce scenario, the disintermediation of the store, replaced by a direct HQ-customer interaction, will mean that the information flow between the two will greatly increase, but it is very unlikely to reach very high levels without creating customer-exhaustion! The main message, therefore, to take from Figure 2.2 is that intra-business and inter-business information flows will remain much greater than that between business and customer. Moreover, these data flows, both individually and in their interactions, are, in general, much more complex than the eCommerce retail customer interaction. It will also involve chains of interactions that must each be satisfied and all of which must integrate, in order to minimise the expense of stock holding but 'satifise' stock availability. For example, the procurement function ought to integrate with the logistics function, in order to answer such questions as 'How many can we supply?' 'Where are the goods held up in transit?' and so on.

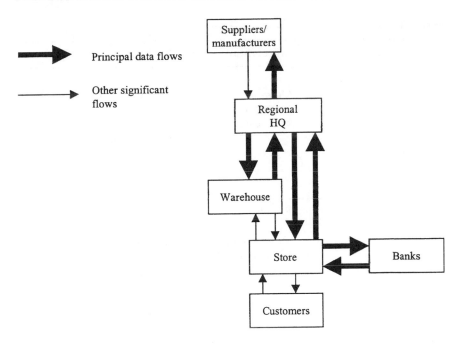

Figure 2.2 Head-office based replenishment retailing information flows

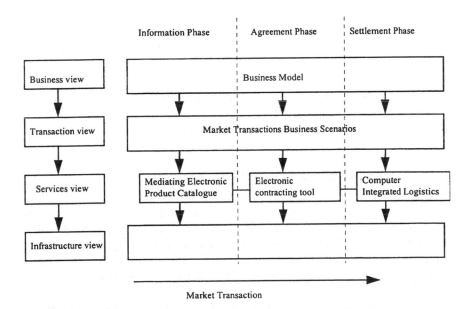

Figure 2.3 Reference model of electronic markets, after B. Schmid and M. Lindemann (p. 116)

Table 2.1   Relating data flow to transaction type

Information	Agreement	Settlement
Between HQ and suppliers		
Sales and sales forecasts	Orders	Remittance advices
Stock	Specifications and designs	Invoices
Product price and availability		Credit notes
Between HQ and outlets (including eCommerce operations)		
Delivery and price information	Recalls	Claims
Shelf space	Order-taking	Receipts
Promotions	Property (or ISP service level) negotiations	Cash banking
Stock information		Credit banking
Sales data		Rents (real and virtual)
Between HQ and warehouses		
Product information	Order notification	Delivery confirmation
Stock levels	Branch allocations	Receipts from suppliers
Batching details		Branch returns and recalls
		Supplier returns

The complexity of the information flows involved and their relationships to business processes has been explicitly defined in an influential business model [116] from which Figure 2.3 has been derived.

We can instantly relate the model's three distinct phases: information, agreement and settlement, to our retail data flows (Table 2.1).

These are only a sample of the many types of transactions involved, but give some idea of the complexity of the process.

2.1   ENTERPRISE RESOURCE PLANNING

Of course, business automation was introduced some time before the ideas of distributed and on-line organisations became high profile. For most parts of supply-chain operations, the relevant generic term was, and still is, *Enterprise Resource Planning (ERP)*. ERP systems were originally turn-key solutions developed for manufacturing operations but have gradually extended their field of operations to include financial control, marketing, business decision support, personnel systems, and others. ERP is largely driven from a data-driven, rather than process-driven

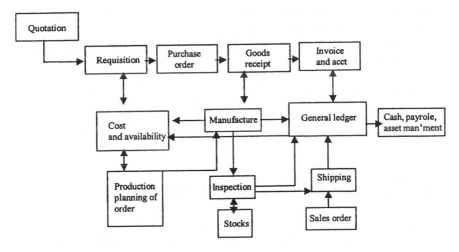

Figure 2.4 Simplified view of ERP business processes.

approach, with users interacting with, and creating entries into, a common *data-warehouse* (see page 183), via input forms and tabulated output, which are generated via custom-designed *views* of the common data. The database is fundamentally an accounting tool, with great emphasis on financial accounting, as this is seen as the main underpinning model for business processes. ERP systems can be very large and their particular feature is that they are designed around sets of data model which are application specific ('marketing', 'production', etc.) but can be considered a subset of the overall accounting model. A highly simplified

Table 2.2 SAP top-level modules

Module name	Examples of function
Financial accounting	Ledger and financial information system
Treasury	Cash management and commitment
Controlling	Direct and overheads costs, profit analysis
Investment management	Programmes and projects
Enterprise controlling	Profit centre accounting and information systems
Logistics	Materiel and supplier data
Sales and distribution	Pricing, shipping, credit management
Materials management	Inventory management, purchasing rules
Quality management	Planning, audits
Plant maintenance	Schedules, records
Production planning	Capacity, orders, materials requirements, routings
Project management	Subtasking
Personnel	Training, development, headcount
Others	Cross-applications, internationalisation, tools

diagram of the typical business processes involved, is given in Figure 2.4. See [117] for a more detailed description.

Note that the accounting basis for ERP is evident, through the central role of the *General Ledger*.

The dominant supplier of ERP software and consultancy is *SAP* [118]. Taking SAP as the 'exemplar' ERP software, it is instructive to note the structure of the top level modules (Table 2.2).

Again the accounting basis is reflected in the many references to financial controls.

2.2 OUTLOOK FOR ERP SOFTWARE

ERP belongs originally to the generation of software products that arrived to support the vision of *business process re-engineering, (BPR)*, which was very fashionable some years ago. Like all visions, it has since its creation come in for some criticism, in this case on grounds of inflexibility. BPR and ERP vigorously promote the need for the explicit creation of an enterprise business model, and the construction of an infrastructure tailored to that model. Subsequently however, some analysts have claimed that it is not desirable, perhaps not possible, to constrain 'modern', virtual enterprises to the rather rigid models that inevitably come out of BRP. In return, the ERP proponents would claim that certain functions of organisations – materials handling, finance, human resource, etc. are common across all organisations and re-invention needlessly adds to cost and time. (One has to note, however, that the introduction of large-scale ERP into an organisation is not a trivial task, even using bought-in products [115]. Many person-months, if not years, of consultancy and testing may be required.) Another rather different argument is that ERP systems have been built around proprietary interfaces and are not 'open' to other databases or to the Internet. (This has also been said about groupware products in general.) This claim would appear to have some justification, but ERP vendors are increasingly opening their systems to integration with 'foreign' software and introducing Web-compatible interfaces. That said, there is more than a little truth in the assertion that a complete but flexible business model is not achievable in a single product. Just as 'traditional' database systems must integrate with Web servers and Internet applications, then so will ERP solutions become part of an integration activity. This is a major cultural challenge for the new generation of eCommerce rapid developers as well as for the more traditional database and internal ERP designers. Flexibility does not sit easily with data integrity and security, but achieving it will be the most important step in migrating from simple eCommerce towards end-to-end, high value-add eBusiness.

2.3 THE CORPORATE DIRECTORY

Not unrelated to the previous discussion on the stand-alone nature of many ERP systems, is the general issue of identifying building blocks at the service and application layer, that are common to all business processes. One of these is undoubtedly a *corporate directory service*. We have seen that details on people's access rights and passwords are critical to security services, the provision of call-centre, eMail and collaborative working applications require consistent addressing and numbering plans, and elsewhere we describe how scheduling and other people-centric processes are critical elements of supply-chain management. In database terms, individual people act as unique *primary keys* to other information which may be spread out across diverse data repositories. A reasonable case can be made for designing virtual organisations around the concept of a single logical database that holds information on everyone – customer, employee, contractor – that is involved in trading. All rights, privileges, skills, preferences payments, etc. should be linked to this data and it should be used as the prime source to bind variables in all business processes – financial approval, targeted marketing, payments, receipts, goods orders, etc. The significance of this, in terms of bought-in ERP, groupware, etc. solutions, is that they should be open enough to be able to take information from this database and use it to bind name variables to their processes, rather than have their own, closed directory structures. A test of a specialised product might be whether it supports an external directory, such as LDAP (see page 138) or whether it cannot be opened up to non-native processes. It is also important that it can be scaled to operate, not just on a LAN, but across a wide area which might require re-defining as partners to the enterprise come and go.

2.4 EXCEPTION HANDLING

At the outset, however, we have to recognise that the theoretical diagrams shown so far are useful for system planning, but also contain hidden dangers. They tend to represent a process in which things tend to go well. This is seldom the case in the real world – or indeed the virtual one. In fact, the move over to on-line trading may well aggravate the situation in some cases. Consider the example of *returns*. This is the name given to goods that are sent back by the customer because they are faulty or just not up to expectation. Any form of distant shopping, as opposed to 'real' shopping (where the goods are available for physical inspection) is plagued with this problem. In some sectors of catalogue shopping, it can lead to as much as one in three items being returned. These returns need to be recorded in the business systems involved:

repayments must be made into financial systems, additions credited to inventory, faults passed into the design database and so on. One part of the solution to effective handling of returns, as we shall see, is in machine-readable packaging labels.

There is also the big issue of confusion: confusion over what is meant by the items in a purchase order, confusion over delivery conditions and whether they were met, confusion regarding where in the delivery path items actually are. (For the former, we need to establish consistent definition of data terms between all parties involved, as described in the discussion on EDI in Part 2, *Managing eBusiness Knowledge*.) To resolve the location of goods problem, we can use package and/or vehicle tracking, as we shall see later in this chapter.

There is often even confusion whether items actually exist in a warehouse. The author was once at a business-case presentation of a proposal for an integrated retail information system. This was laughed out by the retailer who pointed out that the model encouraged HQ to determine the distribution of goods direct to stores, based on calculating warehouse stock-holding as 'goods-in' minus 'goods dispatched'. With typical pilfering rates of 5–10%, this was likely to lead to disappointment. Systems must handle critical data, not just big data-flows, and provision of *exception handling* mechanisms is usually one important example of this.

2.5   DELIVERING SOFTWARE AND ENTERTAINMENT SERVICES

We begin by considering what are probably the easiest products to deliver: software and music. For some time now, a very high percentage of computer software has been downloaded off the Net. So far, these kinds of applications and services have been almost entirely related to providing code for use in conventional PCs or similar home computers. There will be a real possibility in the not too distant future for providing updates to home control systems, and for telemetry applications, some of which we discuss in Part 4, *Service and Support*, where we look at product maintenance. One interesting case is that of entertainment services: games, videos and music. The technical issues surrounding these can be considered to fall into two parts: copyright security, which we discuss in Part 3, *Security*, and the requirements for satisfactory data-delivery performance.

Regarding the latter aspect, we note that the product that has been purchased will be coded digitally and will have to be decoded on the customer's premises at or above a certain minimum speed. There will also have to be some terminal of affordable price and limited size that can do this decoding and, if necessary, provide temporary or long-term storage of the signal. As we have discussed in several parts of this book, the basic requirement for continuous delivery of reasonable quality music is a transmission channel with continuous throughput of several hundred kbits/s. For TV quality video and sound, a figure in excess of 2Mbit/s is required. ADSL technology (Part 1, *Retailing Network Technologies*) begins to make it possible to provide these rates over local telephone connections, as do cable modems and digital TV and radio channels can also achieve this. There are some issues of standardisation. Certainly, MPEG 2 and its MPEG Layer 3 audio component appear to be leading contenders for standardised products, but consumer electronics requires a firm promise for genuinely mass-market take-up before committing itself. There are additional standardisation problems regarding data storage and replay, which involve achieving compatibility for not just off-air (or wire) sourcing but also in physical disk selling. The market for the latter currently vastly outweighs the virtual delivery channel, and thus will call the shots for some time yet.

The exception to this is in TV programming which beams out many megabits of different programming every second. This may mean that digital broadcasting and the TV set-top box, rather than the PC and the Internet, will set the rules for the selling of individual music and video offerings.

It is too early to say for sure.

2.6 DELIVERING 'PHYSICAL' GOODS

Obviously, things are completely different when we come to deal with the delivery of 'real', 'physical' goods: conversion of matter to energy waves and reconstitution at the far-end is either a fantasy or at least something completely beyond our current technology. That said, we note in passing that experiments have been carried out with on-site manufacturing of products, using, for example, laser hardening of epoxy resin according to control systems sent over the network. But this has essentially been limited to the construction of rather crudely formed products such as ashtrays! Nevertheless, on-site manufacture (cooking of food, domestic power tool control, etc.) under remote command, might be considered to be 'quasi-physical' and may become technically and economically feasible before too long.

However, the real place for eBusiness technology in the supply chain is in the transport and processing of information rather than matter transportation. This is not a minority role when seen in terms of cost-benefit and customer service. Of course, supplying goods to customers would not be a problem if inventory stocks cost nothing. We could stock as much as we like and satisfy demand immediately. But *inventory*, that is goods which still remain unsold, bring in no revenue and represent a major part of a company's costs. Moreover, stock reserves exist at many points in the supply chain. Indeed, it is well known that, as we move back in the chain from customer to initial supplier, the fluctuations in inventory holdings increase, owing to managerial second-guessing at each stage, the so-called *Forrester effect* [115]. The best way to reduce this inefficiency cost is to reduce this second-guessing by providing reliable, integrated, end-to-end supply and demand information.

One aim must be to try to dispense with unnecessary and expensive shopping space and, instead, provide on-line access to a virtual store. But we may not want to do so entirely, as stores have a number of advantages over on-line-only operations. So we may have to integrate an eShop fulfilment strategy into our business model, rather than build one in isolation. We may also want to out-source all or part of our fulfilment mechanism. Perhaps we want to distribute goods direct from supplier to customer without any intermediate warehouse. Perhaps we simply want to employ a carrier to select (*pick*) and deliver. In any of these cases we need to maintain control and the way to do this is to maintain data accuracy and control of stock movements. Minimisation of re-keying and regular tracking of product are vital. We need to schedule – to plan the movement of bulk loads and the availability of competent people. Exception handling is, as we said earlier, a major issue.

A further point to consider is that these processes do not run in places where physical conditions are good. Any equipment used in these surroundings must also be robust, perhaps able to run on batteries, work in dim lighting, dust, damp, cold and heat, and be foolproof and reliable. Also in these environments, checks and controls on honesty and professionalism are not easy to police. Opportunity for theft, hi-jack and accidental damage are all high. A certain amount of surveillance is required.

2.7 INTEGRATING SALES AND REPLENISHING SYSTEMS

As this book is about eBusinesses, it may not seem appropriate to spend too much time discussing real stores. However, quite a lot of the technology that has been developed for them, is also applicable to storeless delivery and its characteristics may have been originally determined by the store environment. Moreover, some retail companies, notably in food retailing, are using their existing stores also as warehouses where their *pickers* can load cartons for dispatch as fulfilment for on-line orders. So a little diversion on store-trading may be worthwhile.

It may be quite surprising to those of us who shop, rather than work at the other side of the counter, to discover that it is only very recently that many stores have successfully integrated their sales and replenishment processes. This is true despite the obvious automation of electronic tills and other check-out equipment that comprise, to all intents and purposes, a computer system. There are still quite a large number of food stores belonging to major supermarket chains that rely on manual checking of shelves to alert them to the need for restocking.

The obvious alternative is to monitor the reduction in stock levels by counting the goods that leave the store, using *electronic point-of-sale* equipment, based on bar-code reading at the checkout. What has been lacking, is on-line, wide-area networking. Traditionally, information was collected, batched, and then transmitted over telephone line or by satellite link, ultimately arriving at the warehouse. Increasingly, the requests are sent in smaller and smaller batches, thus approaching a real-time scenario. In this way, almost real-time measurement of demand can be achieved and many stores now have replenishment deliveries two or three times per day. The data from stores is usually sent back to regional headquarters. There, planning programs may modify the immediate demands with forecast requirements, including information from short-term weather forecasts, analyses of current trends and cyclical data from earlier years, before communicating with warehouses and to the transportation fleets that deliver the goods. The warehouses must themselves have ways of accurately knowing what goods they have in stock. For a variety of reasons, of which pilfering is not the least important, this is not as simple as it seems. Goods must then be batched for loading onto fleets and the fleets guided securely to the desired destination. One dream of the distribution business has been *cross docking* wherein there are no, or at least, minimal warehouses. Goods arriving from each supplier to a dock rather like a railway platform, are spilt into their outgoing order sizes and loaded onto lorries, in more or less one operation. By and large, this remains a dream, but is a good conceptual model and an aspiration for a process designer. Direct vending to the customer certainly does not let this become any easier: order quantities are much reduced, whereas order fluctuation probably increases.

2.8   WIRELESS LANS FOR STORE AND WAREHOUSE

Central to the creation of a networked enterprise is the ability to connect up computer terminals with little need to constrain their location. It is relatively easy to wire-up most offices to a network. The layout of desks is usually fixed for a considerable period. Rooms are often quite small or possess low ceilings or under-floor ducting, thus allowing access for power and data connections. But shops and warehouses are much more difficult to wire up. For many years, high-street stores have been aware of this problem in the case of their cash-tills. Often these are run from batteries, which in themselves bring a variety of problems, from early

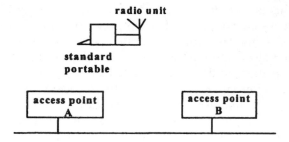

Figure 2.5   Radio LAN

burn-out of switch contacts due to low voltage, high current working, to the obvious problem of replacing and recharging the batteries. The introduction of EPOS into stores and the similar drive to automate warehouses and packing operations and integrate them into HQ data operations have seen these problems increase.

There is little that can be done to get round the powering problem, (other than produce lower power devices) but solutions are emerging for the 'wireless' data-plug, in the form of the *radio LAN*, specified in IEEE Standard 802.11, [119, 120]. The aim is provide convenience rather than exceptional speed, with data rates of the order of 2 Mbit/s, coverage within a building of around 10,000–20,000 square metres and the ability to roam between different radio access points.

As shown in Figure 2.5 the intention is that standard portable equipment will be equipped with plug-in radio units which will communicate with access points on the LAN. The access points create 'microcells', in a scaled-down version of a cellular network (page 37). If the mobile equipment moves out of range of access point A, then it re-registers with point B, and the communication continues. The standard does not lay down how the access points communicate to each other; this is left to the vendor.

Nor does the standard take a view regarding the two competing transmission techniques on offer, *Direct Sequence Spread Spectrum (DSSS)*, and *Frequency Hopping Spread Spectrum (FHSS)*. Both of these techniques are intended to get round the problem of interfering signals that may well occur in the electrically noisy environments of modern buildings, while at the same time trying to keep the signal level of the radio LAN from further contributing to that noise. In the case of DSSS, the signal is 'spread out' across a wider bandwidth than is required to code up the data in a simple, conventional scheme; this makes it possible to recover the original signal, even if parts of the frequency spectrum are too noisy. To understand the basic principles of DSSS, it helps to look at a rather simplified version of how the signal is coded. In Figure 2.6, we see a steam of digital data, which can be fitted into a frequency band A, say.

However, let us instead take the signal and transmit identical copies of

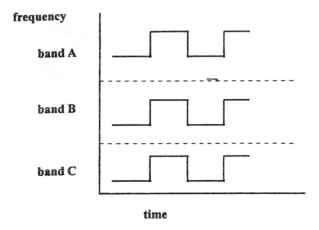

Figure 2.6   Direct sequence spead spectrum (DSSS)

it, within two other frequency bands B and C. We transmit the original and the copies in exact synchronism with each other. When we receive them, we convert them out of their individual bands and now have three identical signals, with exactly the same start and end times. We now add them together, and end up with a signal that is three times as big.

Let us now look at what happens to the background noise. There is a vanishingly small probability that the noise in each of the three bands will be identical. So, when we add the outputs of the three bands together, the noise will not add up to three times its original value, unlike the identical, in phase wanted signal. Sometimes the noise in one band will be large and positive in value while that in another will be large and negative; sometimes they will both be nearly zero. For fully random signals, we can show that the noise will not grow by three times; instead it will grow by the square root of three; that is about 1.7 times. So, by using the extra frequency bands, we have gained a 'noise immunity' of 3/1.7. In general, the spread spectrum gain approaches the square root of the number of bands we use. Typically a spreading ratio of 10 is used, corresponding to a gain in immunity of around 3 times (= square root of 10).

In a real DSSS system, although the noise immunity principle is exactly the same, the data is not sent as multiple frequency bands of the same signal. Instead, each data 'bit' is coded as a set of N bits, each of which is $1/N$ of the duration of the original data bit. This means that the original data bit is 'smeared out' (or 'spread') across a wider frequency range (N times greater) than it would be if sent as a single bit.

Frequency hopping is rather easier to understand (Figure 2.7).

The diagram shows two separate channels of data, corresponding to two users, being transmitted simultaneously, This is done by allowing each channel to 'hop' from frequency band to frequency band according

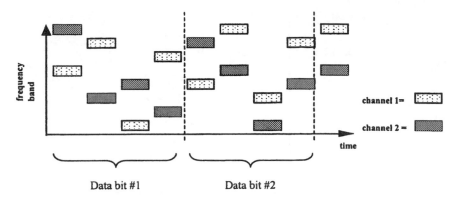

Figure 2.7   Frequency hopping

to a pseudorandom sequence. The sender and receiver are synchronised so that they both know which sequence will be used. Sequences are chosen so that the two channels never hop to the same spot at the same time. One data bit is coded as several frequency hops (4, in the example shown). Thus, if there is any loud narrow band interference, most of the hops can be detected and the data bits they correspond to recovered by averaging the results. Typically, more than 75 frequencies are used and hops occur at a rate of 400 m/s or faster. Since the systems operate at 1–2 Mbit/s, this is equivalent to around 1 Mbit of data per hop.

One of the reasons that standardisation has been attempted across the two radio methods, is that proprietary products were available for both technologies. Consequently, there is vociferous argument between the commercial proponents of DSSS and of FHSS. FHSS is a slightly older technology and may be currently offering more concurrent channels. However, DSSS is seen by some as the longer term solution.

There is also a problem in Europe with the choice of 2.4Ghz as a standard frequency, because this band is already overcrowded. (Although approval has been given in the UK.) Perhaps as a longer term solution to this problem is the emerging HyperLAN standard, which will operate within the microwave bands of 5.7 and 18 GHz. Apart from being less crowded, these frequencies also offer the possibility of 20 Mbit/s performance, although this may not be a key requirement in store or warehouse systems.

2.9   APPLICATIONS FOR WIRELESS LANS

Applications for radio LANS are not limited to connecting the checkout equipment to the corporate network. In some cases they allow the customers (or at least store staff acting as 'personal shoppers' or pickers on their behalf) to walk round the store, accumulating purchases and a running total of their price, using hand-held terminals. Terminals are also providing store managers with the ability to interrogate corporate databases for product information and availability details of items not on display, in front of customers rather than having to retreat to a back-office terminal.

In principle, sales staff could stand anywhere within range of the wireless LAN and locate a product anywhere in the supply chain. With mobile terminals that can read credit cards and issue printed receipts, they could sell the product there and then and schedule delivery. The power requirements of these portable devices is sufficiently low for them to be used all day and plugged in by the sales staff, at a recharging point, thereby obviating the need to supply power within the interior of the store.

Terminals with restricted functionality could also be provided to in-store customers themselves, to allow them to browse the retailers' additional information on their Web servers. Products which really need to be seen 'in the flesh' but which require a level of expert guidance on functionality or finance could be located at stores-in-stores or in the concourses of shopping malls. There they would be on-display but unattended by expert sales staff. Instead, users might employ hand-held terminals to access the required information remotely. These terminals could, of course, also be *WAP*-enabled mobile telephones, or a *Bluetooth* terminal connected over the short range to the LAN. (See Part 1, *Retail Terminals* and *Retailing Network Technologies.*)

It is perhaps in the warehouse that full-power wireless LAN terminals will be preferred to WAP or Bluetooth. Here, *Radio Data Transfer (RDT)* mobile terminals can direct forklift truck drivers to the correct pallet and they can also be used for stock inventory, perhaps combined with bar code readers. Increasingly this technique is fully integrated into supply chain management to allow, for example, a direct communication between the warehouse and the shipping dock, so that delivery orders can be completed and their estimated time of arrival signalled to the loaders.

2.10   BAR CODING

There still remains the problem of checking the contents of the customer's real shopping basket or, equivalently, the contents of the carton that has been *picked* by an agent in a fulfilment house. Visual inspection is the traditional way, but this is fraught with error and the most common way is now to use some form of automated scanning equipment, both at the check-out and in the packing houses that support eCommerce. The earliest system of this type is *bar coding* which was originally developed in the 1950s to serve the needs of American railways in their quest to maintain a record of their freight-cars. Bar-coding appears deceptively simple, but in reality is the subject of many different standards, (one source quotes as many as 225) and requiring some quite tight technologies for its reliable operation.

We mentioned that there are a large number of bar codes at least in theory available. Although some of them can only justify their existence through accident, perhaps through the early adoption of a proprietary standard, there are also genuine reasons why there is no one dominant code. Bar code formats are trade-offs against conflicting requirements, principally that of size versus reliability and/or coding capability. What is possible on the side of a large carton may not be feasible along the side of a small package. Even after choosing an appropriate standard, there are other compromises: one has to consider whether or not use a comprehensive code that uniquely identifies manufacturer, country of origin, etc. or a shorter version using the same coding scheme, but only displaying the part number without further details of origin. This is a data-modelling task which has to take into account possible trading relationships and any overseas expansion plans that the company might have in mind, because uniqueness might be lost if codes are truncated in this way. (One strategy that can be used is to split the coding between the components of the 'pack': for instance part of the code can be on the wooden pallet where the goods are bulk handled and the remainder of the code being on a temporary label applied to the customer's order carton. See, for example [121].)

There are a limited number of ways in which we can use a bar code to represent data (Figure 2.8).

Bearing in mind that scanning of the code can be done manually, at arbitrary speed, and that printing of codes may be done at different scales,

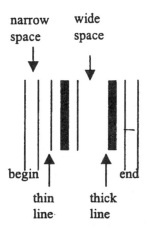

Figure 2.8 Principles of bar coding

it is best to use relative measures to represent the data. This is done in terms of 'thick' and 'thin' lines (and, in some codes, spaces). It is assumed that the scanning speed will be reasonably consistent throughout a single operation so as to make these differences detectable. Some codes have start and stop lines as shown and some use check digits.

Turning to specific standards for bar coding, it is possible to reduce greatly the 225 types mentioned above into a very much smaller number, at least for enterprise-wide trading. The US *UPC Version A bar code standard* [122] was probably the first to achieve wide recognition and is still widely used today. In the mid-1970s, the European Article Numbering Association [123] created an EAN standard, of which *EAN-13* is today's version. Both of these codes are numeric only and have a fixed length for the data, plus check digits for increased reliability. These are the codes that we see every day, for example, on food packaging and are principally used for checkout processing and internal warehouse operations. Whilst this format is compact and easy to agree on and implement, it is not really flexible enough for many other applications such as distribution tracking, packing and inventory control. *Code 39* or *Code 3 of 9* as it is sometimes called, is usually preferred for these purposes. It is of variable length, tolerant of a wide range of size and aspect ratios and has an alphanumeric capability, which makes it useful when the associated printed text has to be interpreted manually.

Where data density of a bar code is a problem, then one solution is to use a two-dimensional bar code. Reference [124] contains a comprehensive description of number of codes and the different techniques used to compress data. Of all the two-dimensional codes, *PDF 417* and *Maxicode* are the most common examples. PDF 417 has an impressive data capacity of up to 1 kbyte of data, in a single coding block. This can be used, for example, to create a label that codes a customer's order, which can be stuck to a carton into which the goods are then picked by an operator who simply scans the label at a picking point. Selection of goods is therefore possible without having to bind a customer reference number to an itemised list, thus removing the need for a network connection between the warehouse and the order department. This has some advantages when fulfilment is out-sourced, (although total isolation of warehouse and order system is clearly not a very good idea).

The assignation of specific code numbers to companies and products is overseen by the relevant number-coding bodies, such as EAN. With some exceptions, (principally where goods are very small and require truncated coding), the assigning of numbers to products is left to the company itself, the regulating body simply assigns a company number and a country code of the assigning body (*not* the company). For more details see [125].

2.11 PICKING, PACKING AND PROCESSING 'RETURNS'

One vital operation in converting the electronic shopping cart to the real thing is to arrange for someone to *pick* the goods that have been ordered and put them into the packaging. Fulfilment houses that do this without electronic support have a bad reputation for wastage and incorrect delivery. The earlier systems often used a *pick-by-light* system, where the goods flowed on a conveyor past the picker, who was instructed by a system of lights as to which goods to pick and when.

Even this semi-automated process was prone to error, as well as being rather expensive.

A cheaper solution, using hand-held bar-code readers can be used to great effect in reducing error virtually to zero. A typical operation might go as follows.

First the order is downloaded into the bar code terminal reader from a system computer which also informs the picker of the size of the carton needed to contain the order. The picker selects an appropriate carton, which may be transported around the warehouse on a trolley. The carton may itself be bar coded and the picker scans this into the terminal. The picker is also supplied with information as to where the goods can be found in the warehouse. Following this route-map, the picker goes to the place where the first item is located. This is called a *slot* and has the item bar code fixed beside it. The picker picks the wanted item and scans the bar code of the slot. A confirmation signal, or a warning, lets the picker know whether they have followed the order properly. There may be repeat items required from that slot. If not, the terminal directs the picker to the next one. Once the pick is complete, the picker seals down the carton, which can, using its own bar code, be sent off into the warehouse distribution system. Because of bar-code standardisation, it is also possible, in the case of business-to-business selling, for the customer to integrate the coding on the received packaging into its goods-inwards logistical system. As we have said earlier, information on theoretical shipments is good and well, but the real proof of the existence of goods within a warehouse has to be something more tangible; bar-code reading of the incoming packages is one such validation.

Some organisations also use bar-coded packaging to send goods out to retail customers. This may seem strange, as the customers are unlikely to scan the codes themselves. However, the reason is simple – how to handle goods returned. Product *returns* are surprisingly high in the catalogue shopping business, in some cases amounting to one in three orders, particularly in fashion goods. This is a serious logistical problem as returned goods need to be reconciled against inventory, money (or credit) has to be paid back to customers, and marketing will almost certainly want to know why they were sent back.

Thus, it is quite convenient to request that customers return the packaging along with the unwanted goods. Customers usually oblige, as this ensures they get their money back. In planning such as system it is important to consider the fact that quite often it will only be a part of the order that is returned, and it will be necessary to label the packaging with an order number, rather than a list of goods. Manual intervention will probably also be required in order to key-in details of which items have been returned and perhaps the reason. Alternatively, we could use a more extensive bar coding of the package, detailing the contents one-by-one, and this would reduce the need for keying and hence errors.

Returnable tickets can also be used, where the customer is asked to tick items returned, together with ticking boxes containing the reason: 'faulty', 'wrong size', 'appearance', 'quality', etc. These can be read by standard *optical character recognition (OCR)* software and fed into a marketing database. After all, if one has the incurred cost of returns, one might as well maximise the customer analysis data thus gained.

2.12 RADIO TAGGING AND 'SMART LABELS'

Bar codes have a number of disadvantages: they are quite difficult to print on irregular-shaped or floppy packages, they need line-of-sight reading, and, of course, once printed they cannot easily be modified. In recent years, an alternative approach has been to use *radio tags* or *smart labels* as they are sometimes known. Originally designed for freight car labelling, the same technology has been adapted for palletised goods and large items such as cars on production lines. Until recently the radio tagging of smaller goods has not been possible because of cost and size limitations. However, the development of systems operating in the 13.56 MHz range now means that tags can be produced thin enough to be laminated between layers of paper or plastic and capable of incorporation within stick-on labels. The quest for a really low-cost, small integrated tag is one of the Holy Grails of the retail industry, as it would significantly cut the costs of both packing and charging for shopping basket contents. The operating principle of radio tags is shown in Figure 2.9.

Although it is possible to use active tags which contain a power source and some way of electrically modulating a radio wave, it is often more convenient to use a passive *transponder*. This is a technique whereby the tag contains a tuned radio circuit which selectively absorbs and re-radiates radio signals of a particular radio spectrum sent by the transmitter. The effect is not unlike that of using a mine detector. Pure transponders of this type usually do not require their own batteries. Some systems allow simultaneous reading of several tags. Tags are assembled from a set of

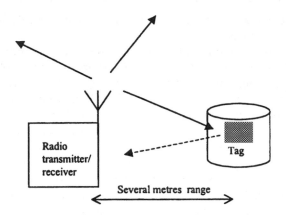

Figure 2.9   Radio tagging of goods

components which together provide a unique absorption characteristic for the tag. The simplest and cheapest tags can only be read, but more complex tags are also capable of alteration via the radio link. This is often of advantage for complex products, such as automobiles, which undergo a number of processes as they pass along an assembly line. The tag can be used to hold the process history. Capacity is typically up to a few hundred bytes of data. Tags can then be re-programmed for reuse. Because communication between the tag and the control unit is by radio, rather than by direct contact or optical reading, tags can be read from, and written to, through a range of dirty environments, such as paint shops, and through wooden or other non-metallic packaging. They are generally able to withstand harsh environments of temperature, humidity and pressure. Data encryption methods can be used to 'scramble' the data so that it can only be read or written to by authorised means.

One of the simplest uses of radio tags is for *Electronic Article Surveillance*, (yet another reminder that the real world of commerce is fraught with theft and malicious damage). Although mainly used for retail applications within-store, surveillance tags can also be used for warehouse anti-theft measures. Labels can be integrated into the product or its shrink-wrapped packaging when it comes off the production line. Tags can be as small as one centimetre square and can operate at a variety of frequencies, in particular, 58 kHz for reliable detection when concealed in liquids or metal foil, 13.56 MHz for general applications and 2.45 GHz where higher speed and volume data transfer are required. See [126] for one vendor's discussion of pros and cons.

As well as their use in security systems, RF tags and smart labels have been introduced in some warehouses and packing centres as replacements for bar codes. Apart from the fact that they can be read automatically when they come within proximity to a reader, they also have the advantage that they can be modified. For instance, a tag can be fitted to a carton and be 'filled in' as the carton is filled with the goods. When the picker places the goods in the carton, they either wand a bar code on the goods or has them automatically read, and a supporting computer sends updating information to the carton tag.

Radio tagging does not need to be restricted to goods; it can also be used on people: according to [127], pickers for Streamline, an American food goods fulfilment house, 'wear tiny portable computers on their wrists that receive individual orders from the central computer via radio frequency. As the worker's pick an item from the shelves, they pass it by a ring scanner they wear on their fingers and their wrist computer confirms that it's the correct item.'.

2.13 DISTRIBUTION AND FLEET MANAGEMENT

Tagging, whether by bar code or by radio means also is of value when goods are on the road or have arrived at distribution points. They make en-route inventory much easier and more reliable to carry out and it is often possible to give customers virtually on-line and real-time information on the whereabouts of important packages. Because of the vagaries of international travel in particular, it is often recommended that tags be read as frequently as possible along the route and this information made available to customers via an on-line server. This also makes the detection of highjacking and the capture of criminals much more feasible on longhaul, international routes.

Of course, any system that places much reliance on tagging and tracking should bear in mind that equipment, such as bar code readers and radio installations is operating in a rather more hostile environment than the average office. Ruggedisation of the laser systems needed to read bar codes, for instance, has improved over the years but they are still liable to damage if thrown into the back of trucks or dropped from more than about one metre onto hard surfaces. Steps may have to be taken to provide tethered units or spares.

In-cab radio systems for the control of fleet transports are becoming more common. The so-called *Trunked data radio systems* include a range of non-public services that allow organisations to create virtual private radio networks between members of their organisation. Market predictions indicate that demand for private radio services will reach 5 million European corporate subscribers by the year 2000. Specifically targeted to meet the needs of fleet management, telemetry service companies and government agencies, the proposed TETRA Trunked Radio System provides both packet and circuit-switched data services. Data rates are not high, (28.8 kbit/s maximum), but there are an extensive set of features on the voice and data side, which may be valuable for service management. These include *calling/called line identification, list search calls* where subscribers can be called in sequence according to a list, *priority calls* which can interrupt calls in progress, even an *eavesdropping* function which allows authorised parties to listen-in. Applications for the data service can include security alarms and environmental monitoring of fragile or dangerous cargoes. These systems can also be integrated with GPS (see page 85) or other geographic positioning technologies, to allow destinations to be aware of the likely time of arrival of goods and, again, as a security precaution for real-time tracking of the vehicle. The pragmatic use of mobile phones and of Bluetooth technology will also become significant.

A number of companies offer combined fleet routing and scheduling systems. Typically, these use one or more digital maps, available at different levels of granularity, and possibly with spoken or graphical description of significant landmarks. Many systems integrate information from a number of sources. Some, such as mileage, overtime, fuel burn rates, etc. are measured in-cab. Others also incorporate things such as customer-availability time windows and other constraints. Many of these systems can be run on PCs and can be understood after a few days training.

Some on-board systems allow for automated collection of information. Typically, drivers are given their own magnetic cards, which they insert into the onboard computer at the start of a run and they remove it when finished. The cards are read in the dispatch office, where the day's mileage, etc. can be read off.

2.14 SCHEDULING

It should not be a surprise that companies invest significantly in scheduling systems. People and inventory stock are perhaps the most expensive commodities in a company. Getting them to the right place and in the appropriate situation is therefore one of the most important tasks in an organisation. It may seem rather inhumane to talk about people and goods in the same sentence, but the fact of the matter is, that scheduling either of them, in many – but not all – ways, requires the same technology. In any case, often the actual service to be provided, for example, a maintenance task, requires that both people and goods be simultaneously present and both need to possess the correct qualities for the operation in hand (Table 2.3).

Examining Table 2.3 we indeed see that most situations require planning of both people and goods. What is also significant is the complexity of making the choice. It is seldom that quantity alone is required. For people there is usually skill requirement; for products, the choice must come from the right box. Scheduling is not just about amount, but also quality.

Table 2.3 Resources to be scheduled

	People	Goods
Call-centres	Sufficient number	Authorising/removing access to services
	Right skills	
Packing centres	Sufficient number	Sufficient number
Customer deliveries	Sufficient number of drivers	Correct product
	Skilled installers	
Production lines	Skilled workers	Correct components
Field maintenance	Skilled workers	Correct components
		Sufficient vehicles

In information systems terms, scheduling is usually a specialist function adjunct to the main *Enterprise Resource Planning (ERP)* activity, which we discuss on page 332. As shown in Figure 2.10, the scheduling engine gets much of its basic data, – e.g. employee information, stock availability and customer requirements – from the main ERP databases.

Therefore, it is important, at least in the case of large organisations, to realise that interfacing between the two will be necessary and to purchase compatible solutions. Another aspect is usability. How easy is it to use the system, particularly if it is to be used by people who are not IT trained? Designers should note that workers in the warehouse or field are very intolerant of process equipment and software, which they often feel has been 'foisted' on them and destroys their skill and initiative. This is particularly true when they are being paid on a piecework basis and where any failure or rigidity in the equipment means that they cannot meet their targets.

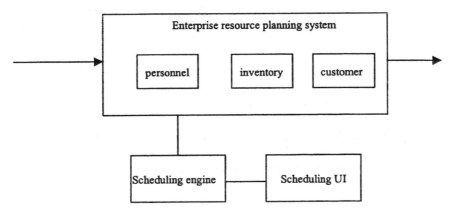

Figure 2.10 Relationship between Scheduler and ERP

2.15   BASIC FEATURES OF SCHEDULING AND SCHEDULING ALGORITHMS

Scheduling is about completing a number of tasks within certain time windows, under a number of constraints. Scheduling has been distinguished from planning, which also concerns itself with the same basic problem, in that the granularity of the time-scales is different [128]. Scheduling is generally a shorter-term phenomenon and thereby also tends to mean that there is less freedom to select the order of carrying out tasks than in the planning case.

Usually the most complex part of the schedule is in dealing with people, because they tend to be the most complex and constrained resource and cannot be duplicated on demand! Therefore, one of the most common inputs to a scheduling engine is the personnel (or 'agent') database. In this will be included against each person's unique key (usually a variant of their name), a list of their skills, their availability, their costing model and any preferences that they have expressed. Most of this seems deceptively simple, but in most cases there are subtleties. For example, it ought to be possible to relate the individual's skill to a wider list that relates to the organisation's process requirements, but this requires a consistent data-model across the enterprise. Consider, for example, the increasing tendency to out-source maintenance work on consumer products. There must be a means whereby a technician's general skills ('wiring', 'electrical testing', etc.) and product-specific experience for any one of the multi-sourced products can be recorded and recalled.

Employees also operate under certain constraints: part-time staff cannot be expected to exceed their conditioned hours. There may be fairness conditions on allocation of overtime and, almost certainly, there will be a desire on behalf of management to minimise overtime costs. The first iteration of shift planning is sometimes carried out anonymously, in order to achieve these constraints. Other conditions then enter into system and so it can be seen that rostering is an iterative task which requires a good development environment if it is to be successfully automated. Usually a highly graphical and forms-based interface is required. In any case, the final tuning of the system will be done with pragmatic managerial input, to meet with sudden changes in demand or staff sickness.

Workforce planning is not restricted to job rostering, of course. More complex systems such as ILOG's Dispatcher [129] can also optimise vehicle routing as well as managing the crews. Routing problems can be incorporated into the model and contingency plans developed.

2.16 FORECASTING

Once or twice we have mentioned *forecasting* as part of the decision-making process regarding stock holding and stock movement. Improvement in affordable processing power and, to a certain extent, in theoretical methods such as neural networks (page 196) have meant that organisations can be more accurate in determining future demand. However, the main improvements have probably come from the availability of more data [130] or from the improved collection of EPOS information from customer-facing systems rather than from warehouse data. Two lessons from this, therefore: firstly it is important and urgent to get one's data model into a form suitable for capturing this data; second, the message, yet again, that data-collection systems must integrate across the whole supply chain. How this is done will largely depend on existing databases, ERP systems and any data warehouse initiatives. First, the causual variables that drive the demand have to be considered quite carefully: not just the obvious factors such as time of year, weather, etc. must be included, but also whether products have been sold upon marketing campaigns, and if so, how does one quantify the strength of the campaign? Not unnaturally, a number of retailers buy-in their forecasting systems, some of which may be available from their ERP suppliers, but this should not be seen as a substitute for trying to understand the underlying model.

A further complication will arise if the forecasting operation is to be extended across an enterprise, involving perhaps suppliers as well as the retailer. Systems must integrate, share a common data model, generate trust and openness whilst keeping some of the data secret.

2.17 PURCHASING

It has been claimed that most businesses spend more than 30% of their income on purchasing goods and services [131]. If this is anywhere near the true figure, then it is clear that a potential must exist for improving the cost of such transactions as well as their quality.

But first we should lay a ghost: there is a naïve belief that on-line information technology inevitably leads to an open, opportunist market place, wherein suppliers offer goods and services 'promiscuously' over a wider market and purchasers can, correspondingly, buy, on whim, beyond their traditional suppliers. Empirical evidence has shown that the opposite appears to be the case. Analysis such as carried out by the late Professor Geoff Lockett, at Leeds Business School [private communication], reveals that companies are trading with fewer suppliers, not more. The reasons for this are not clear, but at least two factors are suspected: firstly, in the area of critical component purchase, businesses prefer to pay a slight premium for the safety of doing business with someone they know and trust. According to Lockett, more than two thirds of purchases are made between organisations that consider themselves to have worked together for mutual gain, and will continue to do so. (We discuss issues of trust in more detail in Part 3, *Trust*.) Secondly, the theoretical freedom offered by IT is often nullified by the practical implementation using proprietary standards which mean it is very difficult to trade outside the relationship without extensive re-coding and capital purchase.

Nevertheless, in some markets, particularly where the goods on offer are commoditised and trust levels are uniform, flexible on-line trading does occur. Basic commodities such as oil have for some time been purchased this way on *spot markets*, often by telephone or telex. More recently, telecommunications circuit spare capacity has been traded in this way and consortia of suppliers and manufacturers in the aircraft, motor car, etc. industries have opened Internet portal sites with the same intention. Transport fleet services are also an obvious area where spare capacity can be traded. These arrangements are, however, only really semi-public. The consumer market, which generally involves simpler purchases without complex contractual agreements, has opened up more readily, with on-line auctions being quite successful. The technology for running these is not particularly complex.

2.18   AUTONOMOUS TRADING AGENTS

There is often talk of on-line trading systems being run using artificial intelligent agents that will negotiate between supplier and consumer. CORBA and other similarly distributed processing environments have explicitly revealed computing models where centralised control is no longer the dominant paradigm: processes run where the system decides they can run the best and this decision is taken via the trading of processing power, rather than its central assignment (see Part 2, *e-Business Systems Architecture*). There is, in principle, no reason why the decision-making powers should be restricted to choice of computing platform or algorithm: we could extend the principle to that of allowing autonomous processes to make purchasing decisions regarding real goods. That is, we carry the concepts of independent but co-operative interaction between the middleware components in DCOM, CORBA, etc. into the application layer, where processes sharing common descriptions of the application can operate on different platforms, working towards an agreed resolution of that application. The processes usually communicate and may involve the transmission of self-contained modules of code from one platform to others. These processes are sometimes called *autonomous* or *intelligent agents,* and, where they can 'roam' about an enterprise network, *distributed*or *mobile agents.*

There is something rather seductive in the concept these agents: sometimes they appear to be possessed with super-human (or certainly super-computing) powers of negotiation. One simply 'fires' them into the network, armed with the right to negotiate the lowest price deal for a product, or the best terms for a supply contract, and this is what they return with. The reality is somewhat different, but trading agents are more than just an empty concept and they may become very important components of eBusiness strategy, in due course.

First of all, what *is* a trading agent? Indeed, what is an agent? Attempts have been made at definition, for example [132], but the safest answer is that there is no simple answer. The term has been used in so many different, inconsistent ways that one person's definition is sure to offend someone else. In fear of adding to the debate, but in the interests of moving things on, we at least try to give a general impression of what a trading/negotiating agent is and does:

- It is a piece of code that interacts with other pieces of code, in order to try to complete a set of transactions satisfactorily, or terminates the action without system failure if this cannot be done.
- Satisfactory completion of a transaction is achieved if a set of *goals* within the agent and in the other processes with which it interacts are achieved.
- The code contains a *strategy* whereby it operates to achieve its goals.
- In trying to achieve a goal, or at the very least, a sub-goal, an agent is allowed to make some decisions, perhaps sharing them with 'foreign' processes, without being controlled by any external, centralised, control function.
- The agent may modify its strategy and perhaps its goals in the light of experience.

Given these very human characteristics, it is easy to see why software agents have been anthropomorphised. We have to be very careful not to confuse the wish list with reality.

There are a number of architectural aspects which primarily differentiates agent-based solutions from other solution mechanisms:

- The design of the agent module is formulated parochially – there are no detailed assumptions about the global architecture (platform vendor, task-optimisation at the other business sites, etc.)
- The interaction between one agent and the complementary process or agent with which it interacts, is based on negotiation, rather than master-slave operation.
- The negotiation is based on interface rules and formats that are known to all the interacting processes.

- The internal details of how these negotiations are processed are not necessarily disclosed.
- In many cases, instantiations of agents may be created without limit, each instantiated agent thus created is *isolated* from the others, its only interaction with others being by a prior agreed messaging protocol.
- Because the way that any one specific instance of task resolution usually depends on the complex interaction of several agent instantiations, each operating towards goals which are based on their current environment, it is not usually possible (or at least very difficult) to specify a compact global algorithm that they are trying to solve. Instead, the result is an example of *emergent behaviour* which results from their interaction. Sometimes the terms *self-organising system* or *connectionism* are used to describe this phenomenon.
- Although the ability to adapt behaviour in the light of past experience is not exclusive to agent architectures, (and some agents are non-adaptive) the methods for achieving this through modification of individual agent goals, is rather specific to them.

It may be apparent, from the above description that agent-based systems do indeed bear a great similarity to object-based component design for distributed, heterogeneous transactions. They interact through specified interfaces, they are not too inquisitive regarding each other's internal mechanisms, they permit multiple instantiation and isolation of these instances and so on. Indeed, object-component methods and languages are often used for their design.

At a more detailed level, it sometimes appears that classifying agent types and applications is a research field in itself. Reference [133] provides one useful taxonomy. According to this, and most other schemes, eBusiness agents are members of a class of *co-operative agents* and may conveniently be divided by function into *requester, provider or broker* agent, depending on whether their prime function is to seek/consume information or services, advertise and carry out these services, or 'matchmake' or 'mediate' between the other two classes. It is interesting to observe that this taxonomy implies an asymmetric relationship will always exist, probably mirroring the laws of contracts. Perhaps it is instructive, therefore, to think of agents as processes which negotiate resources, either directly or via other agents. If this is the case, then we can look for business processes which have this 'contractual' nature, and consider them as obvious candidates for agent-based systems. Amongst these are some obvious candidates, including single purchase by a customer, inter-business contract-bargaining for supply and call-off, and, in a rather different area, work-scheduling in a multi-product, multi-production-line environment.

2.19 SINGLE-PURCHASE EXAMPLE

Although this is really a retail issue rather than a business-to-business one, it is an instructive example to start with. The specification is simple: design a *shopping agent* which can find the 'best' deal on a specific product, by searching on the Web.

That is how the problem is usually specified in the superficial statements made about the miraculous powers of automated agents, and it is often claimed that they will achieve this capability 'soon'. It is of course a terrible specification, that no human, let alone a machine, should ever accept. Nor is it one that can be fairly assessed. First of all, the *goal*, is not sufficiently well defined. What do we mean by 'best'? Do we mean lowest price? Quickest delivery? How soon do we want an answer? (Are we really to search the whole Web?) and so on. We see that even this very simple example requires considerable analysis of the business policy of shopping for the best deal, (e.g. 'get the cheapest!'). We must also specify the business process involved ('Go to a selected list of Web sites for price information; return the result before lunch time', etc.).

Already we have had to put a large amount of human input into what is supposed to be an automated process. But suppose we have been successful in determining what we want and under what conditions. Next we have to specify this in sets of standard terms. We define the shopping process in terms of a relatively uncontroversial elements: finding specifications, checking terms and conditions, checking credit record, giving payment and delivery details, etc. It is possible that these elements can, in turn, be built up from sets of *common business objects (CBOs)*. This is one of the objectives of the Object Management Group (OMG – see page 149) which is trying to achieve standardisation in this area. There is no point in trying to minimise 'price' if the term has not been defined, either as an object itself or as part of another one. We have to reserve judgement as to when, indeed, whether the OMG will be fully successful in identifying all the necessary CBOs. However, if agents created by different parties are to be able to negotiate at all, there will have to be at least some common definition of the significant 'things', (most probably, 'objects'), that are involved in the transaction. What is also required of any attempt to impose an object model at the business level, is that it map sensibly onto the middleware components in legacy systems. It is likely that the *Extensible Mark-up Language (XML)*, will play an important part in defining the things that go together to make up an on-line trading process. As discussed on page 179, XML provides a simple and scaleable way to define terms and to publish this definition on-line.

In its very simplest form, we could imagine a group of suppliers, either privately or through a trade association, setting out an agreed definition of terms such as 'manufacturer', 'model', 'price', 'quantity', 'delivery date', and so on. They agree to publish this set of definitions on-line, as, for example, an XML *Document Type Definition (DTD)* which will list these names and, perhaps the units in which they are expressed. Then, we could imagine they will include these terms as *meta tags* (again, see page 179), somewhere on their on-line catalogue pages. Vendor A might have:

<skis>

<manufacturer> IceKing

<model> CarverKing

<price> £500 </price>

<delivery> 30 </delivery>

</model>

</manufacturer>

</skis>

This (actually rather simplified over the real thing) tells us that the CarverKing model is available at £500, within 30 days.

On the other hand, vendor B might offer:

<skis>

<manufacturer> IceKing

<model> CarverKing

<price> £550 </price>

<delivery> 10 </delivery>

</model>

</manufacturer>

</skis>

That is, dearer but quicker delivery.

We now begin to see how a very simple purchasing agent could be designed to check-out these sites for the 'best' deal. We need to program the agent according to the DTD format, so that it can 'understand' what is meant by 'model', etc. We then need to have an interface which allows the user to type in information according to these terms. For example, one field of a form would request 'model' information. We need to tell the agent where to go to look for pages where it will find meta tags that it can read. To do this, we must define a *strategy*. Again, the parameters for this are human-set. One possibility would simply be to tell it to 'go to Altavista and search for whatever the user had put in under <model> on the agent's on-line form, (in this case, 'CarverKing'). Altavista will return the URLs of a number of hits. The agent is now programmed to visit each one of these hits in turn and look for evidence of the use of <manu-facturer>, <model>, etc. meta tags and, possibly, reference to the DTD. This evidence will tell the agent that it is indeed looking at a page where goods are described in a language and format it can understand. It can therefore reliably extract information from the pages.

Now it has to make the purchasing recommendation. Its simple *goal* is to get the 'best' deal. But how does it define best? Actually, human inter-vention is again required. There has to have been a definition of 'best' supplied to the agent, by the person wanting to buy the goods. The agent is 'only human', in this regard at least! 'Best' could be defined as 'cheap-est', in which only the <price> information would be used. Perhaps a more complex algorithm is preferred: a balance between price and deliv-ery might be defined by the user. (A weighted sum of the price and delivery date might be calculated, perhaps also with overall limits on either.) This algorithm would be programmed in as a *sub-strategy*.

What we want to make clear in all of this, is the non-magical nature of the process. Firstly, there is no way that the agents can second-guess the user's purchasing goals, even though, as we shall see, they can learn from experience. Secondly, the agent's choice will only be as good as the strat-egy allows, and most strategies are relatively naïve. None can be guaran-teed to find the 'best buy', even given a clear definition of the goal.

Obviously though, agents can have more intelligent strategies than that of the example described above. This did have the advantage that, using the XML specifications, it was unlikely to get its facts wrong. But, it might well have missed out on better offers that were not programmed in XML. Much agent research is involved in building agents that can make good attempts at reading free text, rather than that which has been deliberately meta tagged according to a set plan. This will produce more answers and, possibly, better bargains, but at the expense of reliability. The issue is similar to the knowledge management discussions in Part 2, *Managing eBusiness Knowledge*.

2.20 ADAPTIVE AGENTS

Our shopping agent described above carried out a simple task, which was, moreover, fixed. Its interaction with the world was as shown in Figure 2.11.

The agent found what it wanted and told the user. But suppose the user

Figure 2.11   Agent with fixed operation

did not like what they saw? Or perhaps the user then purchased the goods and had a bad experience with supply from the particular vendor? Clearly there ought to be a way to modify the agent's behaviour, in order to improve future performance, in response to feedback from the outside world (the customer, in this case) (Figure 2.12).

The goal has remained the same, but the strategy now includes an adaptation capability which responds to the outside environment. Adaptive agents of this type are in use for a number of tasks, principally in selective retrieval of information, under the descriptions of *personal profiling, preferencing*, etc. They are intended to learn progressively the preferences of individuals and supply only that information which meets a user profile based on previous choice. The learning mechanism can be initiated by the individual user, who passes opinions on previous information delivered by the agent or as a result of collective decisions gathered by a central agent and distributed to individual user agents as feedback from the outside world.

Explaining the latter in more detail: imagine collecting information on the film choices of a number of users. Statistical analysis will almost certainly reveal that a group of users who like one film more than do the average population, will also tend to share a common opinion of

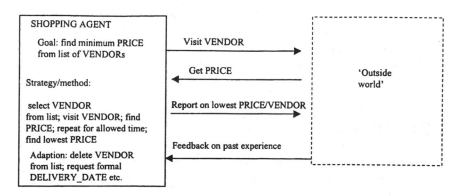

Figure 2.12   Adaptive agent

another film. So, if my agent has gathered a table of my likes and dislikes, it can offer it to a brokering agent which might hold the preference tables for a large population of viewers. Comparing my likes and dislikes with the other viewers' preferences, it is possible to identify films which like-minded viewers have seen and liked, but which I have not seen. The broker agent can then send this information to my personalised agent saying, 'I think you'll like this'. If I decide to view it, I will afterwards be asked to review it and this will lead to a modification of my preference table. That is, to an adaptive element has entered into my agent's selection strategy. (But not to my goal, which remains, 'Locate enjoyable film'.)

2.21   MULTIPLE AGENT-BASED TRADING

Simple though it is, the single-purchase activity illustrates the basic elements of an agent-based solution. The aim of much of today's research, however, is directed towards the solution of intra- or inter-business activities, which are much more complex. In particular, the agent has much more need to adapt its behaviour, often within a single task, and in many cases there are several agents involved. Most of this work is still highly experimental, and has been trialled only in benign environments: no one seems to be ready to pit their agent against an others in a truly combative commercial negotiation.

In constructing a negotiating agent framework, goals and strategies again need to be clearly defined. They tend to be of much higher 'dimension', in that an optimum solution has to be found by manipulating the values of a number of variables across a number of agents. For each agent we try to define a *cost-function*. This could be a trade-off between completion date, price, inventory holding, etc. The negotiation phase involves each of these agents trying to achieve a result which reduces costs to a satisfactory level in terms of the set goals. Notice that the solution is sought on *satisfying*, rather than minimising the cost function. The mathematics and the resulting computation can be demanding; often the problem has to segmented, parameterised and solved for local optima rather than globally [134]. (This may not necessarily be a disadvantage: the algorithms behind optimised, non-parameterised solutions may be more difficult for one's competitors' agents to decipher. Little work has so far been done on this forensic aspect of agent negotiation.) The whole area of agent design is currently a very active area of research (as opposed to implementation), comprising issues of efficient implementation in heterogeneous distributed environments as well as in the design of negotiation strategies. For an up-to-date review of this work, see for example [135].

2.22   MOBILE AGENTS

So far, we have concentrated on the self-contained and adaptive natures of
software agents, and we may have been guilty of implying that they
'move about' the Web, hunting out bargain sites. In fact, most of them
do no such thing. Many reside simply on a client machine, and simply
pull down Web pages which they analyse locally. However, some agents
also possess the property of mobility. This means that chunks of executable code, representing instances of agents, can be launched from, say, a
client, to a server, or vice versa, in order to run some process on the latter.
It is also possible to launch several agents into a computer network, each
one charged with solving the same task or variants on it. An obvious
application for this might be in trying to find a price for a short-lived
commodity – a financial services deal or an airline seat, for example.
An example of the latter is given in [136]. The agent then runs on the
destination host, perhaps negotiating with other agents and either sending a simple 'completion' message to its originator, or firing off either to
the original host or to another computer, a projection of itself bearing
information for further processing. Issues here are clearly of network
congestion and response to its failure or delays, performance issues of
running multiple agents on 'foreign' machines and security. To address
these problems, a number of implementations have been produced [137].
They all impose constraints on the freedom of action of the agent within
the host system, perhaps assigning it to specific *virtual places* in memory,
constraining its access rights to certain places in the file server and terminating it if CPU anomalies occur. Good systems also make heavy use of
security, based on passing-over cryptographic credentials ('certificates')
which can be used to invoke permissions for the agent (file-access, maximum CPU time, etc.). Performance and reliability aspects have been treated in a number of different ways: some systems prefer to treat each agent
as a single thread within a single, multi-threaded process, thereby speeding up inter-agent communication, at the expense of complexity; others
run each agent as a separate process in a multi-tasking environment.
Systems can be entirely distributed or involve a central server that maintains oversight of each interaction. Most systems are interpreted, rather
than compiled, with JAVA being widely used. This leads to rather slow
performance, compared with compiled solutions.

2.23   SECURITY OF MOBILE AGENT SYSTEMS

Not unnaturally, there have been a number of concerned raised about the
security of mobile code solutions, particularly where negotiations involving orders and money are concerned. We discuss security issues in

general in Part 3, but it is appropriate to discuss this specific issue here. What is the problem? In fact, there are two main issues: the damage that mobile code can do to the system it accesses, and the reciprocal possibility of the a system maliciously corrupting the agent. Current research, e.g. [138] would tend to indicate that the latter problem is potentially more dangerous. Using sand-box techniques (page 60) to restrict the agent to a particular part of the receiving host's memory is a strong defence; as described in the reference quoted, it is also possible to equip agents with their own digital certificates and signatures. There are still problems with denial of service attacks, by the malicious spawning of agents which consume receiving host processing power, but this is already a threat with non-agent attacks.

It is more difficult to prevent an attack on a mobile agent, for example, changing the offered or acceptable prices for a negotiating strategy. The agent has to 'open itself up' more than does a host; therefore hosts can see inside to a greater degree. However, some protection is possible: agents can be duplicated and the one that can be kept within a protected environment compared with the agent at risk.

These comments only scratch the surface of mobile agent reliability and security. The subject is still in its early stages and more research has to be done. This, combined with the unknown performance of mobile agents as hard-bargaining negotiators is likely to mean that fully automated commercial negotiations (except between trusting parties over secure networks) is something that will grow only slowly over the next few years. Automated agents are more likely to be used as assistants to human purchasing agents, perhaps as 'finders' for parties wishing to negotiate, for a while yet.

3

Electronic Marketing

Although eCommerce opens up a new channel to customers, it is not particularly difficult to fit it into the traditional models of marketing. Indeed, marketing is a science, with a respectable body of theory and practice behind it, and, as technologists, we should try to understand it, not try to re-invent it. Nevertheless, traditional marketeers can be frightened or deceived by the new technology and one important task for the technologist should be to assist them in spotting new angles for existing concepts. This chapter looks at some of those issues.

3.1 CREATING THE ON-LINE SHOP

To many traditional organisations, opening their on-line shop has been an experimental activity, one perhaps not even really approved of by the main-steam sales and marketing organisation. It often looks like that too! Many sites have been designed by someone in the IT department, often in spare time, without much graphics design skill or consideration being given as to how to integrate what is often a display only of a sample of products into the main catalogue. 'Going on-line' is too important to be left to part-time, non-specialists. It is a job for the marketing team. The eShop should be a representative of the company's *brand*, the 'intangible benefits best interpreted as inspiring customers' confidence' [139]. A badly designed, unreliable, patchy Web site therefore becomes symbolic of a company with a similar ethos and performance.

The process of creating an eCommerce site that instead projects a positive image about the organisation is quite complex. It requires a proper mixture of skills and people. The reason for the complexity is that one has to achieve a consistent understanding and balance between user needs and expectations and the limits of the technology. Of course, this is not a unique case. Almost all software development encounters this requirement: an eCommerce site is 'just' one more example of an interactive multimedia design project. These are known to be difficult! A colleague once described such activities as 'trying to get the Bolshoi Ballet (the graphic designers) and the Red Army (the software developers) to co-operate, under the control of the KGB (the project director)'.

Indeed, the skill probably does lie with the project managers and most of this comes with practice. It is possible to use design studios, but they will always require a 'brief' up-front or developed with the organisation as part of the cost of the contract. Figure 3.1, which is adapted from [140] outlines the standard method for creating professional designs. Market-eers will notice that it is similar to the method employed for designing most marketing campaigns. Inevitably, it will make extensive use of story-boarding.

The storyboards provide a common medium for technical developers, stylists and commissioning marketing people to conceptualise, and refine, the appearance and functionality of the design. This may be carried out in a 'workshop' environment, using white boards and flip-charts, but will ultimately result in a *concept demonstrator*, coded up in some prototyping language (e.g. *Visual Basic*) or perhaps directly into HTML, after which

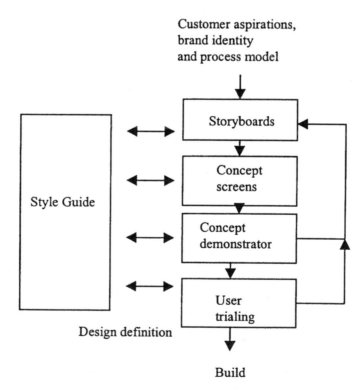

Figure 3.1 Creative process for on-line store design

there will be a major review. Running concurrently with this activity is the creation of a living document, the *application style guide*, which records the styling and branding, as well as the functional model aspects of navigation, etc. The guide is completed when the design phase is over and acts as a reference manual for developers and the client.

The process described is a good way to get all parties thinking alike, with the minimum of misunderstanding. Points to watch out for, however, are the creation of a common misunderstanding about performance and situation, as well as an over-optimistic view of the end-customer's enthusiasm. Performance can be deceptive: remember that the demonstrator is probably operating for a single user, sitting locally. Consider how it might behave when slowed down by multiple accesses and by the limitations of a modem connection. The design studio will also, inevitably, have tried to demonstrate the prototype on a large, high quality monitor, and it will use a browser that is compatible with the design environment. Make sure trials are carried out using the full range of access platforms and transmission environments.

All this will test the *usability* of the system; it is much more difficult to trial its *utility* (usefulness), unless you are going to back up the demonstration with realistic pilots involving delivery of realistic services. This is a good idea, for several reasons. Not only will you be able to start asking questions of your customers regarding the price they would really pay for such a service, it will begin to focus your mind on the true cost of providing it. The emphasis may then begin to shift from the glitz of the front-end multimedia design, to the hard facts of the supply-chain. Notice that one of the really great advantages of on-line service delivery is the opportunity to ask your trial customers for on-line feedback, perhaps through a simple form. It is important, of course, not to overdo it, but it does mean that you can carry out an assessment of immediate customer response in a much more focused way than through traditional market research. Even if it is decided not to ask for customer response, it is still possible to gather statistics on which pages were most popular (see page 379).

The stylistic aspects of Web design are really an issue for artistic designers rather than a technical issue and many textbooks and courses on this subject are beginning to emerge. However, the styling must be carried out within the constraints imposed on it by the technology. In what follows, we shall look at some of these technical issues and their impact.

3.2 INTRODUCTION OF FRAMES

Initially, Web pages each consisted of a single document downloaded from a location on a server, perhaps then pasted into with a number of graphics. Later generations of browser developed this to allow for a

Figure 3.2 Use of frames

screen to display simultaneously a number of *frames*: Web pages, each one fully independent, but with its relative size and position dictated by a single *frame-set* page which controls the display (Figure 3.2).

One very convenient use for this is in displaying a toolbar, perhaps heavily branded with the company logo, whilst allowing users also to view a succession of items in a catalogue presented in another frame. Figure 3.2 shows two screen views rendered in this fashion. On the left, we see the *welcome* screen, which consists of two frames, a menu frame and a welcome frame. Selecting the *our products* link, results in the two frames in the right hand of the figure. The extreme right leads on to a further selection, but the other frame continues to show the same menu as before, thus assisting with navigation.

There is however, an issue about the appropriate use of frames and the debate on the subject is probably the most common argument about Web design that non-specialists will come across. There are some people who believe that frames should never be used; there are others who believe that they can be useful in some circumstances. It is certainly true that it is easily possible to overdo the use of frames and this can lead to consider-able user confusion. Users are not always sure how the two parts of the picture relate to each other, especially if the backgrounds are identical and one part of the picture (i.e. one frame) can be scrolled independently of the other. It is also possible to end up with an incredibly complicated screen when links are provided to third-party sites also making use of frames. (We discuss this further on page 370.) The second problem occurs when a vendor wishes to display a third-party site within its own. This could be the case where a company creates a 'shopping mall' which comprises a number of retailer on-line sites, surrounded by a set of frames of its own design, probably acting as navigators for the mall. If any retailer also has a framed site, an untidy and confusing set of frames-within-frames results. Nor is it usually possible to get round the problem by simply linking to the retailer's information-bearing pages. Almost always the retailer's frame-set will include a toolbar, which will have to replaced by something provided by the mall owner, causing increased expense and a significant maintenance overhead. In any case, it is perhaps not always a good idea to allow users to get direct links to third-party pages.

3.3 'STICKY SITES'

'Don't you realise that real shops are designed to be very easy to get into, but hard to get out of?' This was the remark made by an expert retailing strategist, when asked to view a multimedia shopping terminal, which had a prominently displayed 'exit' button on the screen.

Yet more evidence that the shopping experience does not just happen; it is created. We are still learning how to recreate these effects in virtual shops. Where one has full control of the user interface, this is easier to do: kiosks and interactive TV may be able to achieve this more readily. However, the weightless nature of the Web, particularly the way that HTML links allow users easily to jump from one server to another with 'memory-less' (*stateless*) detachment, means that it is more difficult for on-line sites to hold their customers than it is in a real store. Sites that do manage to a degree to retain customers are known as *sticky*. Most stickiness is a matter of design and marketing flair, making them easy and progressive to move through, with the user feeling that new experiences and offerings are made with every click within the site, but there also some technical aspects that can help to retain customers. This is particularly the case when a site owner is offering service brokering, rather than pure product retail or supply. By this we mean that the site points out to other sites. For example, an on-line travel agency might provide a search service over a number of airlines and car-hire companies, matching dates to availability and lowest prices. Although it may need to let its users have access to the information on third-party airline and car-hire sites, it would rather keep the users on its sites, either to deal through it for commission or because it wants to sell other services – insurance, maps, etc.

One thing it might try to do is to get customers to register, at the earliest possibility, as described earlier, and it might also try to control the session with the customer through the use of cookies. But suppose this is not considered feasible. The simple approach is to use a multi-framed design. Take the example of a travel broker that needs to display the site of the airline for which it wants to sell tickets (Figure 3.3).

Here, 'Best-Travel' is the broker and has embedded the airline ('Europlanes') Web pages in a frame-set. The strategy is to ensure that the user will navigate within the Europlanes site, but return for any further services, to the 'Best-Travel' site. An especially prominent back-button

Figure 3.3 Multiframed 'sticky page'

has been supplied to facilitate this. The button returns the user to the appropriate place on the broker's (not the airline's) site.

A problem arises when you try to embed 'fussy' sites which have complex control bars and multiple frames. In this case, it is perhaps preferable to put these in a separate browser, which can be automatically *spawned* when the user clicks on the relevant button. As shown in Figure 3.4, the user has made a search and then chosen to view one of the results (Europlanes). This appears as a floating window in a separate browser, the window deliberately designed to be smaller than the main Best-Travel one, and certainly not so large it covers the whole screen, confuses the user and loses the BestTravel brand. It is clear it is a separate browser, because it has the full browser surround and toolbars. Note that it also has a back-button, which has again been programmed to return to the Best-Travel (broker) site.

3.4 ANIMATION AND SOUND

The quest for customer seduction is competitive and never-ending. Just as paper catalogues began with text and progressed via line drawings and black-and-white photographs to full-colour, so have Web pages. Seduc-

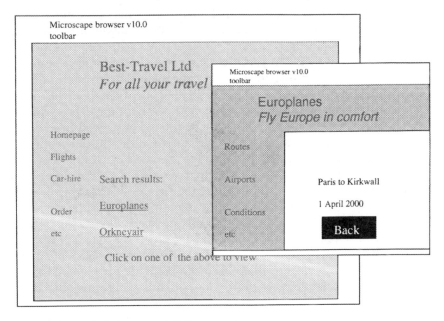

Figure 3.4 Multiple browser 'sticky page'

tion does not appear to be satisfied by the status quo and is probably benchmarked against whatever is the latest innovation of TV and the cinema. Until 3D becomes a matter of every-day fact (which is not likely for several years yet), the target is therefore high quality sound and moving images.

Immediately we run into the issue of delivery bit-rate. With digital TV there is no problem; 2 Mbit/s is a given. With kiosks or other specialist delivery channels it is probably possible either to store the multimedia material locally, or to deliver it via a business-quality high-speed connection (tied cable or radio link). But with most Internet connections using the telephone network, there are serious problems. In Part 1, we discussed the bit-rate requirements for audio and video and also the capability of a number of systems for delivering these rates. Quite simply, it is not currently possible to send continuous streams of high quality audio or moving video, to the vast majority of homes, across the Internet. This is not to say that acoustic and moving image cannot in any circumstances play a part in Internet presentations. Some customers will have higher speed connections, either through ADSL or cable modems or because they are connected to a Local Area Network (LAN), as might be the case with business customers. The important thing is to know one's market. Short audio and video clips, of high quality, can be sent as files, perhaps associated with Java applets that can play them as part of the presentation. Lower quality, continuous sound is possible, using a number of proprietary formats, and a succession of still, high quality, images is often considered by the advertising profession to be better than low-quality shot and/or replayed video. It is also usually considerably cheaper to produce.

Although it is difficult to send 'real' scenes across domestic Internet connections, the continuing increase in processing power within affordable terminals, makes it increasingly likely that client-based *virtual reality* image construction will play a part in on-line shopping. Three-dimensional images (in the sense that they can be rotated or viewed from different angles rather than standing out with apparent depth) can readily be created and manipulated using languages such as *Virtual Reality Modelling Language (VRML)*. If these are partly constructed on the client, then data rates from the server can be quite modest.

Experimental work progresses in the area of creating artificial characters, *avatars*, for example in the form of talking heads, which can be used as naturalistic shopping guides, film and TV presenters, etc. In examples which so far only exist in research laboratories, users can select the image and voice of their choice (or the choice can be made for them, based on profiling). The technology is quite close to full realisation; the major uncertainty will be resolved with market testing.

One thing to note, regarding all seductive designs: novelty attracts but does not necessarily retain. There is good evidence that intriguing interfaces can not only create initial interest but also encourage people to learn how to use services they might otherwise have been frightened of [141]. However, once users have become familiar with an interface, they usually seek out short-cuts and any 'fanciness' that gets in the way begins to turn them off. There is a saying in graphics design, 'Two [colours] to tell, and four to sell', which makes it clear that information-seekers are after the meat, not the colouring.

3.5   ISSUES OF GLOBALISATION

It has often been said that there is no such a thing as global marketing, simply regional marketing carried out on a World-wide scale. Every campaign has to be 'nativised' to meet local conditions. How do we square this with the fact that access to the on-line community is global? There are many issues; here are some:

*   *Language choice*: there may be a need to generate pages which appear in more than one language. In the newer HTTP standards there is provision for the inclusion of a language indicator, which is automatically used by up-to-date browsers (i.e. *not all* browsers) when they communicate with a server. The server in turn can use this information to select a page written in the correct language. When this is not considered an appropriate solution, then the welcome page for a site

can offer a click-through to a number of different language versions. Sometimes this is done using a set of national flags to represent the languages. This is not a good idea. Not only does this often take up too much room and may slow down retrieval of the page, it can also be very politically dangerous, as a German-speaking, Austrian acquaintance once pointed out. The better solution is to use a text menu. But if you want to give a nativised impression, remember that using English descriptions of languages, 'German', 'Finnish', rather than 'Deutch', 'suomea', etc. can give exactly the opposite impression. You should use the target language's name for itself or make use of the standard abbreviations that exist for all of the world's major languages.

- *Language translation*: artificial translation tools are *not* good enough for translating sales information. You could offer them a button to such services, for example Altavista's *Babelfish*, but a disclaimer should be displayed. People who select to use them to translate your pages do so at their own risk. The only safe way is to get material translated by a professional service that uses skilled humans. In Table 3.1, for fun as much as anything, is an extract from a machine translation from Italian into English, of a promotion for cheese.
- *Alphabet issues*: as well as problems with languages, there are often issues of scripts. Many languages have diacritical marks or even full characters which cannot be represented using ASCII Roman text. There exists an established standard, *Unicode* [142], which should be used for coding such text. As described in the reference, not only is ASCII insufficient to meet the needs of all languages, it is also used in an incompatible manner on different platforms. Unicode provides a unique number for every character. The Unicode Standard, which is compatible with the related ISO/IEC 10646-1:2000 standard, has been adopted by many industry leaders and is a requirement for other Web standards, including XML and CORBA, etc. In its most common applications, it uses a 16-bit mode that allows for unique coding of 49,194 alphabet-related characters, but there is even an extension mechanism (*UTF-16*) which can encode a million more. This covers all historic scripts of the world.

Table 3.1   Example of automatic translation

The cheese of Fabrizio MIl of the Sardinia in Sardinia, in spite of the event of new technologies, the operation of the cheese in those second ones place continues has always passed on the rules

The latter ones come scaldato and the rennet helped to is associated to how much is condensed

The beginning Sardinian and constituted of the flower from the crude glue that comes is filled up in on within on smoke of the traditional procedures

3.6 DESIGNING FOR PEOPLE WITH SPECIAL NEEDS

Although the world of eCommerce is supposedly one of cut-throat entre-preunership, perhaps we might be permitted to put in a short plea for consideration for the less fortunate? It is said that on-line services can be a way of helping reduce the isolation felt by people with disabilities such as infirmity or blindness. This can be true, but it is also the case that many of the more sophisticated multimedia designs are increasingly removing some of the advantage. For instance, many people with impaired vision have used text-to-speech converters to read Web pages. With the incor-poration of graphical buttons rather than text ones, it becomes impossible to use that solution satisfactorily. Simply replacing text with loudspeaker messages may not always be appropriate, either. Consider kiosks, in-flight services, or other examples of public place equipment. There may be a need to provide sockets or inductive couplers, rather than broadcast the dialogue to the world.

Many companies express a desire to be seen as 'ethical'; one way they could achieve this would be to include design checks to see if their pages could be used by these communities. It may even be prudent to do so. After all, there is now legislation in many countries regarding access to physical stores; perhaps this may also come to pass for virtual ones. There are a number of initiatives concerned with providing better access to people with disabilities. See, for example [143]. A search on google.com, using the string 'Web design disability' will also reveal a number of other online resources.

3.7 ON-LINE ADVERTISING AND 'AD-STRIPPING'

The nature of Web page design makes it very easy for companies to sell advertising on their sites to other companies; alternatively they can them-selves place advertisements on the sites of others. The impact of these ads can also be measured as part of the monitoring functions carried out on the Web site, as we discuss later. There are a large number of intermediary companies which offer services whereby they obtain advertising space for clients and pay out revenues to the displaying site. Payment is generally computed in terms of so many cents per *click-though* to a link to the advertiser's server, via some intermediary routine that counts the number of times the ad is clicked-on.

An even simpler approach is to pay to display an ad, without any active click-through, on another organisation's page, perhaps as brand reinfor-cement campaign. In deciding to accept such advertisements, the accept-ing site owner should be aware that this could cause problems. Where customers are using slow-speed access, for example via a modem, then there can be excessive waiting time for the down-loading of any images. This is a particular problem with third-party ads. It can be lucrative, but also slow and is aggravated by the way that ad value is computed. Advertisers want to pay on the basis of *impressions*, the number of times an ad is viewed. Although browsers can cache pages on the client, reducing the need to down-load them repeatedly, it is more usual to design composite pages such that banner ads are usually reloaded every time a user revisits a page already viewed, thus letting the server count it as a further impression. The result of this approach is a heavier load on the server and, probably more significant, a needless, frustrating wait for the customer while the usually bulky advertisement is downloaded. It is technically possible with modern browsers to get round this problem: a short message can be sent from the browser to the server each time a cached item is recalled to view. But there is no standardisation yet in place and it will not apply to older versions. As reported in [144], Netscape and Microsoft have incorporated reporting features in their later browsers.

Whilst on the subject of advertising, we perhaps should take note of an interesting development, directly hostile to banner advertising, *ad-stripping* software. This in theory will allow customers to remove advertisements from sites that they view, by running a piece of software on their terminals. There are number of such systems which vary in complexity and 'intelligence'. Some rely on hand-crafted rules but others, increasingly, can learn the conventional structure of advertisements and replace them by blanks. A typical ad-stripper [145] scans a Web page and identifies a number of features generally indicative of advertisements, such as images with links that go to different servers and which have captions that contain words such as 'sponsor', 'funded by', 'special offers', etc. How popular these services may become is difficult to say, but site designers ought to be aware of their existence.

3.8   GETTING A SITE NOTICED

The estate agents cry, 'Location, location, location!' also applies to the retail trade. Strategic positions in the High Street, or close location to a key store, such as a major department store, are seen as a way of maximising footfall. What are the equivalent on-line technologies for achieving this? We need to bear in mind that on-line sites are significantly different from physical outlets, particularly their so-called *weightlessness* which is simply a way of saying that they are largely unconstrained by geographical position and the laws of physical distance do not apply.

But although on-line customers do not really walk the High Street, they do possess a location-point for their travels in cyberspace, by virtue of the URLs they insert in their browsers. So, when they go shopping, they must start out from somewhere. For many, the starting point is the home page of their browser, which is usually the default option of the Internet Service Provider (ISP) from whom that they get service. How do they first get connected to their ISP? There are two principle options: on-line, or from a CD-ROM. It requires slightly more computing knowledge to set-up an ISP via an on-line connection, than from a CD-ROM. The latter usually has self-starting software and all the user has to do is insert the disk and carry out a series of simple operations, with on-screen prompts. This is attractive to many customers. It also is of advantage to the company that gives away the CD-ROM, whether it is an ISP or a retailer. The point is this: the user is automatically connected through to the CD-ROM issuer's site, whenever they start up a browser. It is, of course, possible to change the default page for the browser, but this is often not done. Thus, there are a precious few seconds where the user is exposed to the CD provider's seductions, before they type in a new URL, of the site really wanted. Effectively, the retailer's site becomes a main entrance point or *Web portal*. We shall discuss other aspects of portals later in this chapter.

This process of supplying CD-ROMS may be too expensive or difficult to administer for some retailers; others may simply not be in a business where a retail portal is appropriate. In these cases, the company may rely on being brought to the customer's attention because it has been found through a search engine such as *Altavista* or *Yahoo*. The development of search-engine solutions such as these is a ceaseless, continuing race with the ever-growing content of the Web. It is one that they are probably losing, in the sense that the percentage of the Web that they index and catalogue appears to be falling. Thus it cannot be assumed that a site will become listed by them without some effort on behalf of the site owner. This needs some understanding of how sites are indexed. As mentioned, the process is one of continuous development, and the secrets are not always revealed. However, some general principles are worth noting.

Large-scale search engines do not carry out exhaustive searches over the entire content of a Web page or site, every time a query is made. Instead, they are continuously trawling the Web for sites to *index*. This index is a summary or digest of the contents of the page or site, created according to the search engine's strategy. Clearly they will choose to begin indexing with popular sites and those with which they have some agreements. So, paying to have your page advertised on a popular site may mean you will not only get additional hits off that page, but also may be more likely to get indexed by a search engine. Commonly, indexing software commonly also collects URLs on the page it started with and goes to their respective pages and examines them too. Obviously, this would be a mammoth trail to follow if it were done exhaustively. Therefore the indexer operates only to a certain *depth* or *level*, say, the first, next and third set of pages. Thus it is important to get your URL within about three links of the top-level page.

Here, the continual development of more sophisticated page construction techniques can be a problem. Temporary pages, which act as *welcome screens* before giving way in a few seconds to permanent home pages, are often skipped by search engines. Similarly, a frame-set page which itself does not contain much information (except for URLS of the 'meaty' pages) can act as a barrier between the search engine and the meat. Even more of a barrier are *active pages* and *form* structures which are themselves only formulae for the composition of pages. An on-line search page for a product catalogue does not itself contain any information on the contents of the catalogue, at least from the point of view of a search engine. (It is not going to key-in the necessary words into the search box.)

Once pages have been selected, then the search engine will index them. The search engine may only look at a fraction of the page, usually the beginning. The message is clear: get your key message, company name, list of products, etc. into the first few lines. As well as looking at body text on a page and at URLs linked off it, search engines can also look for specific areas on the page, usually in the *page header*, which contain potted information put there by site designers specifically to make it easy for machines to understand what the page is 'about'.

In Part 3, *Managing eBusiness Knowledge*, we discuss in some detail the meaning of the term 'about' as applied to information retrieval and also mark-up techniques in languages such as XML. But the basic principle can, and is often, applied in the simple case of an HTML page. The method used is to employ the *meta tag* capability within HTML. This allows use to specify a pair of terms or *name-value pairs*, within a frame-work of the form.

<META NAME = "anything you like to put here" CONTENT = "and here" >

Table 3.2 shows how a set of code-lines can be inserted in the HTML header portion of a Web page, to advertise a set of services providing flowers and/or chocolate for 'special occasions'. The *name* portion is a form of data *type*, in the HTML case, an arbitrary type assigned by the designer, not necessarily with any special significance. Table 3.2 shows two examples, 'abstract' and 'keywords'. It is now possible to give values or *content*, in HTML jargon, to instances of data types. We see that 'abstract' has been given the value 'Special occasion presents', and 'keywords' has been assigned values, 'Worldwide flower deliveries' and 'Chocolate'.

Now, we said that names are arbitrary and at the whim of the designer. However, there is semi-official code of practice that has grown up regard-

Table 3.2   Meta tagging to help search engines

<META NAME = "abstract" CONTENT = "Special occasion presents" >
<META NAME = "keywords" CONTENT = "Worldwide flower deliveries" >
<META NAME = "keywords" CONTENT = "Chocolate">

---

ing some specific names. Among these are indeed our two examples, 'abstract', taken to mean a short description of what is on the page or what it is for, and 'keywords', not unnaturally consisting of a comma-separated list of terms that the designer thinks adequately defines it further. Search engines can make use of these conventions.

Specifically, the search engine will parse the page, looking for meta tag blocks such as the example given in Table 3.2. It has been programmed to look out for the appearance of 'abstract', 'keywords' and a number of other data types, ('author' is a common type in library material), and include their corresponding values in its index. Note that this relies on convention, not standard. There is not even agreement as to how many times a data type should be re-valued within a document: in the example in Table 3.2, two separate lines of code have been used to set values 'Worldwide flower deliveries' and 'Chocolate', to keywords. This could just as easily been written in one line with commas separating the values. Most search engines will correctly index the latter approach, but some do not accept the former. With mark-up languages such as XML which are more structured in their handling of data typing, there is more opportunity to work towards standardisation.

There does not seem to be any generally accepted strategy for how much to say in the meta tags. Some companies are content to give their name and a very brief description – 'bookshop', say. These probably consider themselves to be brand leaders. Others give fairly exhaustive lists (Table 3.3).

Notice that, not only are a large number of products listed, but also they are given as several variants: 'Baseball, Baseballs, Baseball Bats, Baseball Gloves, Baseball Equipment', although some search engines with good *stemming* (page 201) capabilities would probably generate hits for them all simply on 'baseball'. Incidentally, there is in practice a limit to how much you can successfully put into the keyword field. Although one might feel the more you had, the more times you will be 'hit', some search engines

Table 3.3 One retailer's exhaustive list of products

<META NAME = "keywords" CONTENT = "The Sport Company, Sporting Goods, Sports Equipment, Sporting Goods, Sporting Goods Equipment, Sporting Goods Manufacturer, Baseball, Baseballs, Baseball Bats, Baseball Gloves, Baseball Equipment, Basketball, Basketballs, Football, Footballs, Football Equipment, Footwear, Shoes, Tennis Shoes, Walking, Walking Shoes, Golf, Golf Clubs, Irons, Wedges, Putters, Woods, Golf Gloves, Golf Bags, Golf Balls, Racquetball, Raquetball Racquets, Squash, Squash Racquets, Soccer, Soccer Balls, Tennis, Tennis Raquet, Tennis Raquets, Tennis Rackets, Tennis Balls, Tennis Strings, Tennis Equipment, Softball, Softballs, Softball Bats, Softball Gloves, Softball Equipment, Uniforms, Baseball Uniforms, Softball Uniforms, Football Uniforms, Basketball Uniforms, Volleyball, Volleyballs, Youth ">

refuse entries from more blatant attempts to use the entire dictionary. (Nevertheless, it is surprising how many sites that have nothing to do with aardvark or zoogeography do pop up when these terms are used in searches!)

It is important to keep checking ones listing on search engines and trying to understand what can be done if it is seldom found.

3.9 MONITORING SITE USAGE

The ultimate measure of the success of an on-line store is the volume and value of sales it makes, for a given cost. However, this measure alone cannot tell you why the store has been successful, nor ensure that this success continues. Just as with traditional stores, we need to look at shopping patterns and the customer behaviour leading up to the sale. We must try to obtain information and form theories regarding the attractiveness of our site to our potential customers. To help us do this, there is a wide range of monitoring and analysis products available for integration with our servers, that can answer questions such as 'Where do our customers come from?' 'How many do we have?', 'Which are their favourite pages?' and so on. A typical monitoring package, which can be applied to a small merchant server installation, without imposing too large a data storage or speed penalty on the server will provide the following information, for any defined period:

- Number of 'hits'.
- Number of 'accesses'.
- Number of 'unique visits' by 'unique hosts'.
- Average and peak number of hits/hour.
- Mbytes of data served during the period.
- Histograms of usage against time of day, day of month, etc.
- Pages visited, their frequency and from which page they were visited.
- URLS of common visitors and their frequency of visiting.

In addition, some information on real-time performance, such as time-to-service a request may be given if you host your service on an ISP's platform. Beware, servers in overload can generate false data, counting repeated attempts as new ones (see page 379).

Indeed, all of these statistics need to be treated with caution. For instance, everyone quotes number of *hits*. This is not surprising, as it is the easiest way to record impressive numbers. But a hit is equal, not to the number of visits made to your site, but only an approximate measure of the number of HTML files or images retrieved from your server. A typical page contains several such items and thus calling down one page may well give rise to several hits.

The term *access* is sometimes also used: this is a rather more accurate estimate of the number of 'real' pages pulled down by users. However, even this can be misleading: a set of framed pages, viewed by users as an apparent single item, will actually result in several accesses.

The term *visit* is an attempt to remove some of this over-optimism. As we pointed out elsewhere, the Internet protocols do not contain any real concept of a 'session' – each HTTP request stands on its own, in a memoryless (*stateless*) environment. A visit is really an attempt to guess whether or not a single IP-addressed client is making continued use of the server, by placing a time-out window around any block of requests. One can also use cookies as a way of generating reasonably defined sessions by a customer, but these too are often arranged to time-out. Thus, whatever way you look at it, Web statistics need to be treated with caution.

It is worthwhile, however, gathering information on repeated behaviour. After all, people may visit your site once, purely out of curiosity or even by accident, but, if they come back at a later date, chance is that they are more than casually interested. Thus if you count repeat visits from each IP address accessing the server, separated by a larger interval, say a day, a week, a month you may get a surer indication of success. You can also collect information on which country name is producing repeat visits or whether you are hitting the.ac or.com community most.

3.10   HOW MANY PROSPECTIVE CUSTOMERS DO YOU HAVE?

In the end, perhaps this is most sought-after information. Here too there is more than a hint of ambiguity. A prospective customer could be defined as a *client* (in terms of a unique PC, for instance) visiting our site, but how do we find out this information? In the case of Internet/Web access, the requesting client sends an HTTP message which contains the unique IP address of the requesting *host*, but this is not necessarily the same thing as a unique client identity.

Figure 3.5 shows an example of a large number of PCs connected to a manufacturing company's corporate intranet and screened from the outside world via a firewall. Irrespective of which one of those PCs accesses an external Web server (say, a component supplier's merchant server), the server will only receive one IP address – that of the firewall. As far as the Internet is concerned, the whole corporate network is represented as one host. (As indeed it should be, for security reasons.) The consequences of this may or may not seriously distort your statistics; it depends on the nature of your business. If you are selling widely to a number of individuals, then firewalls may not be a problem, but if you are dealing in products likely to be purchased from multiple sources within large corporations – stationery, office software, etc. – then you will have no

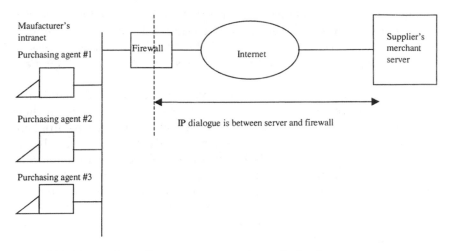

Figure 3.5 How a firewall can obscure customer behaviour

way of estimating your success within them. You would not be able, for instance, to find out which purchasing agents actively sought out your company's products and which avoided them.

There is a similar problem when you are dealing with customers who dial into the Internet and access your site via an ISP – unfortunately this is probably the most common route for retail customers. As shown in Figure 3.6, these customers do not possess (for the purposes of these transactions anyway) such a thing as a permanent and unique IP address. Strictly, the Internet TCP/IP dialogue is between the merchant server and the ISP. The latter effectively extends it to the customer usually by means of the *Point-to-point Protocol (PPP)*, which assigns to their a temporary address for the period whilst logged into the ISP, which is IP address that you will detect at your server. This can change each time the customer logs on again with the ISP.

One other class of customer may mislead you: the automated retrieval inquiry. A number of public search-engines regularly crawl the Net for

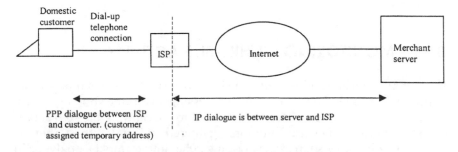

Figure 3.6 How customer identity is obscured by dial-up access

sites to index and include in their repository of Web sites. Of course it is good to be 'hit' by such a service: in the end it will mean more genuine accesses, but do not be deceived into believing that all things accessing your site are people looking to buy your products, not even if they make repeated visits.

The most reliable way to ensure that you are dealing with identifiable customers is to require them to log-on. The simplest way is to let them define a password, which is stored in the server. There are two problems with this approach, which need managing. Firstly, people are very bad at remembering passwords. Calls to computer help desks are overwhelmingly dominated by 'Please reset my password' requests. This is not helped by exhortations to use non-memorable ones and not to write them down, which is eminently sensible in a security context, but acts against you when all you want is to get the customer to identify themselves. One useful variant is to ask the customer to use an eMail address as a reference. This has obvious advantages: it is usually much easier for the customer to remember and has fewer overtones of security; it also gives you the potential to use the information for out-bound eMailing. This prospect may put some customers off and has to be used with discretion.

Another adverse aspect of password control is that it subtly alters the relationship between customer and vendor: in free-access shopping (whether real or virtual), the customer is in control and can choose to enter stores of their choice and reveal or conceal whatever personal information they choose. Once an element of access control is introduced, this freedom is diminished. Common sense would suggest that the customer would expect something in return. Some on-line stores use access control as a way of trying to establish a loyalty relationship, under such slogans as 'In order to serve you better, we need to collect some details from you...'. Others which offer physical delivery of goods, (notably groceries), ask for what is essentially user identification details as part of acquiring the delivery address. Information of this type which tells something about individual customers is valuable from a marketing point of view and we discuss it later in this chapter in relation to customer profiling and notification services.

3.11 MONITORING WITHIN-SITE ACTIVITY

It is not just repeat business from one IP address that is worth monitoring; almost all customer behaviour whilst accessing an on-line shop or service can be of value. Although we have tended to concentrate on Web server statistics, because they are commonly gathered and understood, there is no reason why other sources of shopping behaviour cannot be analysed in the same way. User behaviour can also be monitored on the client (the user's terminal) for instance. If the interaction between the client and the server is mediated, for example, via an ActiveX or JAVA interface running on the client and collecting statistics as it does so, then a large amount of information can be gathered. One specific non-Web example comes to mind: a company sold kitchen units through its smaller branches via an in-store multimedia kiosk, when space was too limited to show physically the entire colour range for each type. Users navigated through the catalogue and, as they did so, the choices they looked at, and the speed with which they moved off them, were recorded. Some rapid rejections of some designs were recorded and, in consequence, they could be withdrawn from crowded 'real' displays.

Capturing customer behaviour patterns in this way can be used to create either a database of individual customer behaviour, as part of a targeted marketing campaign, or for use in a pooled database for product analysis purposes. In both cases, but particularly the former, where individual names and addresses are associated with individual behaviour patterns, great care will be needed to ensure data privacy and rights issues are addressed. The whole process of deciding what to do with individual customer data, whether collected via on-line or at the checkout is, of course, still in its infancy.

As on-line sales and out-bound selling become a more regular part of retailing, it is also possible that customer interaction with the on-line store will be monitored in real-time by intelligent software which will be able to offer the chance to speak to a human assistant. The technical issues surrounding this are discussed on page 316. In service terms, this could mean, for example, the offer of assistance being made when the customer has progressively narrowed their navigation into a focused range of products. Indecision might be detected by the switching back and forward between one or two of them. At this point, the intervention of a trained assistant might be all that is required to clinch the sale. (One is reminded of the dictum, 'Financial services are sold, not bought'.)

3.12 PORTALS AND CHANNEL MASTERS

We have already mentioned the concept of the *portal*. In general terms, a portal is the first or main point of entry that a user makes when going on-line. It could be the welcome page of an ISP; it could be the home page of a major provider of a wide range of on-line information, for example Altavista or Yahoo. Portals which dominate the market for delivering one or more fields of customer interest, are known as *channel masters*. Examples of these include the major search engines and a number of on-line news channels. The main goal of a portal is to get in front of as many potential customers as soon as possible in their online session. To users, it fulfils the equivalent role to the entrance and store guide to a physical shopping mall. To vendors it provides a good way to get high *footfall*. Although many vendors try to publicise the URL of their home page, with a view to getting prospective customers to key this into their browsers, it is nevertheless a fact that many customers, perhaps the majority, access a site via someone else's portal. This is true to such an extent that many eBusinesses make a large proportion of their revenue (perhaps, all of it), not by selling products, but by providing access to the products of others. Technically, portals are just rather complex Web sites with high functionality, much of which may also be relevant to individual vendor sites.

3.13   THE CORPORATE PORTAL

A variant of the portal concept is the *corporate portal*. Rather than being the unique port of call (rather, the unique URL) for all information on a particular area of customer interest, it is the dominant access URL for all information on a specific company, its products and its environment. As such, it may not only cover customer information, but also equally provide information to people internal to the company and to its suppliers. Some define the corporate portal as being entirely within the corporate firewall. Under this definition we could have put this section wholly into Part 3, *Managing eBusiness Knowledge*, where we discuss corporate knowledge management, but this would not have highlighted what is possibly the most exciting and radical aspect: the way that corporate knowledge and activities can be driven directly from customer experience. In this approach, customers, suppliers and the intermediating company which sets up the portal are not treated as separate entities, but as partners (albeit, with qualified rights) in a unified enterprise. That, at least, is the theory. In practice too, there are some gains to be got from some degree of 'externalising' the corporate knowledge base: driving it from customer preference profiles, allowing customers to comment on-line regarding current products, product-reuse opportunities and product improvements, etc. Particularly in the case of companies which sell to corporate customers, an information library of *White Papers*, is a good way of indoctrinating designers into the merits of one's product. (As anyone who has written a technical book will know!) Moreover, the Portal does not just inform customers; it can be a way of sharing knowledge within the company as well. Technically, this is not easy to do: it requires integration at the basic wires and data format levels and it also requires creation of global and local models for corporate knowledge, (sometimes known as the *knowledge map* [146]).

A corporate portal will look rather less like an on-line shop and more like an information gateway such as Altavista, Yahoo, etc. except that it will concentrate on the company's products and services, although perhaps not exclusively. (A pharmaceutical retailer, for example, may

Figure 3.7 A corporate portal

have links out to other health-related sites.) The layout varies, but broadly follows the outline given in Figure 3.7.

Clearly it is supporting the primary function of the organisation by offering a shopping service for those who have identified a specific need and possibly a product that meets that need. It is also selling via more subtle channels: it offers news services which may be product related, perhaps announcing something new or at discount. But it is more subtle yet: it offers a number of 'information services', less directly related to immediate sales opportunities. These can include sports events and charities, sponsored by the company, diets, health information, links to related external sites (perhaps displayed in a new browser window as discussed on page 370). There may be a 'tell-us' facility that allows users to send messages commenting, for example, on products.

Probably there will also be a reserved area, open only to trading partners who have obtained access rights and passwords. There may also be an on-line entry system to allow validated intra-enterprise users to contribute content.

Business-to-business sites will, in general, probably offer more complex negotiation facilities, for the purpose of selling goods, since business contracts are often more than a simple, single transaction. This means that the seller must provide a more complex on-line catalogue. The catalogue will contain meta information surrounding the basic product data, that can be used by purchaser processes, to identify appropriate suppliers. As discussed in Part 3, *Managing eBusiness Knowledge*, there are no generally agreed rules for such meta information apart from that agreed on an ad hoc basis by groups of suppliers and purchasers. However, these groupings are beginning to be quite common and it is likely that industry-wide standards will emerge in the next few years.

3.14 SEARCH FACILITIES

It is very important to make it easy for customers to find their way about your site. Some of this can be achieved by good design logic, by good style and by providing simple site-maps. One approach is to introduce *categorisation*, the splitting up of information into categories that have separate appeal. *Yahoo* is one example of a portal which adopts this approach. But if the site is large and complex, then some form of search engine also needs to be provided, whether the site is for retail or for business-to-business and whether it is meta tagged or not. Effective searching through large volumes of information is a complex process and a very active area of research. In Part 3, *Managing eBusiness Knowledge*, we have discussed some of the theory, but particularly in relation to corporate knowledge management. We can extend this discussion to cover approaches particularly relevant to customer-centred searching.

The simplest case arises with limited amounts of data made available on static Web pages. Here all that is required is to submit a search query (usually as a form) to a CGI-initiated process which runs a comparison between the string of data keyed in by the user and the text on the Web pages. Very often these CGI programs are written in the Perl language, which has good string-handling capabilities and an extensive library of common operations. It is relatively easy to provide quite complex facilities, for example the ability to find strings where several individual words occur within a certain distance of each other, and so on. Also possible, with rather more programming effort, is the provision of *stemming*, searches which not only cover a specific string but also variants of it; for example, searching for 'custom-car' could yield items on 'customcar', 'custom car' and even 'customise cars', or 'customizing autos'. To program these into the system, the programmer has to decide whether to set up simple lists of words which are considered synonyms – meta data, meta-data, meta information – or to introduce general grammar rules such as:

Rule: word pairs X Y are synonymous with hyphenated word X–Y, for any words X and Y.

There is no general rule for getting the balance between a synonym-based and rule-based approach. If the site is big enough, it might be cost-effective to buy an off-the-shelf search engine. There are a number of these available and reviews are published on the Web, for example [147].

If the product catalogue is hosted on a modern database, such as Microsoft's SQL version 7 server, then it is possible to set-up an indexing operation across the pages stored in the database. Typical indexing operations can result in the creation of word-frequency counts for each page stored (that can automatically exclude 'common words' – the 'the's and 'and's that probably do not give any information on the 'aboutness' of the page), lists of URLs, etc. It is easy to programme a search engine to examine the index, rather than the full-text version, thus speeding things up. Stemming and other intelligent operations are also becoming available as standard components of databases.

Finally, we refer back to our discussion on page 200 regarding the difficulty of simultaneously getting all the information you want (*recall*) whilst being able to reject unwanted material (*precision*). This will always be a problem which has no analytical solution and the performance of a site can only be found by regular testing.

3.15 CUSTOMER PROFILING

The relevance of material on a Web site or elsewhere is not an absolute thing: it is highly subjective to the individual user. Basic to the need to provide a searching service is the realisation that your customers are only interested in a subsection, usually a small one, of the entire data on your site. Customers who use a search engine are saying something about themselves – their interests, their location, price points, etc. Rather than present them with a generic search facility, it is desirable to customise one around their specific interests. We can consider the act of gathering these interests as customer-initiated market segmentation. If we can get customers to register with the site before they start a search, or get them to complete an on-line form, we can then create and hold in a database, a personalised profile which can be used to help them find what they want in future. Better still, we can be proactive in bringing to their attention any new developments of interest to them.

In its simplest form, we could get users to tick boxes on a form that listed frequently changing items of interest to them: current price-list, all products in a particular range, and so on. More complex processing, either manual or using some of the techniques mentioned in Part 3, *Managing eBusiness knowledge*, could also correlate these specific interests with other related products: motor car purchases implying a probable interest in low-cost finance, for example, (taking care not to infringe any cross-marketing legislation). A posted or eMailed notification could then be generated and sent to the customer.

This is a good and simple process for direct selling. Where the purpose is more to inform, rather than sell a product, as is the case with giving corporate pricing information or running an on-line news service, then an acceptable alternative is *channel push*.

3.16   CHANNELS AND PUSH

The portal provider decides on a segmentation of Web information that it thinks will appeal to its user-base. This is an editorial activity, not dissimilar to the way that a newspaper splits its content into sections such as 'news', 'sport', 'finance', 'weather' and so on. A Web portal will offer a link to a 'finance channel', say. This channel will itself be under some level of editorial control. It may be provided as a complete package by an established content provider – Dunn and Bradstreet or Financial Times, for a financial channel, for example – rather than simply a loose collection of Web links. The intention is to give the customer the impression that the added value of professional filtering and presentation has been laid on top of raw Web data, for their benefit. If customers accept that this has been done successfully, then, the theory goes, they will be happy to be given this information by a proactive service, rather than make a conscious effort to access it. This is the concept of channel *push*. It is applicable to news services as well as to corporate portals serving internal and external users.

As shown in Figure 3.8, a number of content providers produce a supply of new information, which is labelled with the channel identity. They do this using a publishing tool supplied by the operator of the push service and they are able to load their information into the service provider's database. The service provider has 'signed-up' a number of users to its service. Signing-up typically involves downloading a browser plug-in whose function is to enable the user's machine to poll the push-server at preset intervals (usually determined at set-up, by the user). In some cases, the plug-in may inform the push server of which channels ('news', 'sport', etc.) the user has an interest. If the polling by the user's machine detects that the server is flagging-up a change to a channel, then the plug-in downloads this information and may alert the user to the arrival of new

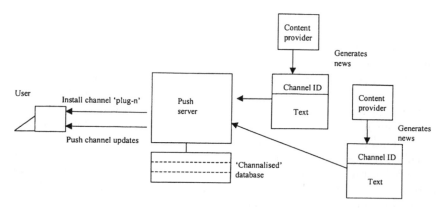

Figure 3.8   Channel 'push technology'

information. There are a number of ways in which the plug-in can notify the user of an update: one approach is to provide a piece of *screen-saver* software, which includes the polling application and which generates a message on the screen, when the system has been inactive for a period of time. An alternative is to modify the browser so that channels that have changed are flagged as such.

We can see from this description that 'push' technology is rather a fraud! The process is actually client *pull*, although it may not appear like this to the user.

Some years ago, when push technology was first promoted, it was rather oversold and the technology itself rather predated the ready availability of Java-enabled applets on standard browsers. Today, much of the proprietary technology of push has been much reduced in novelty value by these applets, which allow alert messages to be easily displayed, often in a way which is less annoyingly obtrusive to users. Channelling, however, in the sense of providing users with professionally edited on-line content, is a growth industry. The channels tend to be on portals rather than as unsolicited push through screen-savers and the like. Thus, from the client end, the push has been largely removed, in favour of HTML pages that the user chooses to view or 'pull'. At the user's end, channel creation is thus a service activity, rather than a client technology. This is not quite so in the case of the server and content provider end; technologies such as XML are clearly relevant to lowering the cost of editorial management and the assigning of information to the appropriate channel. One approach is Microsoft's *Channel Definition Format* which uses XML to define the meta data elements necessary to describe and manage the content. The specification defines a number of major elements but also leaves options open for further extension. The major elements are shown in Table 3.4. Microsoft has produced an XML Document Type Declaration (DTD) for these.

The use of a standardised format such as CDF means that material prepared by a content provider can be shared across a number of channel providers that share this format. It also makes the content generally more acceptable as it can be seen to conform to good-practice and therefore of editorial added value. This is attractive to eBusinesses involved in selling information services (either by subscription from end-users or by site advertising revenue) and it also has some implications for product vendors. In the case of the latter, it may be seen that a well-crafted news-bulletin that is obviously appropriate to a well-publicised channel can be an effective and low cost way of obtaining advertisement, for example for a new product, if it produced according to formats accepted by industry. By advertising via a portal's channel services, vendors can target a wider, more specific audience than they might otherwise get.

So far, portals are mainly geared to handling information 'feeds' from major information providers, in a publishing format that is heavily edited

Table 3.4 Definition of CFD elements

Elements	Description
LastMod	Last modified date for this Web page
Title	Title
Abstract	Short description summarizing the article (200 characters or less recommended)
Author	Author
Publisher	Publisher
Copyright	Copyright
PublicationDate	Publication Date
Logo	Visual Logo for channel
Keywords	Comma delimited keywords that match this channel
Category	A category to which this Web page belongs in. The string value is a URI to a CategoryDef element
Ratings	Rating of the channel by one or more ratings services. (String found in PICS label meta tag)
Schedule	Schedule for keeping channel up to date
UserSchedule	Reference to a client/user specified schedule

using human intervention, but, in the longer term channelling and standardised mark-up lends itself to almost fully automated (therefore low-cost) information distribution. In this regard, it might be useful for vendors to consider how their external publications (product specifications, 'white-papers', etc.) can be integrated with the information available to internal users, on the corporate intranet for example.

3.17 MOBILE 'PUSH'

We should not leave off discussion of push technology without mentioning its particular significance in the case of mobile terminals. Part 1 described the Web mobile phone and the WAP standard. Unlike PC-based systems, where, as we have explained, 'push' is really client-pull, there is a genuine push element about mobile services. Part of the WAP standard is an ability to generate *alert* messages. The WAP designers are aware that mobile communication environments are different from the static case, both in terms of mode of use and purpose of use. A mobile telephone may often be switched off to conserve battery power or to avoid interrupting other activities, thereby it must be told at a later date that a message is waiting. Also, people on the move may need timely alerts to traffic problems or to the nearness of place of interest – car park, shop, etc. Their on-line behaviour is more focussed and immediate; they are unlikely to browse the network so freely as a static user.

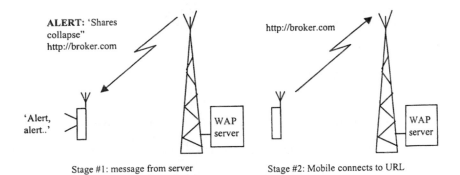

Stage #1: message from server   Stage #2: Mobile connects to URL

Figure 3.9 WAP alert process

Consequently, there is an alert function on WAP which allows the WAP server to send a message to a WAP mobile, telling it that 'something' significant has occurred. Depending on the design of the mobile, it can either store the alert until the user decides to give it attention, or proactively generate a ring or other alarm. Whatever method is chosen, the alert message sent by the server essentially consists of an alert short title and the URL of the location of the full message (Figure 3.9).

Thus, the user's response to the alert is to pull-down the WML page that contains the full message.

3.18 eMAIL 'PUSH'

We mentioned the hype surrounding push plug-in technology and how it had evolved more towards portal channelling. This was not the only evolution that happened. Whilst push was being heralded as tomorrow's technology, a lot of the little guys (and some of the big ones) were quietly achieving the same effect by using eMail. *Direct marketing* by letter mail has been a very cost-effective way of selling for many years [148] and the semi-automated version involving *database marketing* [149] had developed the techniques of profiling and targeting long before on-line selling had taken off.

Post and eMail have one great advantage – you are dealing with a named customer, rather than an IP address; it has the disadvantage that it can be less structured and labour-intensive, particularly if one is handling in-bound messages, whether paper or electronic. As we discussed in Part 4, *Service and Support*, a level of automation and integration with other more structured processes can be achieved, even the intelligent sorting of incoming mail by natural language processing, but one has to accept that the messages will sometimes require manual intervention.

An interesting approach which is increasingly being adopted, is to integrate out-bound mailing with the company's Web site: information,

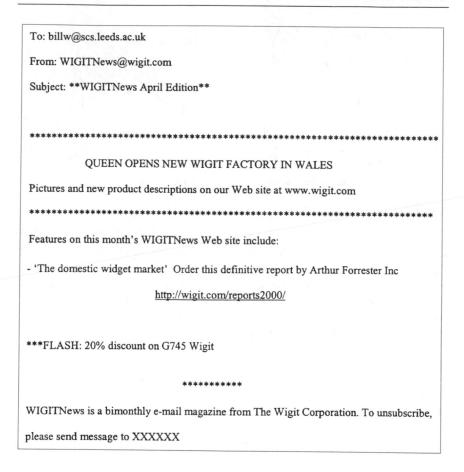

To: billw@scs.leeds.ac.uk

From: WIGITNews@wigit.com

Subject: **WIGITNews April Edition**

****

QUEEN OPENS NEW WIGIT FACTORY IN WALES

Pictures and new product descriptions on our Web site at www.wigit.com

****

Features on this month's WIGITNews Web site include:

- 'The domestic widget market' Order this definitive report by Arthur Forrester Inc

http://wigit.com/reports2000/

***FLASH: 20% discount on G745 Wigit

****

WIGITNews is a bimonthly e-mail magazine from The Wigit Corporation. To unsubscribe,

please send message to XXXXXX

Figure 3.10   A typical eMail push

perhaps customised, is mailed out in summary, with the invitation to visit the Web site for more information (Figure 3.10).

As can be seen, the opportunities for seductive graphics are limited, but this should not be a problem, as the intention is to get a short, snappy 'teaser' onto the customer's PC. Brevity is important, as long messages are often not read and are felt to be an intrusion and consumption of eMail resources.

The idea is to make it as easy as possible for the customer to then get to areas of specific, in-depth interest. Note that Figure 3.10 shows two ways to do this. In the 'headline' (the royal visit) the URL of the main site is simply given in flat text. In the 'CORBA' item, the Web address is given as a hot link, which can be clicked to give immediate access to the browser. This is only possible on some eMail systems. It can also be messy to use the full address of the page and it may be better to link to an intermediate contents page on the server.

Also note the offer of a consultant's report. This is 'something for noth-ing' and so it would be reasonable to direct the customer to a page where they first completed an on-line form, for your marketing purposes. Customers who access your site directly can also be pointed at this form, which is a good way to get them to subscribe to the eMagazine and provide eMail details.

Incidentally, as with paper mail, the key is to get hold of appropriate addresses. Order forms and other input documents can clearly be used to enrol your existing customers. Lists can also be bought. It is important to be very wary about sending bulk *spam* (unsolicited) mail to on-line user-groups, for example. This often causes very vocal and damaging offence and can result in your site being bombarded with eMails of complaint and, perhaps, attempts at hacking. Nor is it advisable to steal lists. Experienced marketeers know that almost all paper-mail lists that are for sale contain one or two dummy names and addresses which are under the control of the list seller. By monitoring mail to these addresses, they know that it is their list that is being used. It would not be excessively suspicious to believe the same thing happens with eMail lists.

3.19   THE PRIMACY OF CONTENT AND OWNERSHIP OF THE CHANNEL

It should hardly need saying that the most effective way to get customers to access a site when they are not driven by an immediate need to purchase something, is to make the site interesting and relevant to them. This also applies when information is pushed to them. Numerous studies show that people do not want to reminded of the status quo; they prefer to be told only when things change. (And only when these changes are 'relevant'.) Thus it becomes important to construct sites so that they can easily incorporate changes and to be generating sufficient items of interest. We have mentioned various channel-creation tools and services. Once the initial novelty of 'being on the Web' has died, the on-line team need to devote considerable energy to assisting the functional departments, particularly marketing, in their need to manage the Web site, so that decisions are business-led rather than left to the Web-master. We see a progressive increase in cost in this area once the Web site becomes at least one of the prime interfaces with customers. Traditional skills such as merchandising will become integrated into the Web-based channel. Therefore, there will need to be tools and packages developed to allow this to happen, using people who are not IT experts. In the same way as functional departments wrenched away many of the powers of centra-lised IT a few years ago, they will take control of the Web channel. Their requirements will be complex and involve end-to-end integration, relating stock holding to sales campaign. They will also want parallel integration of visible presence on the Web (and interactive TV, etc.) with customer service, support and fulfilment facilities.

Thus, in the mature market place, on-line marketing will have adopted many traditional values.

Appendices

Appendix A

ePeople: Choosing the Team

So far, we have discussed hardware and software, platforms and languages, necessary for constructing the eBusiness. In this brief chapter we turn our attention to another aspect, without which nothing will be achieved: *liveware*, the engineers who will build the eEnterprise, and the skills and training that they must have.

Although perhaps chiefly intended for managers of technical teams and for those who help them recruit, it may be also be relevant to experts at the 'byte-face' who may also find it useful to compare their CV with what follows, (or perhaps use it as a training plan?) Organisations might like to carry out a 'skills-audit' across their company, to see whether they measure up to the requirements, before committing to ambitious eBusiness programmes. The examples given may not necessarily be the correct routes to every design, but they represent mainstream thinking in the business and anyone who did not have a basic awareness of a goodly number of them, might not be considered to be completely on top of the subject. Conversely, it is unlikely that many will be found who know everything about them all. Thus any team will include detailed code hackers and requirements-capture people, as well as those concerned with middle-ware systems design and systems architects. The team must also possess skills and attitudes that are a positive contribution to service provision and technological and process change.

A1.1 PLATFORM KNOWLEDGE

This covers a wide area of hardware and software technology, including computing platforms, transmission, mobile, terminals.

Application servers

Competence in application servers is less a skill than a set of experiences, particularly so because it is vendor-specific and depends upon the scale of the operation. Many companies will make use of Internet application service providers, who will run the applications and associated catalogue databases in an operational environment. Application development may, however, be done in-house and, in this case, there may be a need to employ a systems administrator with specialist knowledge of the server and the associated network. The choice of operating system is split between UNIX/LINUX and Microsoft Windows NT products, the hardware platforms include PC-based systems and Sun Solaris as the UNIX leader, and recruitment will presumably follow on this basis.

Networks

Many companies will outsource part or all of their networks to third parties, but where they still retain some network responsibility, standard telecommunication and data network skills are all that is required. There is no special 'e-business network' strand. Watch out for fossilised partisan bias on the telecoms/data divide. Although things have changed for the better, there has been in the past a tendency either to disparage the ability of data networks to ever achieve quality of service, or to have a naïve belief that the telcos should junk all their existing networks and instantly roll-out IP everywhere. There are still areas of genuine uncertainty. Anyone who 'knows' all the answers, should be looked upon with suspicion. In terms of specific technologies, knowledge of TCP/IP must be top of the list. (If this is well understood, then other packet protocols can be picked up.) Where call-centre design or specification is required, then the ability to understand the latest developments in *automated call diverters* is necessary, but also knowledge of how these will migrate into a mixed voice and data or computing architecture (see chapter). Anyone holding a responsible position in wide area network design or selection must know about *ATM* as well as the Internet quality of service techniques of *RSVP, Diffserv, multiprotocol routing, tag-switching, etc.* and be able to form a judgement as to their current status.

Mobile servers, WML and mobile application development

As stated frequently in this book, there is a great interest in mobile Internet applications, particularly, at the present time, those using WAP. The WAP server, or gateway, is simply a piece of software that runs on any

HTML Web server environment and there is probably no need for specialist administration skill. WAP applications are written in *WML*, a language which can easily be mastered by anyone with HTML expertise, since they are very similar. WAP technology and software are immature. Designers must be comfortable with this state of play and possess judgement rather than pure enthusiasm.

Interactive TV, games and other client platforms

These platforms are a minority choice for current eBusiness delivery (the PC dominates.) Because of this, programmers should be capable of avoiding the tendency to assume that PC architecture is necessarily the correct way. The alternatives have special features: they are consumer products, whose specifications are heavily influenced, that is, restricted, by price. On the other hand, they are expected to give an emotionally ('seductive') satisfying performance. Because of the restricted functionality, it is usually not possible to separate hardware from software and programmers may require *embedded systems* design skills and be comfortable at programming in very low level code. At the same time, they must be able to cooperate with graphics design experts, whose task is to create the seductive product.

Delivering to interactive TV may also require understanding of special-purpose server architectures and of streaming file-handling.

A1.2 LANGUAGES AND MIDDLEWARE

In fully integrating all of a company's electronic resources into a functioning eBusiness, there may be a very large skill base required in programming at the basic coding level. There are many legacy systems still in operation as the recent tremendous demand for COBOL programmers to work on Y2K problems should remind us. However, existing staff will probably be trying to forget their old languages; alternatively these can be brought in specially; what is more important to look at is the competence in main-stream or emerging code. So, we concentrate on what staff should be learning, if they do not already know it. It is not always easy to distinguish between a 'language' and a 'script' or a piece of middleware code, but, as a rough and ready guide, treat 'languages' as largely application and vendor independent and sitting above the operating system and below any middleware such as CORBA, etc. As we explained in Part 2, *e-Business Systems Architecture*, there are no indications that specialised operating systems for eBusiness are likely to emerge. The preferred solution is to glue things together using one or more *middleware* products. Which of those a company chooses will largely depend on its legacy and its strategic purchasing policy. This in turn will be determined by the knowledge and experience of its systems architects. Widely used languages and scripting tools include.

HTML

HTML is a moderately complex document formatting language, certainly not as complex or creative as its ancestor, *SGML*. (Skilled SGML code writers will be more than competent members of any complex document management project.) We would expect any Web designer to be competent in HTML, not just in the ability to create basic pages, but also in the use of frames. Someone who claims Web expertise but only knows HTML for static Web pages (whether framed or not) is not really a professional, but might have good (artistic) design skills which would need to be examined separately from their technical competence.

WML

See under 'Platforms'.

Visual interdev, visual basic, active server pages

These are the main bread-and-butter programming and scripting (the different between the two terms is ill-defined) tools used for creating on-line applications. It is relatively easy to pick up basic skills sufficient to create simple applications, but they are also capable of extremely sophisticated and complex use. Thus, someone claiming 'some skill' with them, needs to be tested further to find out the limits of their ability.

C and C++

Most professional programmers will have served at least some of their time writing C and/or C++ programs. These are still very relevant languages for the whole of eBusiness programming. Also worth noting is the fact that many guides to other languages often describe features as being 'similar to' or 'different from' C/C++, thus making it rather more difficult for someone unfamiliar with them to learn the new language. C++ is, of course, the newer of the two, and is object oriented, which would not necessarily be the case with C programs or C programmers.

Perl/CGI

Perl is a very important and flexible language for Web development. Its principal features in this respect, are, firstly, its extremely good text string handling facilities, which make it ideal for writing search engines and data indexing and retrieval applications, secondly, as a very convenient language for applications written behind Web servers and called up by *Common Gateway Interface (CGI)* commands. Perl is probably the most widely used language for CGI-related applications. Perl coders are serious programmers. Whereas it is possible to write a few lines of Perl in order to handle a simple problem, the language in its fullest extent is very flexible, feature-rich, with extensive class libraries (newer versions follow an object oriented model) and capable of similar levels of complexity to C and C++.

Java

Java skills come in at different levels: it is possible to be able to understand and manipulate JavaScript for writing simple Web applications, particularly on clients, but understanding how to use JavaBeans as part of a component architecture probably indicates a higher skill level, if only because the technology is newer and there are fewer people to ask when things go wrong. See also the middleware section for further discussion on Java server-based coding.

A1.3   MIDDLEWARE

Although not a rigid divide, middleware skills are more 'architectural' than are those of pure programming, placing more emphasis on systems design and integration across a number of platforms. In Part 2 we discuss the technical details of various middleware environments; here we mention them briefly. Anyone calling themselves distributed systems designers should have some knowledge of at least one of these and, preferably, the ability to compare and contrast a couple of them.

Microsoft DCOM

If the company is a Microsoft 'shop' then programmers and systems architects need to be on top of DCOM, Microsoft's distributed programming environment. There are arguments regarding how open DCOM can be, but it undoubtedly has a significant share of the market.

CORBA

This is currently DCOM's major rival, and is an open standard, therefore with the strengths and weaknesses thereby implied. Some of the weaknesses have been removed by virtue of CORBA's ability to interoperate with Enterprise JAVA components.

Enterprise Java

This is a relatively new middleware solution, but likely either to become very important in itself or to have a major impact on future middleware direction. Full competence in Enterprise Java for middleware component solutions is probably evidence of 'expert Java programmer' status.

Databases/ERP/transaction processing

Any database-aware person must have some knowledge of the theory of relational and object oriented databases and their relative strengths and weaknesses. Where any serious database application is concerned, for example beyond the realm of a very simple shopping catalogue, then knowledge of SQL, the dominant query language for databases, is a strong asset. The SQL database market is dominated by two vendors: Microsoft with SQL Server (currently at version 7) and ORACLE, but application-specific solutions may use proprietary systems. IBM's own database software, DB2, and Informix, make up most of the remainder of database products. In the rather more specialised area of *enterprise resource planning (ERP)* field, SAP has huge market share.

Transaction processing was, for many years, dominated by mainframe solutions such as IBM CICS (which is still very important), but more recently, the leading on-line, non-mainframe system is BEA's Tuxedo, with IBM, Sybase and Oracle also significant. Microsoft are also coming into the running. Transaction processing is a technology and an attitude of mind and staffing should be considered in this light.

For largish data warehousing, in hardware terms, IBM mainframes still dominate and IBM have a good hold on the smaller platforms, with HP and DEC taking up most of the remainder. Unix is the dominant operating system, with Microsoft NT increasing its popularity but still some way behind.

Directory services/X500/LDAP

We have emphasised in this book the important role that corporate directory structures can have in building an end-to-end enterprise. Although there are a number of proprietary solutions, particularly in the LAN case, increasingly the Wide Area Network solution is *LDAP*. This is a leaner, fitter, version of the rather florid and over-functional X500 (at least in the eyes of computer network designers – telecoms people might disagree!), and thus anyone skilled in X500 and willing to shift to LDAP should have little problem.

A1.4   ABSTRACTION, MODELLING, INFORMATION ENGINEERING

There is a long tradition amongst the computer community of being sceptical about 'architecture', 'abstraction' and other top-down approaches. ('Old men do architecture, when they can no longer cut code!') Some of this is reasonable: every decade a CASE tool or high level support environment is heralded as the answer to the coding bottleneck, only to disappear without trace, often in a very short space of time. However, there are signs that a more rigorous approach to software design is bearing fruit and a useful perhaps suitably modest, set of tools is emerging. Unfortunately, the problem of modelling the dreadfully named 'non-functional aspects' of security, performance, etc. has still not been tackled successfully. (UML uses informal descriptions, *stereotypes*, as an attempt at the problem, and a range of approaches under the general title of *reflective design* is also appearing in the literature.) The common sense of designers will therefore be key to success, in this regard. In many cases, the designers will probably be comfortable in thinking in CORBA/DCOM/Java terms for the implementation. They will also be reasonably favourably disposed to modelling tools.

IDEF

This is a process and data modelling tool which was originally designed for the US DOD, but has had some degree of success in many civil applications. It has the advantage of being a structured approach which has graphical charts which are relatively easy to use as part of requirements capture, with clients who are not IS experts.

UML

The *Unified Modelling Language (UML)* has emerged as a widely used modelling language of choice. It has been selected by the Object Management Group as its standard. It is still a very new tool, first issued in 1996, and therefore, people experienced in using it can be considered to be early adopters.

Meta information: RDF/XML/DTD

Rather than mention XML in the language section, we prefer to include it as part of the higher layer of design abstraction. XML will not be a particularly difficult language for HTML programmers to acquire and, as such, it is something they could be expected to pick-up 'on the job'. However, the more subtle part of XML is the opportunity that it provides for the creation of properly structured data models which can underpin enterprise processes such as invoicing, goods-handling and so on. An XML professional therefore needs to understand the role of *Data Type Definition* and the problems of creating and navigating between different *name spaces* (essentially the different data types used by different applications or organisations). An appreciation of the more formal structuring methods such as *Resource Definition Format (RDF)* might be an advantage and some faint scars left by previous experience of EDI, might be seen as honourable wounds.

A1.5   APPLICATIONS, SERVICES, CULTURAL AND PERSONAL SKILLS

The main body of this book mentions most of the application areas covered by the broad term of eBusiness and it is difficult to comprehensive summary. Possibly it is best here to concentrate on the areas that are likely to be forgotten in the rush to go on-line. In this regard, as well as identifying specific skills such as eMail and security, we need to mention the requirement to integrate large legacy systems, perhaps built on mainframes and running in an insulated environment. Cut-over to a newer on-line approach will require project management skills as well as technical knowledge. (In fact, employing everyone with too much interest in 'the new', could be a disadvantage!)

E-mail, document handling, knowledge management

Even if only modern technology is to be used, it is important to remember that eBusiness is not just about Web servers and Web applications centred on HTTP. Real eBusinesses will trade via eMail (and telephony, which we mention above under *networks*) and eMail systems are a field in their own right. So also are collaborative systems for handling documents, of which *Notes* (formerly *Lotus Notes*), based on IBM's *Domino* server, is the market leader. The market for general knowledge management, on-line analytical processing, intelligent information retrieval is not dominated by any one vendor, and so staff will have to be recruited on the basis of one's own purchasing strategy, if, as is the wisest choice in most cases, a turn-key solution is bought-in. The people who design for the vendors of such systems will have to have a combination of *Artificial Intelligence (AI)* and database skills, with the ability to understand both the abstract concepts of the former with the practical, performance-related issues of the latter.

Security

In Part 3 we told terrible tales of security problems and highlighted the poor state of training in computer security. The problem about recommending security skills is that this is almost a dangerous process: computer engineers can become obsessed with the technology of security, rather than with the much more relevant issue of security *process*. Part 3 does list a number of technical areas, including SSL, IPSEC, encryption, public key infrastructure, firewalls, etc. but note that it is probably better to recruit someone who has operated in a company with a tight security record and knows little more than the basic technology, than to employ someone who understands the issue of weak primes but not how to administer a 'forgotten-password' duty.

Computer security training is surprisingly hard to come by, as it is not adequately taught at undergraduate or Master's levels in most computer courses or business schools in the UK. Mainland Europe students have a slightly better chance and those in the US better still, but overall the position is disappointing. Security problems are often system-specific and it is important not to expect someone experienced in one system to be as adept on another, without proper experience and training.

Cultural and personal

Companies with long histories are likely to be running applications on in-house or proprietary solutions which may not scale in size or inte-grate with newer on-line processes. Obviously you will need to retain (or contract out) a skill-base that can still service the legacy, but you will also eventually need a migration strategy. The people that can do this will require a modicum of new skilling, but the predominant qual-ity to foster is an open mind. In-house software teams are notorious in their 'not invented here' mentality and a desire to write everything from the ground upwards. Similarly, people with IS responsibility in functional units have made their careers in understanding how, for example, an ERP system can be configured to meet a function's require-ments and it is difficult to persuade them to sign-up to a cross-func-tional solution. eBusiness is a new domain, thereby with particular requirements. But there have always been new domains, and the attri-butes to cope with them, are enduring. The most valuable, and most difficult, skill is judgement, particularly when there is little opportunity to acquire it, and many of one's past experiences are of little, or nega-tive value. We should not forget that neither the major telecoms compa-nies nor the major computer ones really foresaw the value of the Internet; in fact, they were later than a large number of non-experts. On the other hand, on-line optimists and cowboys/girls are going bust every day, through failure to appreciate the value of caution, planning and security. One of the major problems in building the eBusiness will be in achieving this mix of processes and more importantly, attitudes. To give one example, database people and Web people do not share the same attitude about reliability versus time-to-market, but if an end-to-end process is to be effected in time to make money, they have to learn to operate together and respect each other's strengths. Strong project management skills are required. Just as important is a senior manage-ment culture that is sympathetic to technology-induced change, whilst preserving sufficient critical faculty, based on confidence preferably built by taking some time to understand the basic principles, (e.g. by dipping into books like this?)

On a lighter note

After that pontification, perhaps a few examples of disfunctional beha-viour might be permitted, in order to highlight the need to restrain the otherwise praiseworthy enthusiasm of eBusiness computer staff. The following, all guaranteed true, happened in my presence:

- Designer, trying to sell his system to the product marketing team: 'It is so complex and clever that I'm the only one who understands it.'
- Designer, to managing director: 'You'll be pretty excited and pleased to know that we have built your [mission-critical] sales and marketing

system on by far the largest, distributed Unix solution in Europe.'

- Managing director: 'Excited, yes. Pleased....'
- Designer, 'No, I did not involve the central computer organisation [whose job it will be to maintain his system] in the design, because they are too stupid and reactionary'.

Appendix B

eBusiness: The Future

As readers might imagine, the relief of completing a text of this size is somewhat tempered by a reviewer's comment that a 'few words on future developments might round off what is now a rather nice book'. This is especially true given the writer's belief that all pundits of the future ought to be suspended above a vat of boiling pitch and required to make three specific, due-dated predictions, whose ultimate truth will determine whether or not the prophet is precipitated. This chapter, which covers my *tentative* predictions, will be short!

A2.1 GENERAL DIRECTION

The eBusiness market has probably reached the end of its beginning. The current shake-up in the dotcom companies is demonstrating the classic signs of market maturity. No longer is it sufficient simply to be an early starter; it is now necessary to demonstrate differentiated added value. This means that technology alone is not sufficient for competitive advantage; it is how effectively it is deployed that will be the making or unmaking of businesses.

We also have to recognise that many of the successes will come about because of superior marketing skill and the best technology may not necessarily win out. The eBusiness industry will therefore now undergo a consolidation phase, rather than revolution, although a number of new home and mobile platforms will become much more visible.

A2.2 ELECTRONIC SHOPS

I do not foresee a revolution in the design of the eShopping experience within the next 5 years at least. The basic shop-front/catalogue/order-taking model is as appropriate now as it has been for hundreds of years. There will be an evolving increase in attractiveness (*seduction*) brought about by the availability of higher speed local networks, probably mainly cable modem or telecoms ADSL. In some ways cable offers an easier technology solution for local delivery, but for an end-to-end delivery across the wide area, the telecoms companies possess better infrastructure and competence.

There is an outside chance that a significant minority of wideband delivery will be by point-to-point radio systems, but there have to be doubts as to whether it will be possible to achieve good radio propagation to all customers. I also expect some problems to arise from large-scale introduction of ADSL, because of cross-talk between wire pairs, as well as in domestic premises (see below).

Given this increased bit rate, what can we expect to see from it? Probably simply the faster retrieval of Web pages, plus a major increase in the amount of moving video. This is not necessarily a good thing! It requires more design skill to create a seductive offering that incorporates moving images. There will be a need for better tools for integrated multimedia design. We are unlikely to see truly immersive 3D real images, in the foreseeable future, for a variety of technical and production reasons, but 3D animation, in the sense of being able to rotate an object, (e.g. the interior of a car), will be possible and used to seduce and inform. Currently, there are experiments underway with images of synthesised characters and artificially generated speech. Soon the more likeable of these will be used operationally.

There will also be serious trials, within a year or so, of intelligent shopping assistants, which will profile customer needs and will push specific products in a hopefully beguiling way. Their underlying intelligence will be modest, as will be their success, within the next 5–10 years.

Simple shopping agents, promoted as a way of bargain-finding, are already with us, and their deployment will increase. They will neither be perfect, because of lack of standardised product descriptions, nor particularly intelligent. In fact, the best ones will probably rely mainly on human programming, again for at least 5 years.

As I say in the next section, TV-based and PC-based shopping will not directly compete. The former will emphasise seduction and entertainment, the latter, exploration and information.

A2.3   HOME PLATFORMS

One big question: will the home platform for eCommerce be a PC or a TV set? Hopefully we explained in Part 1 that this is rather a false alternative, as the answer is, both, and more. Technology convergence and maturity imply product diversity, not restriction. It does look as if some product with the complexity and flexibility of the PC (but with rapidly increasing power and performance) is likely to be with us for several decades and will be used in most homes for serious exploratory activities including expeditionary shopping. However, there will also be entertainment-based equipment which will be easier to use, cheaper and which treats shopping as a more passive entertainment. Although the technologies of TV and of PCs have converged, they will continue to be differentiated at the user interface level. There is, however, a fascinating issue of product churn, TV sets currently churn on a 10-year purchase basis; computers (effectively business ones) churn in less than half that time. What will happen in the domestic market? Will TV churn faster or PC slower? It is unlikely that the difference will remain so large.

Unless, that is, the PC enlarges its function to become a piece of domestic capital equipment – a central-heating controller, a monitor of food-stuff in cupboards, etc. The role of down-loadable upgrades also comes into the equation: the life-time of the hardware becoming almost entirely divorced from the purchasing of relatively low-cost, incremental software upgrades. Because of the availability of high-speed networks, the PC-lite debate will have new strength. I confess to being very uncertain as to which of these countervailing trends will win out.

At a more technical level, there is also an issue regarding the *wired home.* Undoubtedly there is an increasing requirement for a networked architecture within domestic premises. But will this actually be *wired*? It costs several hundred pounds to rewire a house. There are some major questions regarding how multiple ADSL channels, for instance, can be piped through a building on existing telephone wires without creating radio interference. For the same reason, I discount the use of mains wiring as a generally satisfactory way of distributing signals. It is probable that wireless systems will carry a large percentage of telephony and data traffic within domestic buildings, within the next 5 years. I stick my neck out and say that neither optical fibre nor infra-red appear to be anywhere ready to offer serious competition.

So, the emerging solution will be a wide range of computer-based products, from quasi-passive 'interactive' TV to really interactive PC, with wireless-based domestic LAN and wideband, wired connection plus digital broad or narrowcast entertainment/shopping channel over wireless, within the next 5 years.

A2.4   MOBILE PLATFORMS

At the product level, I do *not* see the triple convergence of mobile phone, palm top and mobile computer, despite the marketing propaganda from the respective camps. The ergonomic issues are too complex and there are serious problems with power supplies and display technology, plus the

412 eBUSINESS: THE FUTURE

critical issue of price-pointing. It is possible that hybrid mobile/palmtop and palmtop/computer products will both emerge as high-end models, but mass-markets will continue to see three distinct products.

If these predictions come true, the consequences for eBusiness are significant: the high performance, high price, computer-based products will try to attain full Web functionality, with good multimedia capability. The low end, telephony products will not. This then raises the question of what will be available on the lower-end products and, indeed, whether they will play any major role in eCommerce. In short, will WAP, or any similar stripped-down browser, be sufficiently attractive for home shopping? The jury has to be out on this. If WAP services are to maximise their chances of establishing themselves in the eMarket, they must, as a matter of urgency, get their act together. Mobility may have a natural advantage in that it can offer geographically related services – 'buy it nearby!' – but I believe it to be touch and go for other mass-market eCommerce, with one exception: electronic money. Payment mechanisms do not need a great deal of seductive surround. Nor do they need particularly high bit rate. It is possible that low-end mobile platforms will be an important way of providing any-place, any-time payment services. The SIM chip in GSM mobile phones is, to all intents and purposes, an application smart card and the mobile service provider delivers a reasonably secure and trusted network.

A complication in mobile networking is the rate at which network technology will cut across from the existing telephony architecture to the more flexible and potentially cheaper IP-based services. Wide introduction of WAP is being held up precisely because of the uncertainty of this transition. I expect the high-end, mobile laptop products to force the pace for the introduction of IP networks, thus allowing business users at least be able to get access to mobile moving video of reasonable, rather than premium, quality, within 2–3 years. I am much more doubtful as to whether moving video (or any other 2 mbit/s service) will be a mass-market product within 5 years, on grounds of network costs.

A2.5 SECURITY

I have said a lot on this and will say it again: poor security is the Achilles heel of eBusiness. Actually, this is not strictly true: it is not fatal, but it will severely reduce the pace of introduction of eServices. There is a cultural issue to overcome, both on behalf of customers, who are cautious, and on behalf of the designers, who have given customers the right to be so. More thought and a higher level of professionalism is required from Web designers.

There are also technical and process issues. Public key infrastructures have been introduced slower than once were expected. The timely expiration and renewal of digital certificates has yet to be absorbed into business processes. Systems for reliable and low cost biometric measurements are only just emerging. On the other hand, smart-card standardisation is nearly there. So when will it be possible to say that security in eCommerce is no greater an issue than security in current alternative ways of trading? I'll go for 4–5 years.

This may not apply to mobile services. Combining security and performance is still not completely resolved. One has to remember that the major issue is not confidentiality but theft-of-service. This is difficult to control in a mobile environment and the illicit market is lucrative.

A2.6   SERVER-SIDE

Server-side activities will see a major increase in component architecture development. Vendor policies will be vigorous and much debated. Microsoft will obviously try to 'embrace and extend' open standards and will build on existing software such as NT and DCOM. However, they will face vigorous competition from their competitors and their support for the 'open' solutions of LINUX and especially CORBA and Java. This is too valuable a market to lose, so it will be hard fought and I do not see an early winner.

The critical areas to develop will be in performance and reliability. This will continue to polarise the choice of platform, depending on one's legacy systems. In practice, it is difficult to separate performance from reliability, because the true requirement will not be for mips per buck, but for fault-free months for a given number of simultaneous tasks. Middleware will begin to have to take more interest in system realities. The disgracefully named *non-functional requirements* will be catered for via new approaches such as *reflection*. Expect gradual change, because of size and legacy.

A2.7   WIDE AREA NETWORKS

'IP everywhere'. This may not necessarily be the right solution, but it will come about. All of the major telcos are feverishly building IP networks to replace their existing voice-centred ones. They, and the major equipment manufacturers such as Cisco, are the only people who can hope to solve the quality of service issue which currently makes it difficult to send real-time data, such as voice or video, over IP. There are currently too many solutions and too many unknowns in this area to persuade the established carriers to rip out all they have got and go direct to a native IP solution. For the next 5 years or so, many will run IP over ATM networks which give a good approximation to a traditional, switched network, with guaranteed quality of service. Even using this rather over-engineered approach should allow a significant drop in price per bit carried. Perhaps costs will drop by half or more. In the longer term, probably because of wider deployment of new optical fibre techniques (principally, *dense wavelength division multiplex* and *optical switching*), sufficient bandwidth will be available to allow for ATM to be no longer required and a simpler routing scheme to be introduced. In all events, over the next 10 years, transmission and routing costs should drop by more than an order of magnitude. Vendors, standards bodies and systems architects will enthusiastically embrace these cost reductions, in a race to develop distributed processing solutions and the design of operations such as call-centres.

A2.8   ENTERPRISE INTEGRATION, KNOWLEDGE MANAGEMENT

An enterprise is a collection of stakeholders that combine some of their processes for mutual good. Much has been said by the eConsultants regarding this integration being the most important development in eBusiness. I would not disagree with this opinion. We do need, however, to be tight in our definitions and realistic in our expectations.

At the nuts and bolts and bits level, this is clearly an area where significant progress has been made. We see component architecture rapidly replacing ad hoc remote procedure calls as a way of regularising the interaction between computers belonging to different parts of the enterprise. Scaleable and affordable platforms for distributed transaction processing are now on the market and will be commonplace within the next few years. Wide area directory services are replacing those centred on the restricted domain of the LAN. Even security has improved a little, with the newer breed of virtual private networks. It is now affordable and reasonably safe to effect a quick joining together of two or more partners' computerised operations. The need is not primarily for new technology, rather for experience in its use.

At a higher layer, that of harmonising data semantics and processes, we are less advanced. True, XML and related technologies offer a convenient structure for the coding up of such information, but the problem of its definition remains. This is a process, rather than a technical, issue and is *hard*. It is a problem with *meaning*, and meaning is not fixed, absolute and isolated. Instead, is conventionally assigned, by a group of people, at a certain time, within a certain context. Unfortunately, for completely legitimate reasons, there are numerous competing groups, varying time-scales and a myriad of contexts. One need only look at the laudable but frustrating activities of the Dublin Core team who are trying to standardise the definition of a minimal core set of 'context-independent' (actually 'library-centric') meta data, to see how difficult and protracted it will be to get a definition of a realistic business process. Semantic frameworks for coding inter-business data are emerging between co-operating partners on an ad hoc basis. Even today, the derivation of standards for vertical sectors will produce results within a couple of years and organisations such as the OMG will have some useful definitions of generic processes within that timescale too. But I am much less optimistic about plug-and-play data definitions that will allow any business to deal with any business without significant re-engineering.

The same goes for data analysis tools using *Artificial Intelligence (AI)*. Over the next few years we are going to see a resurgence of claims for AI as a way of analysing customer data and personalising service. Practical and useful ad hoc solutions have been achieved in limited domains. But it is important to note that there is still no general theoretical basis for *machine understanding*, despite several decades of endeavour. There is no reason to suspect we are yet near a solution. Let the buyer beware!

Similarly, activity in the field of trading using artificial agents, is still really at the research stage. I do expect simple agents to be able to conduct negotiations in benign business relationships, within the next year or so. But competitive trading is another thing. I doubt whether businesses would be wise to permit fully autonomous behaviour from such software, for at least 5 years. I do not expect them to do so. Even if the agent's performance is generally superior to that of a human, I expect them to operate in an advisory, rather than executive, role, for that period at least.

I am also sceptical about the claims that distance will cease to have meaning in the enterprise of the future, because electronic connection between workers, and between them and customers and suppliers, will allow virtual co-location to happen. In 10 years time, probably not before, we may have large screens within offices and sufficient, affordable bit rate between them, to allow us to recreate an 'office next door', for those who are really hundreds or thousands of miles away. Whether this will be convincing, is unproven. There has been very research carried out into the social dynamics of such a scenario. There is informal evidence that it does not fully replicate a feeling of nearness.

However, the use of desktop conferencing will grow steadily, as a means of reinforcing, rather than replacing, a feeling of co-locality. A major boost to its usage will be when IP-based networking makes it affordable to leave the connection open all the time across the wide area.

A2.9 LAST WORDS

In the above I may have painted a conservative and unambitious picture of the eBusiness future. I believe it is necessary to do so, for the popular interpreters and the less responsible futurologists have undoubtedly overhyped things. In many respects, we have so far solved the easy things, the lower layers, the local solutions and the basic functional requirements. Performance, reliability, large-scale integration, security and semantic modelling all remain unfinished. But despite all the caveats given above and my reluctance to commit to too many dates and preferred options, I do believe that we are on the edge of a significant quantitative and qualitative shift in the way we go about our business and personal lives, brought about by technology and its integration. Within 5 years, technology will have made major cost savings to the way enterprises carry out their operations, perhaps even to the extent that it is no longer news. Within 10 years, eCommerce will be a critical part of revenue generation of most retail organisations.

Signing off with these words, I sit, sweating, above the boiling vat!

Bibliography

[1] Smith PR, *Marketing Communications: an Integrated Approach*, Kogan Page, London, 1995.

[2] Whyte B (ed.), *Multimedia Telecommunications*, Chapman & Hall, London, 1997.

[3] Whyte WS, *Networked Futures*, John Wiley & Sons Ltd, Chichester, 1999.

[4] http://www.davic.org.

[5] Held G, *Data Communications Networking Devices*, fourth edition. John Wiley & Sons Ltd, Chichester, 1999.

[6] Eberspacher J & Vogel H-J, *GSM, Switching, Services and Protocols*, John Wiley & Sons, Chichester, 1999.

[7] www.trillium.com/whats-new/wp_gprs.html.

[8] Walke BH, *Mobile Radio Networks – Networking and Protocols*, John Wiley & Sons Ltd, Chichester, 1998.

[9] Swain N, Will new standards allow users room to roam? Global telecoms business, *Euromoney*, Dec 1998/Jan 1999.

[10] Richardson KW, UMTS overview, *Electronics and Communication Engineering Journal*, June, 2000.

[11] Solomon JD, *Mobile IP – the Internet Unplugged*, Prentice-Hall, Englewood Cliffs, NJ, 1998.

[12] *Bluetooth Protocol Architecture Version 1*, White Papers. http://www.nokia.com.

[13] Stone M, Davies D & Bond A, *Direct Hit: Direct Marketing With a Winning Edge*, Pitman, London, 1995.

[14] Midwinter T, Desktop multimedia, in Whyte B (ed.), *Multimedia Telecommunications*, Chapman & Hall, London, 1997.

[15] E-commerce: the guide. *Euromoney and Forrester Research*, 1999.

[16] Loshin P, Tune in, turn on the Web, *Byte*, February, 1997.

[17] *The Effects of Network Downtime on Profits and Productivity*. White paper from Performance Technologies Inc.

[18] Hatton, Software failures, *IEE Review*, March, 1997.

[19] *IDC Executive Insights*, November, 1997.

[20] Bissell RA & Eales A, The set-top box for interactive services, in Whyte B (ed.), *Multimedia Telecommunications*, Chapman & Hall, London, 1997.

[21] Borko F et al., Design issues for interactive television systems, *IEEE Computer*, May, 1995.

[22] Thompson J, Interactive multimedia: from couch potato to nerd? *Electronics Communication Engineering Journal*, October, 1998.

[23] Perrier V, Adapting Java for embedded development, *IEE Review*, May, 2000.

[24] http://www.wapforum.org.

[25] http://www.symbian.com.

[26] Goldman et al., *Client/Server Information Systems – a Business-Oriented Approach*, John Wiley & Sons Inc., New York, 1999.

[27] http://www.w3.org.

[28] Homer A, *Professional Active Server Pages*, Wrox Press Ltd, 1997.

[29] Feiler J, *Database-Driven Web Sites*, Morgan Kaufmann Publishers Inc., San Francisco, CA, 1999.

[30] Bhatti N & Friedrich R, Web server support for tiered services, *IEEE Network*, September/October, 1999.

[31] http://www.oucs.ox.ac.uk/hcdt/publications/audio.html.

[32] http://www.cselt.it/mpeg/.

[33] Sanker S, Constrained-latency storage access, *IEEE Computer*, March, 1993.

[34] Kerr GW, Servers for BT's interactive TV services, in Whyte B (ed.), *Multimedia Telecommunications*, Chapman & Hall, London, 1997.

[35] Gemmell DJ, Multimedia storage servers: a tutorial, *IEEE Computer*, May, 1995.

[36] Wijegunartne I & Fernandez G, *Distributed Applications Engineering*, Springer-Verlag, London, 1998.

[37] http://msdn.microsoft.com/library/techart/dw1intro.htm.

[38] West EJW, Using the Internet for business, *Computer Networks and ISDN Systems*, November, 1997.

[39] http://www-4.ibm.com/software/ebusiness.

[40] Izzo P, *Gigabit Networks*, John Wiley & Sons Inc., New York, 2000.

[41] Huston G, *Internet Performance Survival Guide*, John Wiley & Sons Inc., New York, 2000.

[42] Kosiur D, *Building and Managing Virtual Private Networks*, John Wiley & Sons Ltd, Chichester, 1998.

[43] http://www.opengroup.org/directory/.

[44] Hunt R, CCITT X500 directories – principles and applications, *Computer Communications*, December, 1992.

[45] DCE CDS & LDAP, *Integration Via Directory/Dovetail for LDAP and Directory/Segue for LDAP*, White Paper, Triangulum Software, Inc., 1998, www.triangulum.com/wpaper.html.

[46] http://info.internet.isi.edu/in-notes/rfc/files/rfc1777.txt.

[47] Hummel R, Which OS? *Byte*, February, 1997.

[48] http://www-4.ibm.com/software/network/dce.

[49] http://www.asptoday.com.articles/19991118.htm.

[50] Xiaoping J, *Object-Oriented Software Development Using Java*, Addison Wesley Longman Inc., Reading, MA, 2000.

[51] Pritchard J, *COM and CORBA Side by Side*, Addison Wesley Longman Inc., Reading, MA, 2000.

[52] Sheshadri G, *Enterprise Java Computing*, SIGS,Cambridge University Press, Cambridge, MA, 1999.

[53] *Enterprise JavaBeans Technology – a Business Benefits Analysis*, Zona Research Inc, 1999, currently available on the Java.sun.com site.

[54] http://archive.infoworld.com/cgi-bin/displayTC.pl?/971020analysis.htm.

[55] Chung P et al., http://www.cs.wustl.edu/~schmidt/submit/Paper.html.

[56] Subrahmanyam A, at http://www.subrahmanyam.com.articles/transactions?NutsAndBoltsOfTP.html.

[57] Panko R, What we know about spreadsheet errors. *Journal of End User Computing, Special Issue on Scaling Up End User Development*, 10(2) Spring, 1998.

[58] www.ibm.com.

[59] Udell J, *Practical Internet Groupware*, O'Reilly & Associates, Sebastopol, 1999.

[60] Roberts B, Groupwar strategies, *Byte*, July, 1996.

[61] www.staffware.com.

[62] Webber D, Introducing XML/EDI frameworks, *Electronic Markets* 8(1) 1998.

[63] http://www.editeur.org.

[64] Watson H & Haley B, Managerial considerations of data warehousing, *Communications of the ACM*, September, 1998.

[65] Sen A & Jacob V, Industrial strength data warehousing, *Communications of the ACM*, September, 1998.

[66] Bontempo C & Zagelow G, The IBM data warehouse architecture, *Communications of the ACM*, September, 1998.

[67] http://pasific.www.hummingbird.com].

[68] *Customers, Relationships and Households Derived from Account-Oriented Data*, White Paper by Vality Technology Corp.

[69] *Surveying Decision Support: New Realms of Analysis, Information*, Discovery Ltd., January, 1998.

[70] Rockart J & DeLong D, *Executive Support Systems*, Business One, Irwin, IL, 1988.

[71] *Integrated Knowledge Management – the Vision, the Overview*, White Paper. Pcdocs/fulcrum at http://wwwpcdoc.com.

[72] *E-business Makes New Demands of Search Technology*. White Paper from Vality Corp.

[73] Lau L & Whyte B, *Technological Aspects of TRUST in Virtual Organisa-*

tions, Report 98.27 School of Computer Studies, University of Leeds, 1998.

[74] http://www.truste.org.

[75] http://www.verisign.com.

[76] http://www.w3.org.P3P.

[77] Kramer R & Tyler T (eds.), *Trust in Organizations: Frontiers of Theory and Research*, Imprint, Thousand Oaks, Sage Publications, CA, London, 1996.

[78] Anderson R, Why cryptosystems fail, *Communications of the ACM*, November, 1994.

[79] Bauer F, *Decrypted Secrets*, Springer-Verlag, Berlin, 1997.

[80] Grobler P, *Attack?* http://www.geocities.com/ResearchTriangle/Lab/1578/attack.

[81] Dean D et al., *JAVA Security: from HotJava to Netscape and Beyond*. IEEE Symposium on Security and Privacy, Oakland, CA, May 6–8, 1996.

[82] Boyd C, Modern data encryption, *Electronics and Communication Engineering Journal*, October, 1993.

[83] Kaliski B, Emerging standards for public-key cryptography, in Damgard I (ed.) *Lectures on Data Security: Modern Cryptology in Theory and Practice*, Springer-Verlag, Berlin, 1998.

[84] *The Orange Book*, US Department of Defense.

[85] *Towards a European Framework for Digital Signatures and Encryption*, http://www.ispo.cec.be/eif/policy/97503.html.

[86] Puhakainen P, *Certification Authority in a X.509 Public Key Infrastructure*, http://www.hut.fi/~puhis/#_Toc460815746.

[87] Maher, *Requirements and Approaches for Electronic Licenses* http://dimacs.rutgers.edu/Management/.

[88] Crispo, *Untrusted third parties*, http://dimacs.ritgers.edu/Management.

[89] http://www.pgp.com.

[90] http://isp.webopedia.com/TERM/d/digital_signature.html.

[91] http://www.ietf.org/html.charters/ipsec-charter.html.

[92] http://www.rixsoft.com/Knowbuddy/gnutellafaq.html.

[93] Barnett R, Digital watermarking: applications, techniques and challenges, *Electronics and Communication Engineering Journal*, August, 1999.

[94] Special Issue on Image Security, *IEEE Computer Graphics*, 19 (1), Jan/Feb 1999.

[95] Trask N & Meyerstein M, Smart cards in electronic commerce, *British Telecommunications Technology Journal*, July, 1999.

[96] http://www.smartcardsys.com.

[97] Timmers P, *Electronic Commerce: Strategies and Models for Business-to-Business Trading*, John Wiley and Sons Ltd, Chichester, 1999.

[98] Norris M et al., *E-Business Essentials*, John Wiley and Sons Ltd, Chichester, 2000.

[99] Treese G & Stewart L, *Designing Systems for Internet Commerce*, Addison-Wesley, Reading, MA, 1998.

[100] *IEEE Spectrum Special Edition on Electronic Money*, February, 1997.

[101] http://www.w3.org.

[102] Morris S, Electronic payments: crime and prevention, a treasury viewpoint, *IEEE Spectrum*, February, 1997.

[103] http;//mondex.com.

[104] http://www.w3.org/TR/WD-Micropayment-Markup.

[105] http://www.setco.org.

[106] http://www.inetmi.com.

[107] *Secure Computing*, August, 2000.

[108] Sevcik P & Forbath T, *The Call Centre Revolution*. Northeast Consulting, http://www.3com.co/technology/tech_net/white_papers/500685.html.

[109] http://www.mitel.co.uk.

[110] Silling J, CTI, piece by piece, *Byte*, February, 1997.

[111] http://www.thebradygroup.com/KSBAWhitePaper.shtml.

[112] *Financial Times Case Study*, 2 October, 1996, page VIII.

[113] http://notes.net.

[114] Roussos K, *WAP Enabled Telemetry*, MSc dissertation, University of Leeds School of Computing, 2000.

[115] *Information and Technology in the Supply Chain*, PricewaterhouseCoopers, Euromoney Publications, 1999.

[116] Schmid B & Lindemann M, *Elements of a Reference Model for Electronic Markets*, Proceedings of the 38th HICCS Conference, 1998.

[117] Miranda R, SAP R/3 Overview, www.originet.com.br/users/miranda/overview.htm, 1997.

[118] http://www.sap.com.

[119] http://www.proxim.com/wireless/whiteppr/whatwlan.shtml.

[120] http://www.ndclan.com/Wireless/wlanW1.htm.

[121] http://www.elogistics101.com.

[122] http://www.uc-council.org.

[123] http://www.ean.be.

[124] http://www.adams1.com/pub/russadam/stack.html.

[125] http://ean.eurosecure.net.

[126] http://www.sensormatic.com.

[127] http://www.foodlogistics.com.

[128] *Manpower* White Paper, Ilog.com, April, 1997.

[129] http:// www.ilog.fr.

[130] http:// WWW.ris.com, Retailers see forecasting in their future, *RIS News*, Febriary 1998 Issue.

[131] Gebauer J & Segev A, *CALS Expo International & 21st Century Commerce*, Long Beach, CA, 1998, also in virtual-organisation.net Newsletter V2, No 3

[132] Foner L, *What's an Agent, Anyway? A Sociological Case Study*, Agents Memo 93-01, Agents Group, MIT Media Lab, foner@media.mit.edu.

[133] Klusch M (Ed.), *Intelligent Information Agents*, Springer, Berlin, 1998.

[134] Zeng et al., Dynamic supply chain structuring for electronic commerce

among agents, in Klusch M (ed.), *Intelligent Information Agents*, Springer, Berlin, 1998.

[135] Zhang C & Soo V-W (eds.), *Design and Applications of Intelligent Agents*, Springer, Berlin, 2000.

[136] Merlat W, An agent-based multiservice negotiation for e-commerce, *BT Technology Journal*, October, 1999.

[137] Brewington B et al., Mobile agents for distributed information retrieval, in Klusch M (ed.), *Intelligent Information Agents*, Springer, Berlin, 1998.

[138] Tschudin C, Mobile agent security, in Klusch M (ed.), *Intelligent Information Agents*, Springer, Berlin, 1998.

[139] Kotler P et al., *Principles of Marketing*, Prentice Hall, London, 1996.

[140] Smith G & Jones W, Design of multimedia services, in Whyte B (ed.), *Multimedia Telecommunications*, Chapman & Hall, London, 1997.

[141] Igbaria M et al., The respective roles of perceived usefulness and perceived fun in the acceptance of microcomputer technology, *Behaviour and Information Technology*, October, 1995.

[142] http://www.unicode.org.

[143] http://www.w3.org/WAI/.

[144] http://www.wired.com/news/topstories/0,1287,3500,00.html.

[145] Kushmerick N, *Learning to Remove Internet Advertisements*, ACM Conference on Autonomous Agents, Seattle, WA, 1999.

[146] *An Enterprise Portal Bridge to E-business*, Deplhi Group, Hummingbird White Paper, 617-247-1511, 2000.

[147] http://searchenginewatch.com.

[148] Stone M et al., *Direct Hit – Direct Marketing With a Winning Edge*, Pitman, London, 1995.

[149] Linton I, *Database Marketing*, Pitman, London, 1995.

Index

Notes:

Where an index entry constitutes a major topic of a chapter, it is referenced in **bold** to the Part and chapter, e.g. Knowledge management **2.2**.

Computing and telecommunications terms abound with abbreviations which are more widely used than their expansions; consequently this index uses abbreviations where they are the more commonly used form.

It would not be generally useful to index the terms *eCommerce* and *eBusiness* since they are all pervasive in this book. Consequently no references are given.